AF885739

Research and Application of Cable-Driven and Rigid Parallel Robots

Xiaoqiang Tang · Zhufeng Shao · Rui Yao

Research and Application of Cable-Driven and Rigid Parallel Robots

Development of the 40-meter Scale Model of the FAST (China Sky Eye) Feed Support System

清華大學出版社
TSINGHUA UNIVERSITY PRESS

Springer

Xiaoqiang Tang
Department of Mechanical Engineering
Tsinghua University
Beijing, China

Zhufeng Shao
Department of Mechanical Engineering
Tsinghua University
Beijing, China

Rui Yao
Chinese Academy of Sciences
Beijing, China

ISBN 978-981-99-7451-1 ISBN 978-981-99-7452-8 (eBook)
https://doi.org/10.1007/978-981-99-7452-8

Jointly published with Tsinghua University Press
The print edition is not for sale in China (Mainland). Customers from China (Mainland) please order the print book from: Tsinghua University Press.
ISBN of the Co-Publisher's edition: 978-730-25-5497-4

© Tsinghua University Press 2024

This work is subject to copyright. All rights are solely and exclusively licensed by the Publisher, whether the whole or part of the material is concerned, specifically the rights of reprinting, reuse of illustrations, recitation, broadcasting, reproduction on microfilms or in any other physical way, and transmission or information storage and retrieval, electronic adaptation, computer software, or by similar or dissimilar methodology now known or hereafter developed.
The use of general descriptive names, registered names, trademarks, service marks, etc. in this publication does not imply, even in the absence of a specific statement, that such names are exempt from the relevant protective laws and regulations and therefore free for general use.
The publishers, the authors, and the editors are safe to assume that the advice and information in this book are believed to be true and accurate at the date of publication. Neither the publishers nor the authors or the editors give a warranty, expressed or implied, with respect to the material contained herein or for any errors or omissions that may have been made. The publishers remain neutral with regard to jurisdictional claims in published maps and institutional affiliations.

This Springer imprint is published by the registered company Springer Nature Singapore Pte Ltd.
The registered company address is: 152 Beach Road, #21-01/04 Gateway East, Singapore 189721, Singapore

Paper in this product is recyclable.

Foreword by Zheng You

The Five-hundred-meter Aperture Spherical radio Telescope (FAST), known as "China Sky Eye", is the world's largest and most sensitive single-dish radio telescope with Chinese independent intellectual property rights. Its completion and operation are of great significance for China to achieve major original breakthroughs in the scientific frontier and accelerate innovation-driven development. With the ultra-high sensitivity of the FAST, China has achieved a zero breakthrough in pulsar discovery and increased the timing accuracy of pulsars by an order of magnitude. It is expected to implement the detection capability of primordial gravitational waves. The FAST project has taken 20 years to complete from launch in 1999 to national acceptance in 2020, which embodies the wisdom and efforts of a large number of scientific researchers and engineers.

Tsinghua University has always kept in mind its responsibility and mission to serve the major national strategic demands. Several research teams represented by the authors of this book have participated in the research and design of the FAST, undertaking a lot of core work and playing a decisive role in its construction. The National Astronomical Observatories of the Chinese Academy of Sciences (NAOC) fully affirms the important contribution of Tsinghua University to the FAST project and names an asteroid as "Tsinghua University Star" (with a permanent number of 16982). The authors of this book have overcome a series of key theoretical and technical problems of the cable-driven and rigid parallel robots mechanism, completed the research and development of the 40-meter scale model of the feed support system, and successfully realized the astronomical observation function, which has laid a solid theoretical and experimental foundation for the implementation of the FAST project.

This book summarizes the theoretical and experimental research on the final scale model of the FAST feed support system and systematically describes the key technologies, focusing on the cable-driven and rigid parallel robots. The content covers the accuracy assurance and optimization design of the cable-driven parallel mechanism, the dynamic modeling and inertia matching of the rigid parallel mechanism, and the efficient modeling method and vibration control of the flexibly supported parallel mechanism. This book has a clear engineering background and emphasizes the combination of theory and practice. Engineering problems are studied at the theoretical level for in-depth research, and new theories are established considering the practice, which are applied to the engineering practice to promote technological breakthroughs and innovations. As a result, this book is not only a professional theoretical work on the cable-driven and rigid parallel robots but also a reference for engineering scientists and technicians to inspire their scientific research methods and ideas.

The cable-driven and rigid parallel robots possess outstanding features and advantages. It is expected that the authors of this book can further promote the innovation of related robotic equipment based on the solid research foundation and propose original solutions for such working conditions as high speed, heavy load, and large workspace to advance the development and deep integration of robotics and intelligent manufacturing.

March 2020

Zheng You
Professor and Vice President of Tsinghua University
Academician of the Chinese Academy of Engineering
Head of Intelligent Manufacturing Standardization
Advisory Group of Experts
Member of National Manufacturing Strategy Advisory
Committee

Foreword by Qiming Wang

FAST is the world's largest single-dish radio telescope built in Pingtang county, Guizhou province, which is proposed by astronomers represented by Nang Rentong in the 1990s, breaking the dimension limits of 100 m in traditional telescope engineering. Its main performance indexes have achieved the international leading level, which is of great significance to promoting original breakthroughs in Chinese astronomy.

FAST has three major independent innovations: taking advantage of the giant karst depression in Guizhou province as the telescope site; the active reflector that can form a 300-meter aperture instantaneous paraboloid in the observation direction, correcting the spherical aberration, improving the convergence capability of electromagnetic wave, and achieving broadband as well as full polarization; the two-level feed adjustment mechanism (feed support system) with the largest span in the world that consists of a cable-driven parallel robot and a rigid parallel robot, achieving high-precision positioning of the receivers at 10 mm root-mean-square. Tsinghua University has been involved in the research of the active reflector and the feed support system since the 1990s, made significant contributions, and trained a group of researchers for the FAST project, covering key content of scheme design, optimization analysis, motion control, and the building of scale models.

Tsinghua University has proposed several schemes for the large-span feed support system, and scale physical models are established for each scheme. The 40-meter scale model of the feed support system illustrated in this book is the final set of large outdoor model before the construction of FAST prototype. Its main significance is as follows:

(a) This final 40-meter scale model is the closest to the prototype. The configuration composed of "6-cable parallel robot + A-B rotator + Stewart parallel robot" is the same as the prototype, and the feasibility of the final scheme has been fully demonstrated from theoretical research to experimental data. Effective astronomical observations have also been implemented. The final 40-meter scale model provides an important theoretical and experimental foundation for the FAST prototype.

(b) The research on the final scheme has solved many key theoretical and engineering problems, such as modeling of large parallel robots, optimization design methods, dynamic characteristic analysis, and vibration control strategies, which not only provides direct support for the subsequent design and engineering construction of FAST but also guides the research in related fields.

The content of this book fully reflects the difficult and important points of the FAST feed support system in the design phase, and the research ideas and experimental design could give readers inspiration and reference. The detailed derivation of equations provides help for readers who are interested in related research and prompts a more comprehensive understanding of the research process. Robotics develops rapidly, and robots are increasingly used. The authors of this book work hard to advance the research on cable-driven and rigid parallel robots. It is expected that they can make more breakthroughs and innovations in theory and application and further contribute to the development of robotic equipment.

March 2020

Qiming Wang
Researcher of the NAOC
Chief Engineer of the FAST Project

Preface

A core part of the 13th Five-Year Plan on national scientific and technological innovation is to enhance original innovation, strengthen basic research, build major scientific and technological innovation bases, and strive to lead the development of the world's scientific frontiers in more fields. Fundamental research is inseparable from the support of scientific engineering and equipment. Mega-science projects are called "national instruments" and are landmark projects of China's scientific and technological power. The Five-hundred-meter Aperture Spherical Radio Telescope (FAST) is a representative project of China's Mega-science engineering and successfully passed the national acceptance in January 2020. To ensure world-class detection capabilities and sensitivity, FAST has to simultaneously achieve a 500-meter ultra-large-scale reflective surface and millimeter-level signal reception accuracy. Its engineering research and construction process faced many challenging theoretical and technical problems.

The authors of this book fully participated in the construction and research of the final scale model of the FAST led by the National Astronomical Observatory of the Chinese Academy of Sciences. Focusing on the feed support system composed of long-span cable-driven and rigid parallel robots, research on optimization design, modeling, simulation, and accuracy assurance was carried out, laying a solid foundation for the high sensitivity of the FAST prototype. The feed support system is one of the three major innovations of the FAST, which directly determines its accuracy and performance.

The parallel robot is a new type of mechanism proposed to address the shortcomings of the series robot. It has the core advantages of high stiffness, good dynamics, and a high load-to-weight ratio with broad application prospects. Parallel spindle heads and high-speed sorting robots have been successfully commercially applied. Cable-Driven Parallel Robots (CDPRs) are a class of high-performance robots that combine the advantages of cable-drive technology and parallel configurations. CDPRs inherit high dynamics and large load-bearing capacity of parallel robots while achieving significant improvements in workspace, lightweight, and efficiency. Reflecting the frontier trend of integrating rigidity and flexibility in robotic equipment and embodying advanced lightweight design concepts, CDPRs have expanded the workspace and load of traditional industrial robots by more than two orders of magnitude and possess ultra-high dynamic characteristics. Cable-driven and rigid parallel robots have become a research hotspot and an international frontier in high-end equipment and robots.

This book systematically describes the main problems and solutions in the development of the FAST feed support system, introducing in detail the optimal design theory, dynamic modeling methods, and vibration control technology of cable-driven and rigid parallel robots. Chapter 1 introduces the FAST project and the parallel robots; Chaps. 2 to 4 study the modeling, static characteristics, and stiffness of the long-span CDPRs; Chap. 5 studies the comprehensive optimization design method of large-span CDPRs; Chaps. 6 and 7 establish the dynamic model of cable-drive and rigid parallel robots, and propose an efficient dynamic modeling method for rigid-flexible series coupling system; Chap. 8 mainly researches vibration control of rigid-flexible series coupling system; Chap. 9 is oriented to engineering applications and proposes the inertia matching index and method for parallel robots to serve transmission matching and motor selection. Chapter 10 completes the development of the final scale model of the FAST feed support system. Chapters 2 to 5 and 10 are written by Tang Xiaoqiang and Yao Rui, and the rest are written by Tang Xiaoqiang and Shao Zhufeng.

This book is the first work to focus on the FAST project and provides a detailed introduction to the research on cable-driven and rigid parallel robots. It can serve as a reference book for scientific and technological workers in related fields of mechanical engineering. We hope this book can provide inspiration and ideas for theoretical research and prototype development of high-end equipment. This book can be used as an auxiliary textbook or reference book for graduate and undergraduate students majoring in mechanical engineering, and it is also suitable for reading by science and technology enthusiasts to understand the research progress of China's FAST Mega-science project and cable-driven and rigid parallel robots.

Due to the limited level of the authors, errors and deficiencies in the book are inevitable. Readers are kindly requested to provide criticism and suggestions. We hope that readers will work with us to promote theoretical research and engineering applications of cable-driven and rigid parallel robots, promote original innovation of high-end equipment, and serve the transformation and upgrading of the manufacturing industry.

Beijing, China
March 2020

Xiaoqiang Tang
Zhufeng Shao
Rui Yao

Contents

Main Parameters of Each Chapter

Chapter 2

$\boldsymbol{l_0}$	Single-cable model: the unstrained length of the cable
$\boldsymbol{\Delta l}$	Single-cable model: the amount of elastic deformation of the cable
$\boldsymbol{T}$	Single-cable model: the tension at the fixed end of the cable
$\boldsymbol{\rho}$	Single-cable model: the linear density of the cable when elastic deformation doesn't occur (initial inelastic deformation)
$\boldsymbol{E}$	Single-cable model: the elastic modulus
$\boldsymbol{A_0}$	Single-cable model: the unstrained cross-sectional area
$\boldsymbol{V}$	Single-cable model: the vertical component of the cable tension
$\boldsymbol{H}$	Single-cable model: the horizontal component of the cable tension
$\boldsymbol{L}$	Single-cable model: the horizontal span
$\boldsymbol{h}$	Single-cable model: the height difference between the two connected points of the cable
$\boldsymbol{B_i}$	Cable-driven parallel mechanism: the connected point of the stationary platform
$\boldsymbol{A_i}$	Cable-driven parallel mechanism: the connected point of the moving platform
$\boldsymbol{O'}$	Cable-driven parallel mechanism: the origin of the moving coordinate system
$\boldsymbol{\sigma_0}$	Cable-driven parallel mechanism: the vector of the cable tension
$\boldsymbol{J_0^T}$	Cable-driven parallel mechanism: the force transfer matrix
$\boldsymbol{F}$	Cable-driven parallel mechanism: the external screw at the center position of the moving platform
$\boldsymbol{l_c}$	The cable length vector deduced by the precise catenary equation
$\boldsymbol{l_p}$	The cable length vector deduced by the parabolic equation
$\boldsymbol{l_l}$	The cable length vector deduced by the linear equation
$\boldsymbol{\varepsilon_p}$	The modeling end errors of the cable-driven parallel mechanism caused by the errors of the parabolic model
$\boldsymbol{\varepsilon_l}$	The modeling end errors of the cable-driven parallel mechanism caused by the errors of the linear model

$\boldsymbol{\varepsilon l_{pi}}$	Errors of the *i*-th cable length of the fitting parabolic model with higher-order polynomials
$\boldsymbol{\varepsilon l_{li}}$	Errors of the *i*-th cable length of the fitting linear model with higher-order polynomials
$\boldsymbol{R}$	Transfer matrix from moving to inertial coordinate systems of the cable-driven parallel mechanism
$\boldsymbol{U}$	The unit vector of $\boldsymbol{l_l}$
$\boldsymbol{r_i}$	The position of the vector representation A_i
$\boldsymbol{r_a}$	Radius of the moving platform of FAST
$\boldsymbol{r_b}$	Radius of the distribution circle of the cable tower of FAST
$\boldsymbol{h_{tower}}$	Height of the cable tower of FAST
$\boldsymbol{d}$	Diameter of the cable of FAST
$\boldsymbol{m_0}$	Mass of the moving platform of FAST
$[\boldsymbol{\varepsilon_0}, \boldsymbol{\varepsilon_{angle}}]^{\mathrm{T}}$	The maximum allowable modeling error of FAST

Chapter 3

$\mathbf{TCI}_{max}$	The difference between the maximum and minimum cable tension
$\mathbf{TCI_{rmax}}$	The ratio between the maximum and minimum cable tension
$\mathbf{GTCI_{max}}$	The difference between the global maximum and minimum cable tension
$\mathbf{GTCI_{rmax}}$	The global ratio between the maximum and minimum cable tension
$\boldsymbol{\delta X}$	The end error of the moving platform of the cable-driven parallel mechanism
$\boldsymbol{\delta OO'}$	The position error of the terminal of the moving platform of the cable-driven parallel mechanism
$\boldsymbol{\delta \Omega}$	The attitude error of the terminal of the moving platform of the cable-driven parallel mechanism
$\boldsymbol{\delta \varepsilon}$	Cable length error after adopting the linear model and its modeling error compensation
$\boldsymbol{\delta O' A_i}$	The error vector of the cable connection points of the moving platform
$\boldsymbol{\delta O B_i}$	The error vector of the cable connection points of the static platform
$\boldsymbol{l_i^{\mathrm{T}}}$	The unit vector of the limb
$\boldsymbol{J}$	The invertible Jacobi matrix of the long-span cable-driven parallel mechanism
$\boldsymbol{E}$	The error transfer matrix of the long-span cable-driven parallel mechanism
$\forall_{\mathbf{err}}$	The allowable error space at the given attitude
$\mathbf{V}_{\forall_{\mathbf{err}}}$	The generalized volume of the allowable error space
$\forall_{\mathbf{control}}$	The allowable error space at the given attitude satisfying the tension conditions
$\mathbf{V}_{\forall_{\mathbf{control}}}$	The controllable generalized volume at the given attitude within the error space

$\boldsymbol{\Theta}_{\forall}$ The ratio of the generalized volume of the cable tension
$\boldsymbol{\sigma}_{\mathbf{min}}$ The minimum cable tension of the cable-driven parallel mechanism
$\boldsymbol{\sigma}_{\mathbf{max}}$ The maximum cable tension of the cable-driven parallel mechanism

Chapter 4

$\boldsymbol{k_0}$ The unit stiffness of the cable
$\boldsymbol{l_{ci}}$ The length of the *i*-th cable derived from the exact model of the cable-driven parallel mechanism
$\boldsymbol{k_i}$ The stiffness of the single cable
$\boldsymbol{K}$ The stiffness matrix of the long-span cable-driven parallel mechanism
$\boldsymbol{\sigma_i}$ The cable tension of the *i*-th cable
$\tilde{\boldsymbol{K}}$ The simplified stiffness matrix established by adopting the modeling error compensation method
$\boldsymbol{l_{li}}$ The cable length deduced based on the linear model
$\boldsymbol{\varepsilon l_{li}}$ The error compensation value of the *i*-th cable of the linear model
$\boldsymbol{\Delta l_{li}}$ The elastic deformation amount of the *i*-th cable
$\boldsymbol{S_{\mathrm{G}}}$ The stiffness index in a single direction in the global workspace
$\boldsymbol{S_{\delta}}$ The index of the variation rate of stiffness in the global workspace
$\boldsymbol{\lambda_g}$ The similarity ratio of gravitational acceleration
$\boldsymbol{\lambda_l}$ The similarity ratio of length
$\boldsymbol{\lambda_d}$ The similarity ratio of diameter
$\boldsymbol{\lambda_A}$ The similarity ratio of area
$\boldsymbol{\lambda_m}$ The similarity ratio of mass
$\boldsymbol{\lambda_\rho}$ The similarity ratio of linear density
$\boldsymbol{\lambda_{\rho_v}}$ The similarity ratio of bulk density
$\boldsymbol{\lambda_\sigma}$ The similarity ratio of force
$\boldsymbol{\lambda_E}$ The similarity ratio of elastic modulus
$\boldsymbol{\lambda_K}$ The stiffness similarity ratio of the cable-driven parallel mechanism

Chapter 5

$\boldsymbol{W}$ The performance index of the long-span cable-driven parallel mechanism
$\boldsymbol{b_1}, \boldsymbol{b_2}, \ldots, \boldsymbol{b_i}$ The dimensional design parameters of the mechanism
$\boldsymbol{S_\Gamma}$ The magnitude of the rate-of-change of the performance of the long-span cable-driven parallel mechanism

Chapter 6

$\{\boldsymbol{G}\}$: O–XYZ	Global coordinate system of FAST
$\{\boldsymbol{C}\}$: $O' - X'Y'Z'$	Coordinate system of cable platform
$\{\boldsymbol{P}\}$: o–xyz	Coordinate system of the moving platform of the fine-tuning Stewart platform
$\{\boldsymbol{B}\}$: $o' - x'y'z'$	Coordinate system of the base platform of the fine-tuning Stewart platform
$\{\boldsymbol{L}\}$	Local coordinate system of the upper parts of the branch chain of the fine-tuning Stewart platform
$\{\boldsymbol{U}\}$	Local coordinate system of the lower parts of the branch chain of the fine-tuning Stewart platform
$\{\boldsymbol{L}'\}$	Non-inertial translational coordinate system that fixes at the origin of $\{\boldsymbol{L}\}$ and is parallel to $\{\boldsymbol{G}\}$
$\{\boldsymbol{P}'\}$	Non-inertial translational coordinate system that fixes at the origin of $\{\boldsymbol{P}\}$ and is parallel to $\{\boldsymbol{G}\}$
${}^{\mathrm{G}}\boldsymbol{R}_{\mathrm{B}}$	Rotation matrix of the base platform of the fine-tuning Stewart platform in $\{\boldsymbol{G}\}$
${}^{\mathrm{G}}\boldsymbol{t}_{\mathrm{B}}$	Position vector of the base platform of the fine-tuning Stewart platform in $\{\boldsymbol{G}\}$
${}^{\mathrm{G}}\boldsymbol{R}_{\mathrm{C}}$	Rotation matrix of the cable platform in $\{\boldsymbol{G}\}$
${}^{\mathrm{G}}\boldsymbol{t}_{\mathrm{C}}$	Position vector of the cable platform in $\{\boldsymbol{G}\}$
${}^{\mathrm{G}}\boldsymbol{\omega}_{\mathrm{C}}$	Angular velocity of the cable platform in $\{\boldsymbol{G}\}$
${}^{\mathrm{G}}\boldsymbol{\varepsilon}_{\mathrm{C}}$	Angular acceleration of the cable platform in $\{\boldsymbol{G}\}$
${}^{\mathrm{C}}\boldsymbol{R}_{\mathrm{B}}$	Rotation matrix of the base platform of the fine-tuning Stewart platform in $\{\boldsymbol{C}\}$
$\boldsymbol{b}_{\mathrm{i}}$	Position vector of the universal joint of the fine-tuning Stewart platform in $\{\boldsymbol{B}\}$
$\boldsymbol{p}_{\mathrm{i}}$	Position vector of the spherical joint of the fine-tuning Stewart platform in $\{\boldsymbol{P}\}$
${}^{\mathrm{B}}\boldsymbol{R}_{\mathrm{P}}$	Rotation matrix of the moving platform of the fine-tuning Stewart platform in $\{\boldsymbol{B}\}$
${}^{\mathrm{B}}\boldsymbol{t}_{\mathrm{P}}$	Position vector of the moving platform of the fine-tuning Stewart platform in $\{\boldsymbol{B}\}$
${}^{\mathrm{B}}\boldsymbol{\omega}_{\mathrm{P}}$	Angular velocity of the moving platform of the fine-tuning Stewart platform in $\{\boldsymbol{B}\}$
${}^{\mathrm{B}}\boldsymbol{\varepsilon}_{\mathrm{P}}$	Angular acceleration of the moving platform of the fine-tuning Stewart platform in $\{\boldsymbol{B}\}$
${}^{\mathrm{G}}\boldsymbol{S}$ ${}^{\mathrm{B}}\boldsymbol{S}$	Vector of the branch chain of the fine-tuning Stewart platform in $\{\boldsymbol{G}\}$ and $\{\boldsymbol{B}\}$
${}^{\mathrm{G}}\boldsymbol{s}$ ${}^{\mathrm{B}}\boldsymbol{s}$	Unit vector of the branch chain of the fine-tuning Stewart platform in $\{\boldsymbol{G}\}$ and $\{\boldsymbol{B}\}$
$\boldsymbol{r}_{\mathrm{Lo}}$	Vector of the centroid of the upper parts of the branch chain of the fine-tuning Stewart platform

$\boldsymbol{r}_{\mathrm{Uo}}$	Vector of the centroid of the lower parts of the branch chain of the fine-tuning Stewart platform
L	Length of the branch chain of the fine-tuning Stewart platform
${}^{G}W$	Angular velocity along the vertical branch chain direction of the fine-tuning Stewart platform in $\{\boldsymbol{G}\}$
${}^{G}A$	Angular acceleration along the vertical branch chain direction of the fine-tuning Stewart platform in $\{\boldsymbol{G}\}$
$\boldsymbol{a}_L$	Acceleration of the upper parts of the branch chain of the fine-tuning Stewart platform in $\{\boldsymbol{L}'\}$
$\boldsymbol{a}_U$	Acceleration of the lower parts of the branch chain of the fine-tuning Stewart platform in $\{\boldsymbol{L}'\}$
$\boldsymbol{a}_{Lr}$	Acceleration of the non-inertial translational coordinate system $\{\boldsymbol{L}'\}$ in $\{\boldsymbol{G}\}$
${}^{\mathrm{G}}\ddot{\boldsymbol{t}}_{\mathrm{B}}$	Translational acceleration of the base platform of the fine-tuning Stewart platform
${}^{\mathrm{G}}\boldsymbol{\omega}_{\mathrm{B}}$	Rotation angular velocity of the base platform of the fine-tuning Stewart platform
${}^{\mathrm{G}}\boldsymbol{\varepsilon}_{\mathrm{B}}$	Angular acceleration of the base platform of the fine-tuning Stewart platform
${}^{\mathrm{G}}\boldsymbol{R}_{\mathrm{B}}$	Attitude of the base platform of the fine-tuning Stewart platform
$\boldsymbol{e}_{po}$	Vector of the centroid position of the moving platform of the fine-tuning Stewart platform in $\{\boldsymbol{P}\}$
${}^{G}\boldsymbol{e}_{po}$	Vector of the centroid position of the moving platform of the fine-tuning Stewart platform in $\{\boldsymbol{G}\}$
${}^{G}\boldsymbol{\omega}_{p}$	Angular velocity of the fine-tuning Stewart platform in $\{\boldsymbol{G}\}$
${}^{G}\boldsymbol{\varepsilon}_{p}$	Angular acceleration of the fine-tuning Stewart platform in $\{\boldsymbol{G}\}$
$\boldsymbol{a}_{\mathrm{Pr}}$	Convective acceleration of the non-inertial translational coordinate system $\left\{\boldsymbol{P}'\right\}$ of the fine-tuning Stewart platform
$\boldsymbol{I}_{Lo}$	Rotation vector of the upper parts of the branch chain of the fine-tuning Stewart platform in $\{\boldsymbol{L}\}$
$\boldsymbol{I}_{L}$	Rotation vector of the upper parts of the branch chain of the fine-tuning Stewart platform in $\{\boldsymbol{G}\}$
$\boldsymbol{I}_{Uo}$	Rotation vector of the lower parts of the branch chain of the fine-tuning Stewart platform in $\{\boldsymbol{L}\}$
$\boldsymbol{I}_{U}$	Rotation vector of the lower parts of the branch chain of the fine-tuning Stewart platform in $\{\boldsymbol{G}\}$
$\boldsymbol{I}_{Po}$	Rotation vector of the moving platform of the fine-tuning Stewart platform in $\{\boldsymbol{P}\}$
$\boldsymbol{I}_{P}$	Rotation vector of the moving platform of the fine-tuning Stewart platform in $\{\boldsymbol{G}\}$
M_U	Modulus applied by the Hooke joint to the branch chain to inhibit its rotation around itself for torque limitation
$\boldsymbol{F}_s$	Force applied by the spherical joint to the branch chain

C_U	Coefficient of friction at the Hooke joint
C_S	Coefficient of friction at the spherical joint
$\boldsymbol{F}_{ext}$	External force acting on the moving platform of the fine-tuning Stewart platform
$\boldsymbol{M}_{ext}$	External moment acting on the moving platform of the fine-tuning Stewart platform
C_P	Coefficient of friction of the prismatic joint
$\boldsymbol{F}_{PJ}$	Driving force of the prismatic joint

Chapter 7

${}^G\boldsymbol{S}_{Cj}$	The position vector of the *j*-th drive cable from the exit point to the anchor point of the cable platform in $\{\boldsymbol{G}\}$
${}^G\boldsymbol{s}_{Cj}$	The unit directional vector of the *j*-th driving cable from the exit point to the anchor point in $\{\boldsymbol{G}\}$
L_{Cj}	The length of the *j*-th driving cable
K_j	The elastic coefficient of the *j*-th cable
A	The cross-sectional area
E	The elastic modulus
C_j	The damping coefficient
L_{C0j}	The initial length in the ideal state
$\boldsymbol{M}$	The mass of the feed cabin
$\boldsymbol{I}$	The inertia matrix relative to the center of mass
$\boldsymbol{F}_e$	The external force acting on the feed cabin
$\boldsymbol{N}_e$	The external moment acting on the feed cabin
${}^G\ddot{\boldsymbol{t}}_C$	The linear accelerations of the cable platform in $\{\boldsymbol{G}\}$
${}^G\ddot{\boldsymbol{\theta}}_C$	The angular accelerations of the cable platform in $\{\boldsymbol{G}\}$
$v_m(z)$	The mean velocity
$v_g(z, t)$	The turbulence velocity
$p(z, t)$	The wind pressure

Chapter 8

φ_{Cable}	The natural dip angle of the cable platform (feed cabin)
$\varphi_{\mathrm{Receiver}}$	The observation attitude angle of the receiver
${}^G\boldsymbol{X}_P$	The target pose of the fine-tuning Stewart mechanism
${}^G\boldsymbol{X}_C$	The pose of cable-driven parallel robot
$\boldsymbol{\theta}$	Angles of A-B rotator
${}^C\boldsymbol{t}_M$	The position vector of the origin of the intermediate coordinate system in $\{\boldsymbol{C}\}$

${}^{B}\boldsymbol{t}_{M}$	The position vector of the origin of the intermediate coordinate system in $\{\boldsymbol{B}\}$
${}^{B}\boldsymbol{R}_{P}$	The identity matrix
${}^{B}\boldsymbol{t}_{P}$	The position vector of the moving platform of Stewart platform in the base coordinate system $\{\boldsymbol{B}\}$ under the initial state
$\boldsymbol{d}_j$	The position vector of the anchor point on the cable platform
$\boldsymbol{c}_j$	The position vector of the cable exit point on the cable tower
η	The deceleration ratio of the reducer
θ_A	The rotation angle of axis A
n_A	The number of cycles of motor rotation
θ_{B0}	The angle between O_1O_2 and O_1O_3 in the initial state of A-B rotators

Chapter 9

$\boldsymbol{\tau}$	The driving force vector of the limb
$\boldsymbol{M}_{\mathrm{P}}(\boldsymbol{X}_{\mathrm{P}})$	The inertia matrix of the imaginary axis space
$V_{\mathrm{P}}(\dot{X}_{\mathrm{P}}, X_{\mathrm{P}})$	The vector of centripetal, Coriolis, and friction forces
$\boldsymbol{G}_{\mathrm{P}}(\boldsymbol{X}_{\mathrm{P}})$	The gravity vector
$\boldsymbol{X}_{\mathrm{P}}$	The attitude vector of the moving platform
$\dot{X}_{\mathrm{P}}$	The velocity vector of the moving platform
$\ddot{X}_{\mathrm{P}}$	The acceleration vector of the moving platform
$\{\boldsymbol{N}_i\}$:O_i - $X_iY_iZ_i$	The local coordinate system that is fixed at the rotation center of the Hooke joint of the i-th limb
$\{\boldsymbol{K}_i\}$:o_i - $x_iy_iz_i$	The local coordinate system that is fixed at the rotation center of the Hooke joint of the i-th limb
$\{\boldsymbol{M}_i\}$	The local coordinate system with the origin fixed at the rotation center of the spherical joint and the direction of the coordinate axes parallel to $\{\boldsymbol{K}_i\}$
$\boldsymbol{X}$	The attitude of the moving platform of the Stewart mechanism
$\boldsymbol{J}_{link}$	The Jacobian matrix of the limb
${}^{K}\boldsymbol{W}_i$	The oscillation velocity of the limb
$\boldsymbol{F}_P$	The combined forces and moments of the moving platform at the geometric center
$\boldsymbol{f}_{\mathrm{E}}$	The external force acting on the moving platform
$\boldsymbol{n}_{\mathrm{E}}$	The external moment acting on the moving platform
m_{P}	The mass of the moving platform
$\boldsymbol{e} = [e_x, e_y, e_z]^{\mathrm{T}}$	The coordinate of the mass center in the coordinate system of the moving platform
${}^{\mathrm{B}}\boldsymbol{I}_{\mathrm{P}}$	The inertia matrix of the moving platform with respect to the geometric center in the global coordinate system $\{\boldsymbol{B}\}$
$\boldsymbol{I}_{\mathrm{PC}}$	The inertia matrix of the moving platform with respect to the mass center in the coordinate system

$\boldsymbol{F}_{Ui}$	The combined force and moment at the upper part limb of the Stewart mechanism at B_i in $\{\boldsymbol{K}_i\}$
m_{U}	The mass of the upper part limb
$\boldsymbol{I}_{\mathrm{U}}$	The rotational inertia matrix of the upper part limb with respect to B_i in $\{\boldsymbol{K}_i\}$
$\boldsymbol{I}_{\mathrm{UC}}$	The inertia matrix of the upper part limb with respect to its mass center in $\{\boldsymbol{K}_i\}$
$\boldsymbol{F}_{Li}$	The combined force and moment at the lower part limb of the Stewart mechanism at B_i in $\{\boldsymbol{K}_i\}$
m_{L}	The mass of the lower part limb
$\boldsymbol{I}_{\mathrm{L}}$	The rotational inertia matrix of the lower part limb with respect to B_i in $\{\boldsymbol{K}_i\}$
$\boldsymbol{I}_{\mathrm{LC}}$	The inertia matrix of the lower part limb with respect to its mass center in $\{\boldsymbol{M}_i\}$
$\boldsymbol{F}$	The driving force of the limb of the Stewart mechanism
$\boldsymbol{I}_E$	The equivalent inertia of the Stewart mechanism
L_M	Motor armature winding inductance
R_M	Motor armature winding resistance
$\boldsymbol{J}_{\mathbf{M}}$	Motor rotor rotational inertia
$\boldsymbol{K}_{\mathbf{E}}$	Motor counter-electromotive force constant
$\boldsymbol{K}_{\mathbf{T}}$	Motor torque constant
$\boldsymbol{K}_{\mathbf{PI}}$	Current loop scaling factor
$\boldsymbol{T}_{\mathbf{II}}$	Current loop integration time constant
$\boldsymbol{K}_{\mathbf{PV}}$	Speed loop scaling factor
$\boldsymbol{T}_{\mathbf{IV}}$	Velocity loop integration time constant
$\boldsymbol{K}_{\mathbf{PP}}$	Position loop scaling factor
$\boldsymbol{T}_{\mathbf{P}}$	Position loop integration time constant

Chapter 1
Overview

This chapter begins with a brief introduction to the history of astronomical telescopes, reviews the preparation and construction process of the Five-hundred-meter Spherical radio Telescope (FAST) that is known as "China Sky Eye", and clarifies its important role in astronomy and deep space exploration. Section 1.2 illustrates the main components, structural properties and operating principles of FAST, and describes its innovations and breakthroughs compared with the Arecibo Observatory's massive radio telescope. As one of the three independent innovations of FAST, the feed support system is a pivotal guarantee system for its sensitivity, which needs to achieve lightweight and high-precision dynamic tracking with the large workspace. Therefore, the FAST feed support system adopts a multi-level serial structure of rigid-flexible coupling, consisting of a cable-driven and a rigid parallel robots. In Sect. 1.3, the parallel mechanism and its complementary relationship with the serial manipulator are explored, in addition, the Z3 spindle head and Delta high-speed parallel robot that have gained large-scale commercial applications are presented. Section 1.4 discusses the cable-driven parallel mechanism, clarifies that the fundamental function is to keep cables tensioned, and describes the performance advantages, application fields and examples. This chapter give readers a preliminary understanding of the radio telescope, FAST, as well as the rigid and cable-driven parallel robots, which is the foundation for the subsequent study of the feed support system.

Main contents of this chapter are:

(1) Origin of the FAST;
(2) Main structure and operating principles of the FAST;
(3) Overview of the parallel robot;
(4) Overview of the cable-driven parallel robot.

© Tsinghua University Press 2024

X. Tang et al., *Research and Application of Cable-Driven and Rigid Parallel Robots*,
https://doi.org/10.1007/978-981-99-7452-8_1

1.1 Origin of the FAST

Astronomical telescopes, the important tool for observing celestial bodies, are the foundation of modern astronomy. The light-gathering ability of a telescope increases with its aperture, and the stronger the light-gathering ability of it, the fainter and more distant objects and the earlier universe can be observed. Figure 1.1 shows the Guo Shoujing Telescope (or Large Sky Area Multi-Object Fiber Spectroscopy Telescope, LAMOST) at the Xinglong Station of the National Astronomical Observatories of the Chinese Academy of Sciences (NAOC) in Chengde City, Hebei Province. With an aperture of four meters and a five-degree field of view, it can track the spectra of 4000 targets over a large sky area (20 square degrees) for a long time, and can observe faint objects of magnitudes 20.5 within 1.5 h of exposure, making it a telescope with the highest spectral acquisition rate in the world.

In 1932, Jansky. K.G. used a radio antenna to detect radio radiation from the center of the Milky Way (in the direction of Sagittarius), opening a window for mankind to observe beyond the conventional optical band. After World War II, radio astronomy stood out. Radio telescopes are the main detection tool of radio astronomy, and the "four discoveries" of astronomy in the 1960s, including pulsars, quasars, cosmic microwave background (CMB) and organic molecules, are all related to it. The basic principles of a radio telescope are similar to that of an optical reflector telescope, which consists of the antenna (reflecting surface) and the receiving system. The antenna collects radio radiation from celestial bodies, and the receiving system processes, converts, records and displays these signals, as shown in Fig. 1.2.

To satisfy the observation requirements and improve the resolution and sensitivity, the antenna aperture should be made large, which is the giant radio telescope. To implement it, there are two approaches [1]. The first one is making a "fully steerable" rotating paraboloid, in which the parabolic antenna and the receiving system rotate together in two degrees of freedom in space, as shown in Fig. 1.2. Due to the effect of gravitational deformation, the maximum aperture of the parabolic antenna can only reach 100 m (the Effelsberg Radio Telescope). The second one is making the antenna

Fig. 1.1 LAMOST

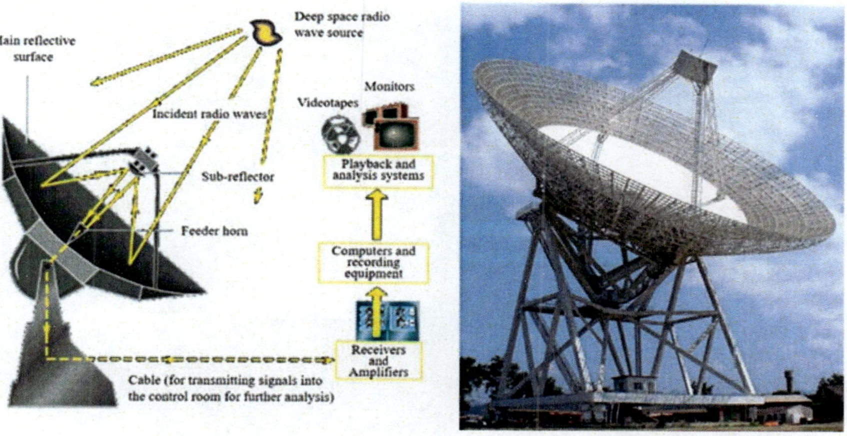

Fig. 1.2 Radio astronomical telescope (fully steerable)

a fixed spherical surface that installed in a recess, and adopting a feed support system to track the radio source converged by the antenna. Since the motion of the antenna is eliminated and the fixed support is employed to avoid the effect of gravitational deformation, the antenna aperture can be made very large. The 305 m spherical reflector radio telescope—Arecibo Radio Telescope was built in the 1960s in a natural sinkhole (Arecibo Valley) on the island of Puerto Rico in Central America (see Fig. 1.3). It has made great achievements for astronomical observations. However, due to the limitations of the technology at that time, its feed support system was bulky and inefficient, with a mass of up to 1000t. The receiver was installed under a curved cantilever and could move along the cantilever. The cantilever was suspended under a triangular frame and could rotate around the frame, which was fixed in the air with multiple sets of steel cables.

In 1993, at the International Radio Science Federation Congress in Kyoto, Japan, radio astronomers from more than 10 countries, including China, Australia, France and the United States, jointly initiated the construction of a new generation of Large radio Telescope (LT) with a receiving area of 1km^2, to develop a radio astronomical telescope with a larger receiving area, improve sensitivity and resolution, and realize the astronomical observation of longer distance. In 1994, Chinese astronomers put forward the concept and preliminary scheme for the construction of the Five-hundred Meter Aperture Spherical radio Telescope (FAST). In March 1999, the "Pre-research on Giant Radio Telescope FAST" was set up as a major project of the Knowledge Innovation Project of the Chinese Academy of Sciences.

In 2007, the FAST project was formally established as a national major science and technology infrastructure, and the site was selected in a karst depression in Pingtang county, Qiannan Buyei and Miao autonomous prefecture, Guizhou province, and was completed on September 25, 2016, after 22 years, as shown in Fig. 1.4 [2, 3]. *Science* reported the progress of FAST four times, and *Nature* listed the completion

Fig. 1.3 Arecibo radio telescope

of FAST as a significant science event that shaped 2016, believing that its scientific achievements in the future are highly desirable [4]. The FAST project, known as "China Sky Eye", is the world's largest single-dish and most sensitive radio telescope with independent intellectual property rights in China. Due to its high sensitivity, 133 high-quality pulsar candidates have been observed and 93 pulsars have been identified by October 2019, which not only achieved a zero breakthrough in pulsar identification in China, but also reached the international leading level in this field. Meanwhile, significant progress has been made in pulsar timing observation, which can improve the pulsar timing accuracy by at least one order of magnitude and is expected to realize the detection capability of primordial gravitational waves in the next five years. However, the theoretical and technical difficulties, as well as the project volume of FAST are rare in the world. This book will focus on the core theoretical and technical research of the FAST feed support system combined with the 40-meter scale model.

Fig. 1.4 The FAST project

1.2 Main Structure and Operating Principles of the FAST

The main structure of the FAST is shown in Fig. 1.5, consisting of the feed support system and the active reflector. The spherical active reflector is built based on the karst topography of Guizhou province, which not only effectively reduces the project volume, but also improves the drainage capacity. The active reflector is based on a cable net structure that is like a large "net pocket", on which reflector panels are installed (see Fig. 1.6a) and at each cable cross node (see Fig. 1.6b) a down-tied cable and an actuator are connected to the foundation pier (see Fig. 1.6c). By controlling the actuator to pull the node of the cable net, the parabolic fitting within 300 m aperture of the reflecting surface is realized, and the microwave radio signal from the target object is reflected and focused to a point on the receiving surface.

The operating principles of the FAST are shown in Fig. 1.7. Due to the rotation of the earth, taking the FAST as the observation point, target objects move from S_1 to S_2 with respect to the reflecting surface, and the corresponding area of the active reflector

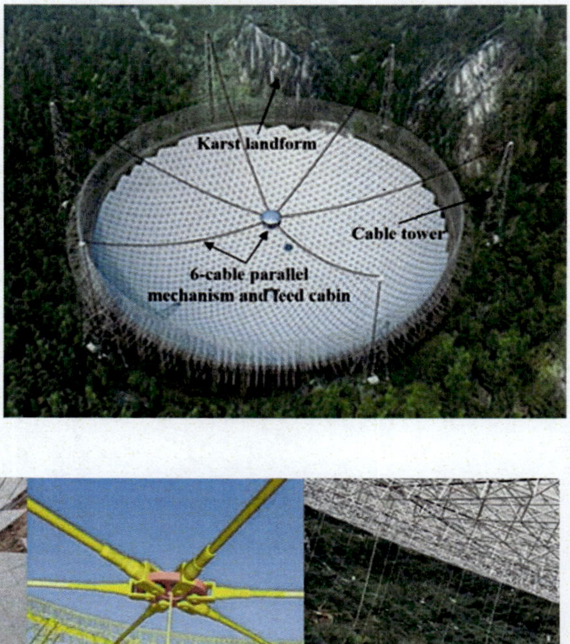

Fig. 1.5 Main structure of the FAST

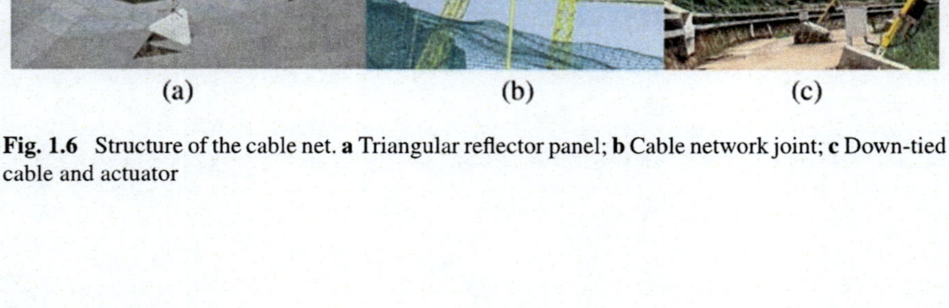

(a) (b) (c)

Fig. 1.6 Structure of the cable net. **a** Triangular reflector panel; **b** Cable network joint; **c** Down-tied cable and actuator

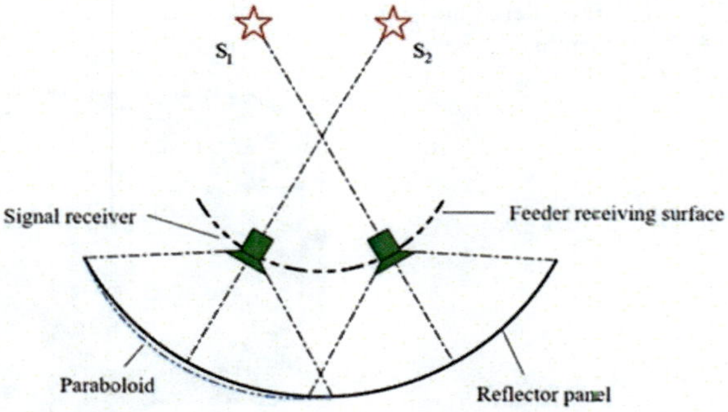

Fig. 1.7 Operating principles of the FAST

starts working in turn, completes parabolic fitting, and converges the microwave signal of the objects to the receiving surface, forming a trajectory of the focal point. The receiver needs to track the trajectory with certain accuracy to realize the signal acquisition. To ensure the observation effect, the position accuracy of the receiver is required to be 10 mm in root-mean-square (RMS) and the pointing accuracy is 0.2° in RMS (described by RPY angle). The large workspace and terminal high precision requirements impose demands on design and control for the feed support system that carries the receiver. Considering the construction cost restraints, the FAST feed support system adopts a three-level serial configuration, which is a macro–micro manipulator system composed of a long-span cable-driven parallel robot (CDPR), an A-B rotator and a rigid Stewart parallel robot.

The CDPR and the A-B rotator form a wide range of motion platform (flexibly support) to ensure the workspace requirements of the receiver; the fine-tuning platform (rigid Stewart parallel robot), is connected in series below the A-B rotator to improve the trajectory accuracy of the terminal receiver. The CDPR, also known as the first-level support system of the FAST feed support system, is composed of six steel cables, the cable tower and the drum mechanism (drive units), which is a complex flexible parallel robot. The span of the CDPR is 600 m, and each steel cable is controlled by a set of servo motors and drums to realize the motion of the feed cabin within 206 m in the air through the cooperative retraction and release of the steel cables. Due to the elastic deformation of the steel cables, the CDPR must be fully closed-loop controlled. Considering the large workspace, it needs to adopt the non-contact measurement equipment with kilometer distance, high accuracy and high sampling frequency.

The A-B rotator and the Stewart parallel robot in the feed cabin constitute the second-level fine-tuning system of the feed support system, as shown in Fig. 1.8. The A-B rotator can be driven by either the branch chain or the servo motor + reducer. The fine-tuning Stewart parallel robot is shown in Fig. 1.9, including the base, the moving platform and six branch chains. One end of each limb is connected to the moving platform by a spherical joint, and the other is connected to the base by a Hooke joint, driven by a set of servo motors and ball screws.

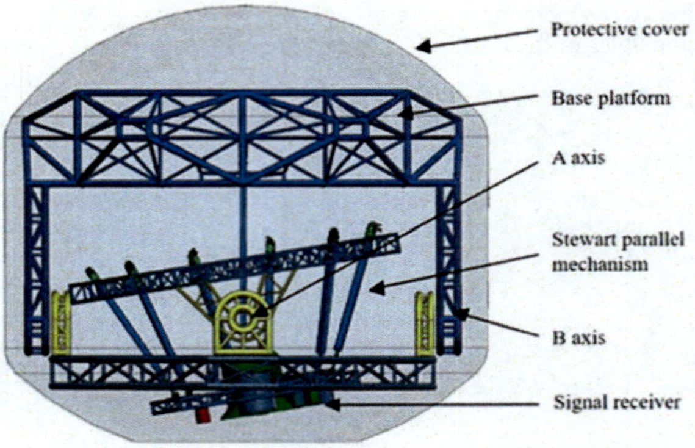

Fig. 1.8 3D model of the FAST feed cabin

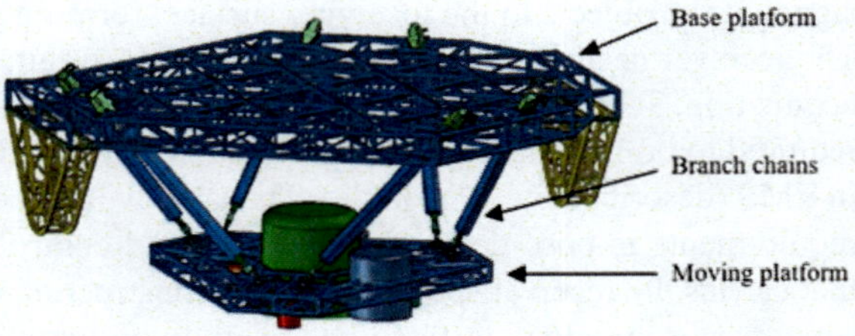

Fig. 1.9 3D model of the fine-tuning platform

The feed support system with rigid-flexible coupling properties is adopted, which is composed of the CDPR and rigid parallel robot. The mass is effectively reduced while dynamic rigidity and accuracy are ensured. In the final design, the mass of the FAST feed support system is only 30t, which is 3% of that of the feed support system of the Arecibo telescope. In addition, the active reflector and three-level feed support system effectively increases the observation range of FAST, enabling the maximum zenith angle to reach 40°.

In summary, the main innovations of the FAST are:

(1) Making full use of the unique karst topography. Building the reflector of the radio telescope based on a natural depression to reduce the amount of excavation. The permeability of karst topography can ensure the continuous infiltration of precipitation into the ground, which can effectively prevent the damage and corrosion on the telescope caused by water.
(2) Active reflector technology. Thousands of reflector panels on the cable net are employed to form a reflecting surface with controlled deformation. The reflector can change continuously with the motion of celestial bodies with respect to the earth, realizing the long-time continuous observation of specific objects and the observation of celestial bodies in different directions.
(3) A macro–micro manipulator system composed of the rigid and cable-driven parallel robots. The large workspace of the cable-driven parallel robot and the high stiffness of the rigid parallel robot are utilized to achieve high-precision

positioning and large observation angle of the receiver, while the mass and construction cost of the feed support system are greatly reduced.

As an important part of the FAST, the feed support system functions as the ciliary muscle of the human eye. The high sensitivity and detection capability of the FAST are achieved through the high precision motion of the feed support system. The following are brief overviews of the parallel robot and the cable-driven parallel robot.

1.3 Parallel Robot

Most common industrial robots adopt a serial mechanism, similar to the human arm, as shown in Fig. 1.10a. There is only one kinematic chain composed of joints and components between the base and the terminal of the serial robot, which is based on the open-loop mechanism, and all joints are active. Although the serial robot is characterized by a large workspace and high flexibility, it has the disadvantages of low stiffness, accuracy and load capacity. The parallel robot is proposed to overcome the drawbacks of the serial mechanism, and its main appearance is characterized by the multi-loop parallel configuration. The parallel robot generally consists of a base platform, a moving platform and multiple kinematic branches connecting the two platforms, and the moving platform has multiple degrees of freedom. The parallel robot is a closed-loop manipulator with a large number of passive joints, as shown in Fig. 1.10b.

As early as 1895, Cauchy proposed the "octahedral manipulator", which is the earliest documented parallel robot. Subsequently, James E. Gwinnett designed a large amusement equipment in 1928, which is considered to be the earliest spatial spherical parallel robot [4]. In 1935, Willard L.V. Pollard designed the first industrial

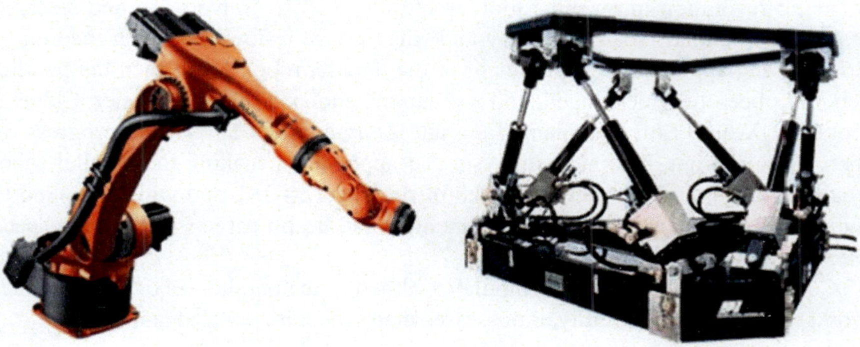

(a) Serial robot (b) Parallel robot

Fig. 1.10 Serial and parallel robots

parallel robot—a 5-degree-of-freedom spray-paint parallel robot [5]. Later, in 1947, Eric Gough, an automotive engineer, inadvertently applied the Hexapod parallel mechanism when designing the tire testing machine [6], but it did not attract attention at that time. In 1965, the scholar Stewart presented a paper on the International Conference on Mechanical Engineering, proposed a similar 6-degree-of-freedom parallel robot—the "Stewart platform", and sparked a heated discussion [7]. Today, the Stewart parallel robot (or Gough platform, as shown in Fig. 1.10b) is still one of the research focus in the field of the parallel robot. In general, before the 1970s, the design and application of the parallel robot were in the exploratory stage, without the complete theoretical basis and systematic analysis method.

The 1970s to the mid-1980s was a critical development period for the theory of the parallel robot. Hunt, an Australian professor of engineering, analyzed the configuration and kinematic properties of the parallel robot in detail in 1978, which laid a foundation for the theoretical research and industrial applications of the parallel robots [8]. Subsequently, a large number of international scholars such as Earl, McDowell, Fichter, Yang, Lee and Duffy [9–13], and Chinese scholars like Huang Zhen, Wang Jinsong and Huang Tian [14–19] have conducted in-depth and systematic researches on the parallel robot, having made great progress in the dimensional design, kinematics and control of the parallel robot. At the same time, a small number of practical parallel robots were invented and started to be applied in industrial production. However, due to the limitations of the manufacturing technology of compound joints and the development level of the electronic computer and servo control system, the parallel robot did not get a lot of practical applications at that time, and mostly was in the stage of theoretical research and prototype experiment.

Since the 1990s, with the popularization of powerful industrial computers and the maturity of servo control technology, the application research of the parallel robot has been developed rapidly. At the International Machine Tool Fair held in Chicago in 1994, Giddings & Lewis and Ingersoll introduced the VARIAX CNC machine tool and Hexapods machining center based on the Stewart parallel robot, which successfully introduced the parallel robot into the field of machine tool and triggered a "major revolution in machine tool structure" [20–23]. Universities and research institutes in China, Japan, Germany and Russia have followed up with theoretical research and application development of the parallel robot. Since then the parallel robot has been fully developed, and a group of scholars such as JP Meler, Clément Gosselin, Xinjun Liu, Xiaoqiang Tang and Zhufeng Shao have made progress on the optimal design, kinematic analysis and application, making the parallel robot gradually become an important branch of robotics [24–28], and widely applied in the fields of mechanical engineering, medical and health care as well as aerospace (see Fig. 1.11).

Compared with the serial manipulator, although the parallel robot has a smaller workspace and less flexibility, it possesses many incomparable advantages:

(1) The parallel robot adopts a closed-loop kinematic chain, thus it has a stable structure, improved stiffness and high load-to-weight ratio;

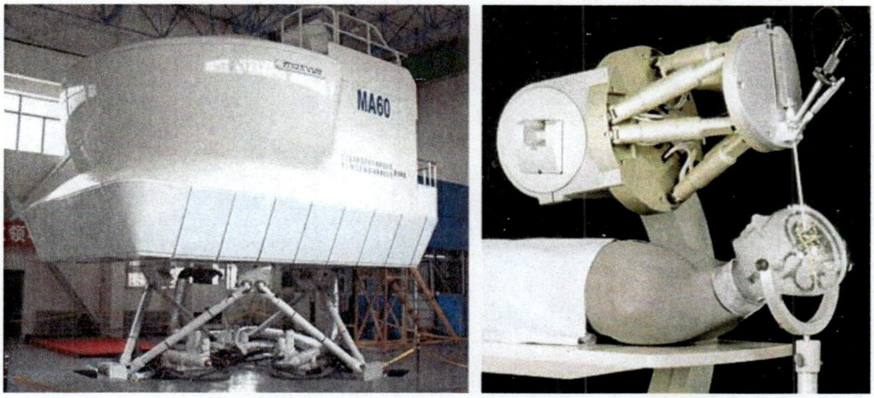

(a) MA60 flight simulator (b) Surgical parallel robot

Fig. 1.11 Examples of parallel robot application

(2) The driving parts of the parallel robot are arranged near the base, which can effectively reduce the mass of the moving parts and easily obtain better dynamic properties to achieve high speed motion with high acceleration;
(3) Due to the similar or even identical kinematic chains, it is easy to standardize the structure and carry out modular design and manufacture.

These advantages of the parallel robot make up for the drawbacks of the serial mechanism and broaden the application range of robots and automation. With the improvement of manufacturing and assembly technology and the emergence of high-precision joints and sensors, the properties of the parallel robot have become more and more obvious, showing broad application prospects in the high-precision, high-speed as well as heavy-load condition. Typical parallel robots that have been applied commercially include the Sprint Z3 spindle head and the high-speed parallel robots represented by the Delta robot.

The Sprint Z3 spindle head is shown in Fig. 1.12a. Its moving and fixed platforms are connected by three chains, each of which consists of a linear guideway (prismatic joint), a rotation joint and a spherical joint. The linear guideway is circumferentially uniformly distributed at an angle of 120°, and the servo motor drives the slicing plate on the guideway to move back and forth through a lead screw, and the sliding plates are connected to the spindle part (moving platform) by a rotational joint, a rod and a spherical joint. If three sliding plates move synchronously, the spindle part moves back and forth in the Z direction. If not, the spindle part swings within ±40° along the A or B coordinates. The conventional *A/C* and *A/B* spindle heads are poor in rigidity and low in acceleration, thus cannot meet the requirements of high removal rate machining of aircraft structural parts. The Sprint Z3 [29] spindle head overcomes these drawbacks and is characterized by good rigidity, high power (80 kW) and large acceleration of 1 g. Based on the Z3 spindle head, a serial manipulator is added to realize the translation motions along X and Y axes, forming the ECOSPEED [30]

Fig. 1.12 Sprint Z3 parallel spindle head and ECOSPEED machining center

(a) Z3 spindle head

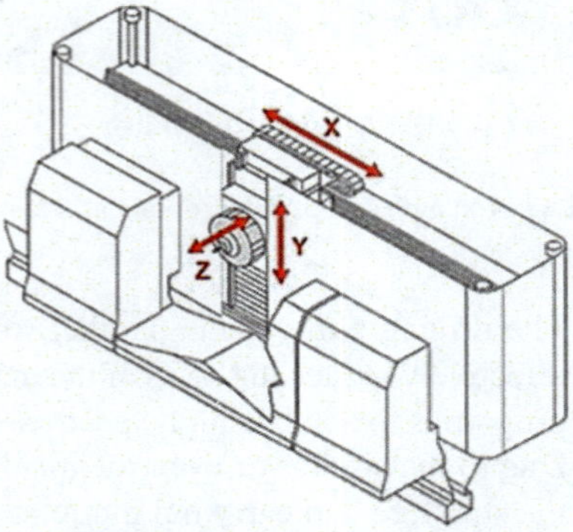

(b) ECOSPEED machining center

machining center (see Fig. 1.12b), which has been widely applied in the field of aerospace structural parts machining with overwhelming advantages of rigid and efficiency.

The Delta parallel robot [31] is shown in Fig. 1.13a, and its moving platform and base are connected with three identical branches, each of which consists of an active pendulum and a passive parallelogram structure. One end of the pendulum is connected to the base through an active rotational joint, and the other end is connected to the parallelogram structure through a rotational joint. The parallelogram structure is connected to the moving platform through a rotational joint. The parallelogram structure consists of four rods connected by Hooke joints, which restrains the rotation of the moving platform. The three branch chains are circumferentially uniformly distributed, and the servo motor drives the pendulum up and down through the reducer. If the three pendulums move synchronously, the moving platform moves up and down in a straight line. If not, the moving platform can be driven to do spatial motion. FAUNC (Japan) and ABB (Sweden) (see Fig. 1.13b) have developed a series of high-speed sorting robots based on the Delta mechanism, which can achieve termianl acceleration of 10 g and repeatable positioning accuracy of ± 0.1 mm. By adding serial rotation manipulator to the terminal, the assembly function of 6 DOFs can be realized, such as FAUNC M-3IA/6A in Fig. 1.13c.

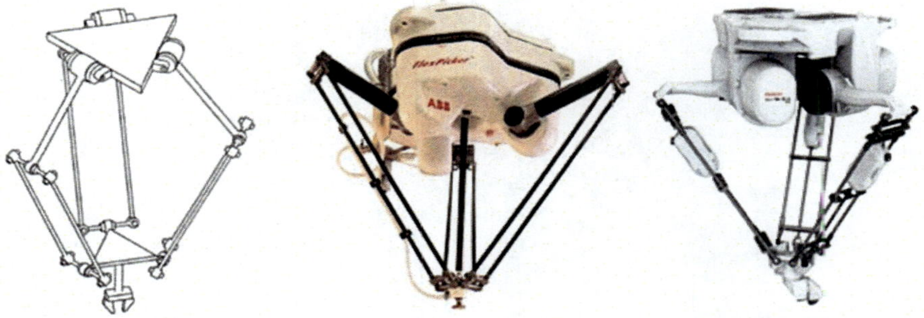

(a) Delta mechanism (b) ABB robot (c) FAUNC robot

Fig. 1.13 Delta mechanism and high-speed robots

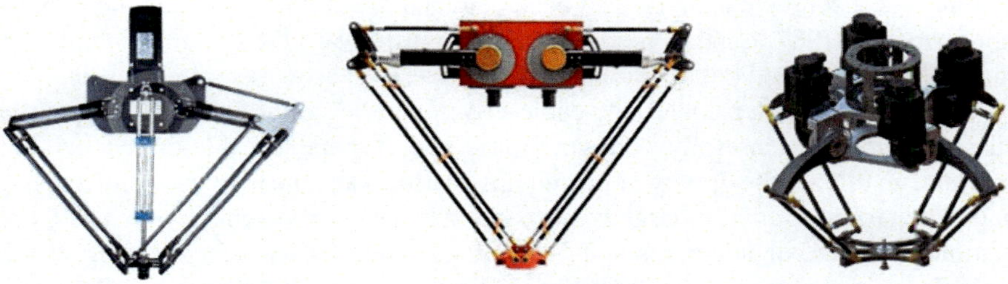

(a) D2-1000 robot (b) Diamond robot (c) X4 robot

Fig. 1.14 Rigid parallel sorting robots

Researchers have developed several high-speed parallel robots with the active pendulum and parallelogram structure for such as D2-1000, Diamond, Unigrabber 2, H4, Par4, and X4 [32–36] to further improve speed and efficiency. Some of them are shown in Fig. 1.14.

1.4 Cable-Driven Parallel Robot

The research and application of the cable-driven parallel robot, which is a special flexible parallel mechanism, originated in the late 1980s. By using cables instead of rigid chains, the CDPR not only possesses the advantages of high load ability of the parallel mechanism, but also is characterized by low weight and inertia, large workspace and simple structure [37]. Thanks to these advantages, the CDPR has attracted much attention. However, the property that the cable can only be pulled imposes restrictions on the configuration design of the CDPR, while keep cable tension is a prerequisite for the operation of the CDPR [38].

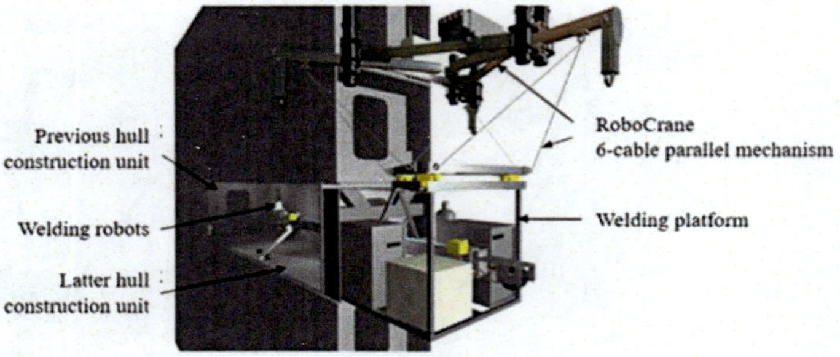

Fig. 1.15 RoboCrane robot

In the initial exploration stage, scholars began to study the basic theory of the CDPR, and theoretical analysis had been conducted, but the systematic theory was not formed In 1987, Landsberge [39, 40] first put forward the design principles of the CDPR and carried out positioning, stiffness as well as workspace analysis.

To ensure cables tensioned, the cable-driven parallel robot generally adopts the suspended or the tensegrity configurations. The suspended cable-driven parallel mechanism utilizes the gravity of the moving platform and the load to tension cables, and is generally applied in large workspace or heavy-load machinery. In 1989, the National Institute of Standards and Technology (NIST) started the research of the RoboCrane project [41, 42]. As shown in Fig. 1.15, the cable-driven parallel robot used in the RoboCrane project is an inverted Stewart platform, which adopts six cables, and the gravity is utilized as an invisible cable to achieve 6 DOFs motion. In the past 20 years, the cable-driven parallel robots have been widely applied in lifting, robot calibration, machining, assembly, submarine salvage, radio telescope and heavy lifting [43, 44]. As shown in Fig. 1.16, Skycam, an American company, has developed a cable-driven parallel robot for camera shooting [45], which realizes the motion of the camera in a large workspace with high speed and acceleration by four cables.

The tensegrity CDPR arranges cables on both sides of the moving platform to make cables pull against each other, keeping cables tensioned. It is generally a redundant drive robot, the number of the driving cables and motors is larger than that of DOFs of the moving platform. Through cables pulling against each other, the CDPR achieves high stiffness and high acceleration. Typical representatives include the CableEndy robot [46] that can achieve a speed of 6 m/s and an acceleration of 10 g, and the FALCON robot [47, 48] that achieves motion control of the end effector through seven cables, with a speed of 13 m/s and a maximum acceleration of 6 g. Other examples are the CableRobot Simulator [49] for the motion simulation, SEGESTA [50] and IPAnema [51] for the pick-and-place operation, the cable-driven high-speed robot TBot [52] and the wind tunnel experimental system [53], as shown in Fig. 1.17.

Fig. 1.16 Skycam robot

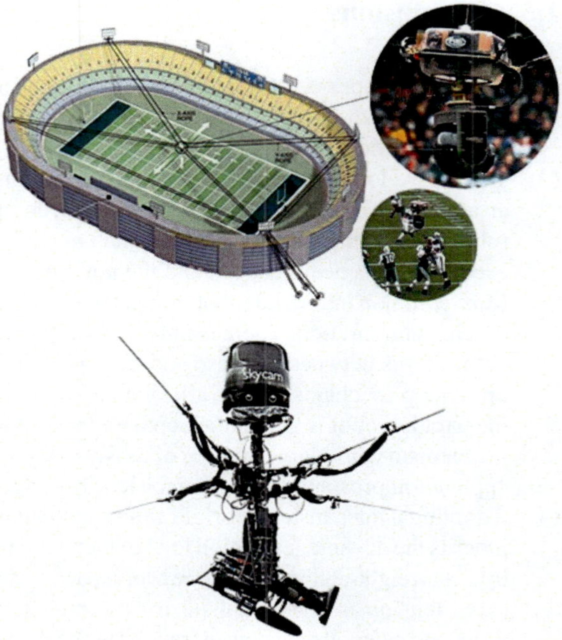

(a) CableRobot-Simulator

(b) CableEndy

(c) SEGESTA

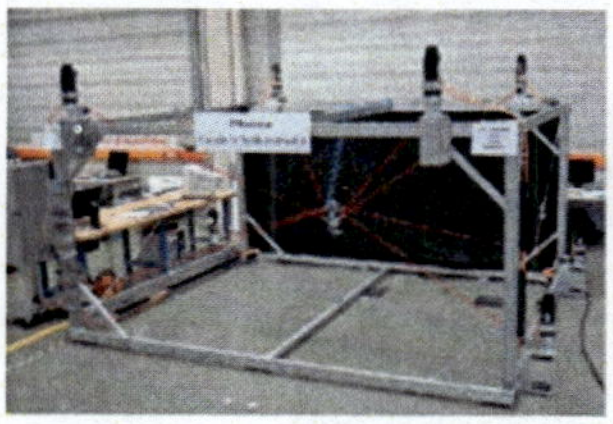

(d) IP Anema-I

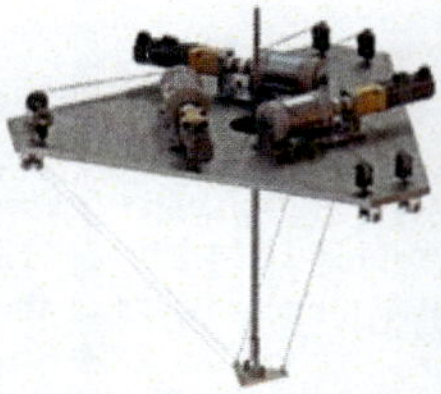

(e) High-speed cable-driven parallel sorting robot

(f) Wind tunnel experimental system

Fig. 1.17 Push–pull cable-driven parallel robots

1.5 Conclusions

(1) Radio telescopes are the main detection tool of radio astronomy. The FAST project, known as the "China Sky Eye", is the largest single-dish and most sensitive radio telescope in the world now.
(2) To achieve long-time continuous high precision observation, the FAST adopts an active reflector and a macro micro system composed of cable-driven and rigid robots. By taking full advantage of the unique karst topography, the amount of excavation has been reduced and the ponding issue has been avoided.
(3) Most common industrial robots adopt the serial mechanism, similar to the human arm in configuration. There is only one kinematic chain composed of joints and components between the base and the terminal of the serial robot, which is the open-loop mechanism with all joints active. The main appearance feature of the parallel robot is the multi-loop parallel configuration, and it is a closed-loop mechanism with a large number of passive joints. The parallel robot shows broad application prospects in high precision, high speed and heavy-load occasions.
(4) Adopting cables instead of rigid limbs, the cable-driven parallel robot not only inherits the advantages of high load of the parallel robot, but also is characterized by low weight and inertia, large workspace and simple structure. Ensuring the cable tension is a prerequisite for the operation of the cable-driven parallel robot and an important consideration for the design. The suspended cable-driven parallel robot utilizes the gravity of the moving platform and the load to keep cables tensioned, and is generally applied in the large workspace or heavy-load machinery.
(5) The tensegrity CDPR makes cables pull against each other to tension cable, which possesses high stiffness and high acceleration.

References

1. Marr JM, Snell RL, Kurtz SE. Fundamentals of radio astronomy: observational methods. CRC Press; 2015. p. 21–4.
2. Tang XQ, Shao ZF. Trajectory generation and tracking control of a multi-level hybrid support manipulator in FAST. Mechatronics. 2013;23(8):1113–22.
3. Jiang P, Yue Y, et al. Commissioning progress of the FAST. Sci China (Phys Mech Astron) 2019;62(05):7–28.
4. Gwinnett JE. Amusement device, January, 20, 1931. US Patent No. 1,789,680.
5. Roselund, HA. Means for moving spray guns or other devices through predetermined paths. March, 14, 1944. US Patent No. 2,344,108.
6. Abdellatif H, Heimann B. Advanced model-based control of a 6-DOF hexapod robot: a case study. IEEE/ASME Trans Mechatron. 2010;15(2):269–79.
7. Shao ZF, Tang X, Chen X, Wang LP. Optimum design of 3–3 Stewart platform considering inertia property. In: Advances in mechanical engineering, vol. 5, p. 249121.
8. Hunt KH. Kinematic geometry of mechanisms. Britain: Oxford University Press; 1978.
9. Earl CF, Rooney J. Some kinematics structures for robot manipulator designs. J Mech Transmissions Autom Des. 1983;105(1):15–22.

10. Fichter EF, McDowell ED. A novel design for a robot arm. In Proceedings of international computer technical conference. San Francisco; 1980. p. 250–55.
11. Lee KM, Shah DK. Kinematic analysis of a three-degree-of-freedom in-parallel actuated manipulator. IEEE J Robot Autom. 1988;4(3):354–60.
12. Yang J, Geng ZJ. Closed form forward kinematics solution to a class of hexapod robots. IEEE Trans Robot Autom. 1998;29(1):1–9.
13. Duffy J. Statics and kinematics with applications to robotics. New York: Cambridge University Press; 1996.
14. Huang Z, Zhao YS, Zhao TS. Advanced spatial mechanics. Beijing: Higher Education Press; 2006.
15. Huang Z, Cao Y. Property identification of the singularity loci of a class of Gough-Stewart manipulators. Int J Robot Res. 2005;24(8):675–85.
16. Huang Z, Li SH, Zuo RG. Feasible instantaneous motions and kinematic characteristics of a special 3-DOF 3-UPU parallel manipulator. Mech Mach Theory. 2004;39(9):957–70.
17. Wang J, Guan L, Wang L, Li T. Research on parallel robot configuration creative design. Chin J Mech Eng. (in Chin) 2004;40(11):7–12.
18. Huang T, Chetwynd DG, Mei JP. Tolerance design of a 2-DOF over-constrained translational parallel robot. IEEE Trans Rob. 2006;21(1):167–72.
19. Wang JS, Huang T. Parallel machine tool–a forthcoming opportunity and challenge to the machine tool industry. China Mech Eng. 1999;10(10):1103–7.
20. Yang JX, Yu DW, Wang LP, et al. Current researches and development of parallel machine tools. Mach Build Autom. 2002;3:10–141.
21. Yan CX. VARIAX, a major innovation in machine tool construction, was launched. World Manuf Technol Equip Mark. 1995;1:16–7.
22. Zhang S, Heisel U. Parallel motion machine tools. Beijing: Machinery Industry Press; 2003.
23. Li JQ, Ding HS, Fu T, et al. The history, current situation and future of parallel kinematic machine. Mach Tool Hydraulics. 2003;3:3–8.
24. Kong X, Gosselin C. Type synthesis of parallel mechanisms. In: Springer tracts in advanced robotics, vol. 33; 2007.
25. Liu XJ, Wang J, Gao F, et al. Mechanism design of a simplified 6-DOF 6-RUS parallel manipulator. Robotica 20(01):81–91.
26. Shao ZF, Mo J, Tang XQ, Wang LP. Transmission index research of parallel manipulators based on matrix orthogonal degree. Chinese J Mech Eng. 30(6):1396–405.
27. Wang LP, Zhang Z, Shao ZF, Tang X. Analysis and optimization of a novel planar 5R parallel mechanism with variable actuation modes. Robot Comput Integr Manuf. 56:178–90.
28. Zhang Z, Wang LP, Shao ZF. Improving the kinematic performance of a planar 3-RRR parallel manipulator through actuation mode conversion. Mech Mach Theory 130:86–108.
29. STARRAG, Ecospeed and axis machining center [EB/OL]. (2017-03-10) [2023-06-08]. https://www.starrag.com/zh-cn/machine/ecospeed-f-1xxx/131.
30. Machine Tool Network, Switzerland starrag 5 axis machining center ECOSPEED F HT 2 [EB/OL]. (2018-09-04) [2023-06-08]. http://www.158jixie.com/news-detail/279/279921.html.
31. Tang BG, Huang T. Kinematic calibration of delta robot. Chin J Mech Eng. (8):59–64.
32. Mei JP, Wu ML, Wang PF. Hardware design of control system and interface development for diamond600 robot. J Mach Des. 2004;21(z1):61–3. https://doi.org/10.3969/j.issn.1001-2354.2004.z1.028.
33. Pierrot F, Marquet F, Company O, et al. H4 parallel robot: modeling, design and preliminary experiments. In: Proceedings 2001 ICRA. IEEE international conference on robotics and automation. IEEE; 2001.
34. Nabat V, Rodriguez MDO, Company O, et al. Par4: Very high speed parallel robot for pick-and-place; 2005.
35. Mo J, Shao ZF, Guan L, Xie F, Tang X. Dynamic performance analysis of the X4 high-speed pick-and-place parallel robot. Robot Comput-Integr Manuf. 46:48–57.
36. Staicu S, Shao ZF, Zhang Z, Tang X, Wang L. Kinematic analysis of the X4 translational–rotational parallel robot. Int J Adv Robot Syst. 15(5):1729881418803849.

37. Zheng YQ, Liu XW. Research survey and development tendency of wire-driven parallel manipulators. China Mech Eng. 2003;14(9):808–10.
38. Barrette G, Gosselin CM. Determination of the dynamic workspace of cable-driven planar parallel mechanisms. J Mech Des. 2005;127(2):242.
39. Landsberger SE. Design and construction of a cable-controller parallel link manipulator. Cambridge, MA, USA: Massachusetts Institute of Technology; 1984. p. 81–8.
40. Landsberger SE, Sheridan TB. A minimal linkage: the tension-compression parallel link manipulator. In: Takampri T, Tsuchiya K, editors. Robotics, mechatronics and manufacturing system. New York, USA: Elsevier; 1993. p. 81–8.
41. Albus JS, Bostelman RV, Dagalakis NG. The NIST Robocrane. J Robot Syst. 1993;10(5):709–24.
42. Albus J, Bostelman R, Dagalakis N. The NIST SPIDER, a robotcrane. J Res Nat Inst Stand Technol. 1992;97(3):373–85.
43. Bostelman, R V, Jacoff A, Proctor FM, et al. Cable-based reconfigurable machines for large scale manufacturing. In: Proceeding of 2000 Japan-USA symposium on flexible automation-international conference on new technological innovation for the 21st century, Ann Arbor, MI; 2000.
44. Pusey J. Design and workspace analysis of a 6–6 cable suspended parallel robot. Mech Mach Theory. 2004;139(7):761–78.
45. Tanaka M, Seguchi Y, Shimada S. Kineto-statics of skycam-type wire transport system. In: Proceedings of USA-Japan Symposium on Flexible Automation, Crossing Bridges: Advances in Flexible Automation and Robotics Minneapolis, Minnesota, ASME; 1988. p. 689–94.
46. Rajnoha A. CableEndy juggler—cable-driven parallel robot. https://www.youtube.com/watch?v=7HNAL8ZKdyM&t=13s
47. Kawamura S, Choe W, Tanaka S, et al. Development of an ultrahigh speed robot FALCON using wire drive system. In: IEEE international conference on robotics and automation. IEEE; 1993.
48. Gu W, Eisenhauer G, Kraemer E, et al. Falcon: on-line monitoring and steering of large-scale parallel programs; 1995.
49. Miermeister P, Lachele M, Boss R, et al. The CableRobot simulator large scale motion platform based on cable robot technology. In: 2016 IEEE/RSJ International Conference on Intelligent Robots and Systems (IROS). IEEE; 2016.
50. Mikelsons L, Bruckmann T, Hiller M, Schramm D. A real-time capable force calculation algorithm for redundant tendon-based parallel manipulators. In Proceedings of the IEEE international conference on robotics and automation (ICRA '08); May 2008. p. 3869–74.
51. Pott A, Mutherich H, Kraus W, et al. Cable-driven parallel robots for industrial applications: the IPAnema system family. In: 2013 44th International Symposium on Robotics (ISR). IEEE; 2013.
52. Zhang Z, Shao ZF, Wang L. Optimization and implementation of a high-speed 3-DOFs translational cable-driven parallel robot. Mech Mach Theory. 145;103693.
53. Zheng YQ, Lin Q, Liu XW, et al. On wire-driven parallel suspension systems for static and dynamic derivatives of aircraft in low-speed wind tunnels. J Aeronaut. 2009;8:195–200.

Chapter 2
Modeling Methods for the Long-Span Cable-Driven Parallel Robot

The FAST feed support system includes the first-level support system and the second-level fine-tuning system. The former (first-level feed support system) is a suspended cable-driven parallel robot (CDPR) driven by six steel cables with a span of 600 m, which is a typical long-span CDPR, and the weight and elastic deformation of cables affect the precise kinematic modeling of the mechanism. Therefore, when conducting kinematic modeling of the long-span CDPR, the effects of the elasticity and weight of cables need to be taken into account. The kinematic and static modeling of the long-span CDPRs are interrelated and need to be considered together. This chapter will focus on the integrated kinematic and static modeling methods for the long-span CDPR.

The present modeling of the long-span CDPR is usually based on the single-cable model, which can be divided into three types, namely, the precise catenary model, the parabolic model and the linear model. The precise catenary model introduces the elastic deformation, but the modeling equations are highly non-linear, require iterative computations, and complex to solve. It is difficult to use for real-time control of the CDPR. The simplified catenary model, i.e., the parabolic and the linear models are convenient to solve with fast deduction speed, but their precision is limited.

In this chapter, a brief analysis of the current researches on the long-span CDPR is firstly presented; Sect. 2.2 describes three modeling methods of single cable, and the modeling method of the long-span CDPR is deduced; In Sect. 2.3, the modeling error of the long-span CDPR based on the simplified catenary model is analyzed, and an error compensation method adopting higher order polynomials is proposed; Sect. 2.4, taking the FAST first-level feed support system as an example, modeling, error analysis and compensation are completed, and the error compensation polynomial is obtained to meet the precision requirement.

This chapter will discuss the integrated kinematic and static modeling of the long-span CDPR, introduce three single-cable models considering the self-weight and elastic deformation, complete the modeling of the long-span CDPR, focus on

© Tsinghua University Press 2024

X. Tang et al., *Research and Application of Cable-Driven and Rigid Parallel Robots*,
https://doi.org/10.1007/978-981-99-7452-8_2

the analysis of the modeling error of the simplified model, study the compensation method of the modeling error, and provide the theoretical foundation for the application of the simplified model in real-time control.

Main contents of this chapter are:

(1) Overview of the long-span CDPR research;
(2) Single-cable models and the long-span CDPR modeling;
(3) Model simplification, error analysis and compensation;
(4) Modeling of FAST first-level feed support system.

2.1 Overview of the Long-Span CDPR Research

Due to the advantages of large workspace, the long-span CDPR has attracted more and more attention. Theoretical and experimental studies have made significant progress, including statics modeling, kinematics, dynamics, stiffness and vibration analysis, etc. [1]. It is important to note that the cable length of the CDPR is directly related to the cable tension, and its static modeling is correlated with the kinematics. Zheng Yaqing mentioned in her Ph.D. thesis [6] that the self-weight of the cable cannot be neglected once its length exceeds 10 m. This type of CDPR is also called the long-span one.

The mass and elastic deformation of the cable affecting the precise modeling of the CDPR. Therefore, more and more scholars take the two factors into account in modeling the long-span CDPR and usually introduce cable modeling methods of the construction and bridge field. Prem [7] introduced the cable mass in the modeling of the single cable, presented an precise modeling method based on the catenary model, and derived parabolic and linear modeling equations; Irivne [8] added elastic deformation to the modeling of the single cable and put forward a precise modeling method based on the catenary model; Kozak [9] pushed forward the idea of introducing elastic deformation into the modeling of the CDPR and adopted Irivne's catenary modeling method to improve modeling precision, establishing the static equations of the fully-constrained CDPR.

The long-span CDPR has features of large workspace, light weight, and low cost, and has brought it into the view of astronomers. As shown in Fig. 2.1, a giant radio telescope was proposed and studied in 1998 at McGill University in Canada, which adopted an eight-cable CDPR to control the attitude of the feed cabin [6]. This telescope has a flat reflecting surface with the aperture of 200 m, and a feed cabin mounted on a balloon. Multiple cables are used to achieve the motion and attitude adjustment of the feed cabin where the receivers are installed. Taghirad [12] studied the kinematics, singularity, dynamics and trajectory planning of this CDPR. Kozak [9] introduced the catenary modeling method of cables into the kinematic modeling of the long-span cable-driven parallel mechanism to improve the modeling precision and studied its stiffness.

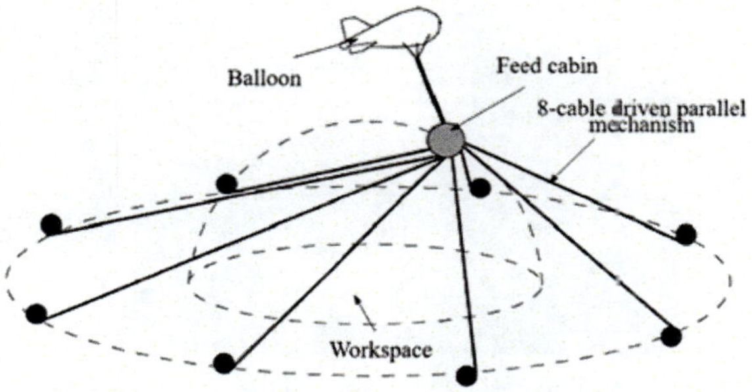

Fig. 2.1 The long-span cable-driven parallel mechanism of a giant radio telescope

Chinese scholars have conducted a lot of research on the feed support system of FAST, and different schemes containing theoretical studies and experimental validations have been proposed.

In terms of theoretical studies, Duan [14] introduced K. Prem's single-cable catenary modeling into the modeling of the cable-driven parallel mechanism, deduced the static equivalent equations. Qu [16] analyzed the effect of cable elastic deformation on the modeling error of the long-span CDPR, and presented error compensation equations on elastic deformation. Li [17] conducted research on the statics of the cable-driven parallel robot of FAST from the finite element perspective.

For the long-span CDPR in FAST, it is difficult to build an experimental model of the same size, therefore the experimental validation of FAST employs small-scale models. Lu [18] from Tsinghua University studied the early conceptual model of FAST, completed the kinematic and static analysis of the four-cable CDPR, and carried out experimental studies on the established scale mode (Fig. 2.2a). Wei [20] from Xi'an University of Electronic Science and Technology investigated the modeling, statics and stiffness of the long-span six-cable CDPR (Fig. 2.2b) and proposed an algorithm of vibration control.

In this chapter, the common problems in the static modeling of the long-span CDPR are analyzed, the effect of cable elastic deformation and weight on the model are discussed, and formula derivation and solving process are presented. Most importantly, this chapter discusses on a balance solution between deduction speed and precision through the compensation for modeling errors.

2.2 Single-Cable Models and the Long-Span CDPR Modeling;

The single-cable model can be divided into three types: the precise catenary model, the parabolic model and the linear model. This section derives and improves the three models, taking the cable elastic deformation into account.

(a) The early four-cable CDPR for the FAST first-level support system

(b) The six-cable CDPR

Fig. 2.2 Scale models of the FAST feed support system

2.2.1 *Single-Cable Catenary Model*

The precise catenary model of the single cable is shown in Fig. 2.3. The coordinate BXZ is established at the fixed point B of the cable, and the following parameters are defined when building the precise catenary model of the single cable:

l_0 is the original length of the cable, and is the cable length without elastic deformation. Δl is the amount of cable elastic deformation. T is the cable tension at the fixed end. ρ is the linear density of the cable without elastic deformation. E is the elastic modulus; A_0 is the cross-sectional area without elastic deformation. V is the vertical component of the cable tension T; H is the horizontal component of the cable tension T; L is the horizontal span of the single-cable model; and h is the height difference of the cable between the two connected points (A and B).

As shown in Fig. 2.3, the precise catenary equation of the single cable is established based on the above symbols and coordinates. Firstly, the single cable needs to satisfy the following equivalent conditions:

$$\sum x = 0, H + dH - H = 0 \tag{2.1}$$

$$\sum y = 0, H(dz/dx) + d(Hdz/dx) + \rho g dl_0 - H(dz/dx) = 0 \tag{2.2}$$

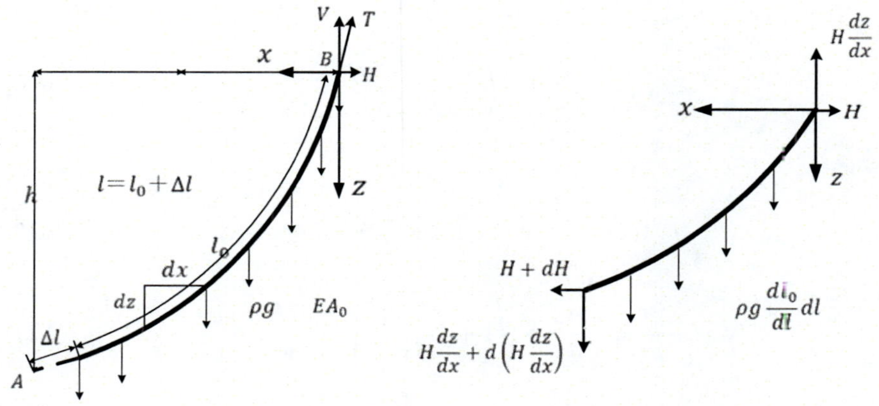

Fig. 2.3 Precise single-cable catenary model

where,

$$\frac{dl}{dl_0} = \frac{T}{EA_0} + 1 \tag{2.3}$$

$$T = H\sqrt{1 + \left(\frac{dz}{dx}\right)^2} \tag{2.4}$$

From $dl = dx\sqrt{1 + (dz/dx)^2}$, Eq. (2.2) can be expressed as:

$$d\left(H\frac{dz}{dx}\right) + \rho g \frac{EA_0}{T + EA_0} dl = 0 \tag{2.5}$$

$$\frac{d^2z}{dx^2} + \frac{\rho g EA_0}{H} \frac{\sqrt{1 + \left(\frac{dz}{dx}\right)^2}}{H\sqrt{1 + \left(\frac{dz}{dx}\right)^2} + EA_0} = 0 \tag{2.6}$$

By replacing dz/dx with p, Eq. (2.6) can be written as:

$$\frac{dp}{dx} + \frac{\rho g EA_0}{H} \frac{\sqrt{1 + p^2}}{H\sqrt{1 + p^2} + EA_0} = 0 \tag{2.7}$$

Then,

$$\frac{dx}{dp} = -\frac{H^2}{\rho g EA_0} - \frac{H}{\rho g\sqrt{1 + p^2}} \tag{2.8}$$

Therefore,

$$x = -\frac{H}{\rho g} sh^{-1}\left(\frac{dz}{dx}\right) - \frac{H^2}{\rho g E A_0}\frac{dz}{dx} + c \tag{2.9}$$

where,

$$sh^{-1}(x) = \ln\left(x + \sqrt{1 + x^2}\right), x \in (-\infty, +\infty) \tag{2.10}$$

$$x = -\frac{H}{\rho g} \ln\left(\frac{dz}{dx} + \sqrt{1 + \left(\frac{dz}{dx}\right)^2}\right) - \frac{H^2}{\rho g E A_0}\frac{dz}{dx} + c \tag{2.11}$$

According to the figure, the boundary conditions of the cable are:

$$x = 0, z = 0;$$

$$x = L, z = h;$$

The length of the cable can be expressed as $l = l_0 + \Delta l$, where

$$l = \int_0^L \sqrt{1 + \left(\frac{dz}{dx}\right)^2} dx \tag{2.12}$$

$$l_0 = \int_l \frac{1}{\frac{T}{EA_0} + 1} dl \tag{2.13}$$

The solution method can be represented by a flow chart, as shown in Fig. 2.4.

2.2.2 *Single-Cable Parabola Model*

When the sag and span of the cable are relatively small, the precise catenary equation of the cable will be approximated as a parabolic equation. If the modeling precision requirement is satisfied, the parabolic equation approximation method can be adopted to speed up the deduction.

The single-cable parabolic model is shown in Fig. 2.5. Considering the cable weight and elastic deformation and, the single-cable equivalent equation can be derived from Eq. (2.6) as:

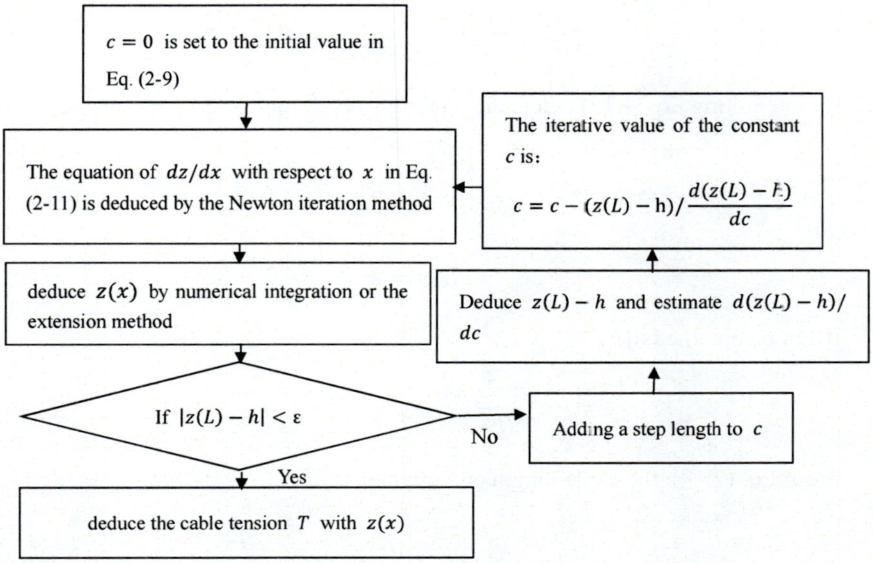

Fig. 2.4 Flow chart of the precise catenary model solution

$$H\frac{d^2z}{dx^2} + \rho g \frac{EA_0\sqrt{1+\left(\frac{dz}{dx}\right)^2}}{H\sqrt{1+\left(\frac{dz}{dx}\right)^2} + EA_0} = 0 \tag{2.14}$$

Since the sag and span are relatively small, by neglecting dz/dx [1], Eq. (2.14) can be written as:

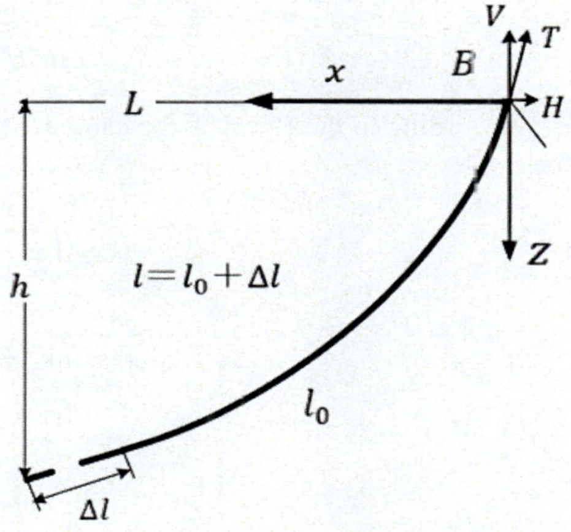

Fig. 2.5 Single-cable parabolic model

$$\frac{d^2z}{dx^2}+\frac{\rho g E A_0}{H}\frac{1}{H+EA_0}=0 \tag{2.15}$$

By integrating Eq. (2.15) twice, the boundary conditions of the cable are deduced as:

$$x=0, z=0;$$

$$x=L, z=h;$$

It can be obtained that:

$$z=\frac{\rho g E A_0 x}{2H(H+EA_0)}(L-x)+\frac{h}{L}x \tag{2.16}$$

From Eq. (2.12), the cable length equation is:

$$l=\int_0^L\sqrt{1+\left(\frac{dz}{dx}\right)^2}dx \tag{2.17}$$

Substituting z of Eq. (2.16) into Eq. (2.17), the Taylor expansion and integral of $\sqrt{1+\left(\frac{dz}{dx}\right)^2}$ gives the equation of the cable length as:

$$l=L\left(1+\frac{(c_0L)^2}{6}+\frac{h^2}{2L^2}\right) \tag{2.18}$$

where,

$$c_0=\frac{\rho g E A_0}{2H(H+EA_0)}$$

According to the Ref. [1], the cable tension T and its vertical component V can be expressed as:

$$V=\frac{\rho g l}{2}+H\frac{h}{L} \tag{2.19}$$

$$T=\left(V^2+H^2\right)^{\frac{1}{2}} \tag{2.20}$$

2.2.3 Single-Cable Linear Model

The single-cable linear model is relatively simple, as shown in Fig. 2.6. Generally, a long-span cable is hard to keep straight under the effect of self-weight and elastic deformation. However, if the tension at both sides of the cable is much larger than its self-weight, the single-cable model can be simplified to a linear. If the modeling precision is acceptable, the linear equation can be adopted to approximate the catenary to speed up the deduction.

According to Fig. 2.6, the linear equation of the single cable based on elastic deformation can be derived as:

$$l = \left(h^2 + L^2\right)^{\frac{1}{2}} \tag{2.21}$$

$$T = \left(V^2 + H^2\right)^{\frac{1}{2}} \tag{2.22}$$

$$L = l\frac{H}{T} \tag{2.23}$$

$$h = l\frac{V}{T} \tag{2.24}$$

$$\Delta l = \frac{Tl}{EA_0} \tag{2.25}$$

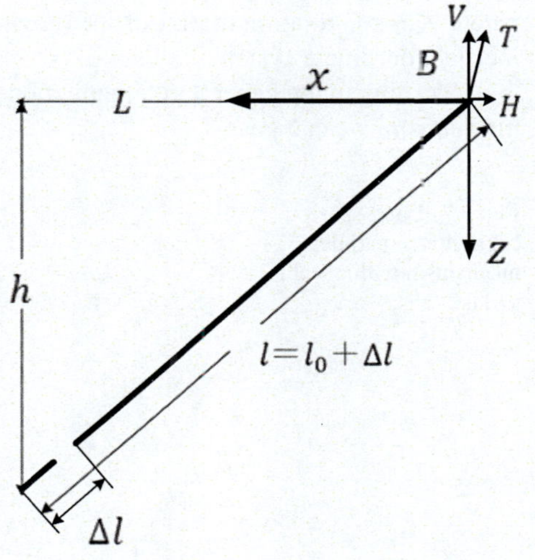

Fig. 2.6 Single-cable linear model

2.2.4 *Modeling Methods for the Long-Span CDPR*

Ideally, the kinematic and static modeling of the CDPR is similar to that of the rigid parallel robots. However, if the self-weight and elastic deformation of the cable are considered, the coupling of the cable tension and length causes the complexity of the kinematic modeling of the long-span CDPR, and the kinematic and static modeling are closely related. In this section, the integrated modeling and solution methods of the long-span CDPR on the basis of three single-cable models are discussed.

Figure 2.7 shows a long-span cable-driven parallel mechanism with n degrees of freedom (DOFs). Two coordinate systems are established: one is the inertial coordinate system $\Re : O - XYZ$, the origin of which is located at the center of the base platform of the cable-driven parallel mechanism with the Z-axis upward; the other is the moving coordinate system $\Re' :' -X'Y'Z'$, the origin of which is located at the center of the moving platform with the Z'-axis upward along the normal of the moving platform. $B_i (i = 1, 2, \ldots, m)$ is the cable exit point of the base platform and $A_i (i = 1, 2, \ldots, m)$ is the cable anchor point of the moving platform.

The symbols involved in the modeling process of the cable-driven parallel mechanism are defined as follows: $\boldsymbol{O}^{'\Re}$ is the position vector of the origin of the moving coordinate system O' in the inertial coordinate system; $\boldsymbol{B}_i^{\Re}$ is the position vector of B_i in the inertial coordinate system; $\boldsymbol{A}_i^{\Re}$ is the position vector of A_i in the inertial coordinate system; $\boldsymbol{A}_i^{\Re'}$ is the position vector of A_i in the moving coordinate system; the subscript i indicates the i-th cable.

According to Fig. 2.7, the cable vector can be expressed as:

$$\boldsymbol{A}_i^{\Re} = \boldsymbol{R}\boldsymbol{A}_i^{\Re'} + \boldsymbol{O}^{'\Re} \tag{2.26}$$

where $\boldsymbol{R}$ is the rotation matrix of the moving coordinate system with respect to the inertial coordinate system.

According to the general kinematic modeling method of the cable-driven parallel mechanism, we can get:

Fig. 2.7 Long-span cable-driven parallel mechanism with m cables and n DOFs

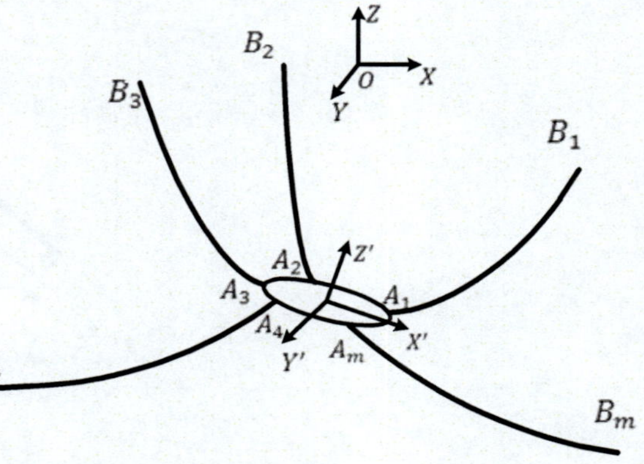

$$\boldsymbol{l}_i = \boldsymbol{B}_i^{\Re} \boldsymbol{A}_i^{\Re} \tag{2.27}$$

$$\boldsymbol{u}_i = \boldsymbol{l}_i / \boldsymbol{l}_i \tag{2.28}$$

$$\boldsymbol{r}_i = \boldsymbol{O}^{'\Re} \boldsymbol{A}_i^{\Re} \tag{2.29}$$

where $\boldsymbol{u}$ is the unit vector of $\boldsymbol{l}_i$; $\boldsymbol{r}_i$ is the position vector of $\boldsymbol{A}_i$.

The static equation of this cable-driven parallel mechanism can be expressed as:

$$\boldsymbol{F} = \boldsymbol{J}_0^T \boldsymbol{\sigma}_0 \tag{2.30}$$

where $\boldsymbol{\sigma}_0$ is the cable tension vector, and the i-th component corresponds to the cable tension of the i-th cable; $\boldsymbol{J}_0^T$ is the force transfer matrix of the cable-driven parallel mechanism; $\boldsymbol{F} \in R^n$ is the external force system. $\boldsymbol{\sigma}_0$ and $\boldsymbol{J}_0^T$ can be expressed as:

$$\boldsymbol{\sigma}_0 = [\sigma_1, \sigma_2, \cdots, \sigma_m]^{\mathrm{T}} \tag{2.31}$$

$$\boldsymbol{J}_0^T = \begin{bmatrix} u_1 & \dots & u_m \\ r_1 \times u_1 & \dots & r_m \times u_m \end{bmatrix}_{n \times m} \tag{2.32}$$

Equations (2.26) to (2.32) describe the kinematic and static models of the cable-driven parallel mechanism under ideal conditions, and the cable self-weight and elastic deformation are neglected and the cable is an ideal linear. In this case, the cable tension is only related to the configuration of the cable-driven parallel mechanism:

(1) For the IRPM (Incompletely Restrained Positioning Mechanism) mechanism ($m = n$), Eq. (2.30) has a unique solution;
(2) For the CRPM (Completely Restrained Positioning Mechanism) mechanism ($m = n + 1$) and RRPM (Redundantly Restrained Positioning Mechanism) mechanism ($m > n + 1$), the number of unknowns cable tension in Eq. (2.30) is larger than that of equations, and there are infinite solutions. The direct solution is turned into an optimization issue by minimizing the norm of cable tensions as the optimization objective.

To obtain precise kinematic and static models, the three single-cable models presented earlier will be adopted in the modeling of the long-span CDPR. In the modeling process, the solution of the ideal model can be used as the initial iterative value to obtain the static equivalent equations based on the three single-cable models:

$$\boldsymbol{F} = \boldsymbol{J}^{\mathrm{T}} \boldsymbol{\sigma} \tag{2.33}$$

where $\boldsymbol{J}^{\mathrm{T}}$ is the force transfer matrix based on the current direction of the cable tension; $\boldsymbol{\sigma}$ is the current tension of the cable-driven parallel mechanism, where the i-th component corresponds to the cable tension of the i-th cable.

To make the cable tension on the moving platform of the CDPR satisfy both the force balance Eq. (2.33) and the force balance equation of the single cable, the solving process of the static equation of the long-span cable-driven parallel mechanism is shown in Fig. 2.8.

2.3 Model Simplification and Error Compensation

The simplified catenary model possesses the advantage of simplicity and fast deduction speed over the precise catenary model, but it also has the disadvantage of larger modeling error. The simplified model is not applicable when high precision of the mechanism is required. To exploit the simplified model for practical control, the modeling error needs to be analyzed, the compensation of the modeling error needs to be completed, and the unification of precision and efficiency needs to be realized.

This section first analyzes the error of the simplified single-cable model and the caused terminal error of the cable-driven parallel mechanism. An error compensation polynomial is proposed, and the genetic algorithm is adopted to optimize the coefficients of this polynomial to ensure that the terminal error of the compensated model satisfies the design requirement in the global workspace.

2.3.1 *Error Analysis of the Single-Cable Models*

Figure 2.9 shows the precise catenary model, the parabolic model and the linear model.

Generally, the cable length obtained from the precise catenary model is the precise value. The difference between the cable length deduced by the parabolic or linear model and that deduced by the precise catenary equation is considered as the modeling error. The single-cable modeling error of the parabola and linear is:

$$\varepsilon l_p = l_p - l_c;\ \varepsilon l_l = l_l - l_c \tag{2.34}$$

where l_c, l_p, l_l denote the values of the cable length deduced by the precise catenary, parabolic and linear equations respectively.

2.3.2 *Model Error Compensation of the Long-Span CDPR*

Assuming that the end position of the long-span cable-driven parallel mechanism is $\boldsymbol{O}' = [x, y, z, \alpha, \beta, \gamma]^{\mathrm{T}}$, the inverse kinematic equation can be obtained as

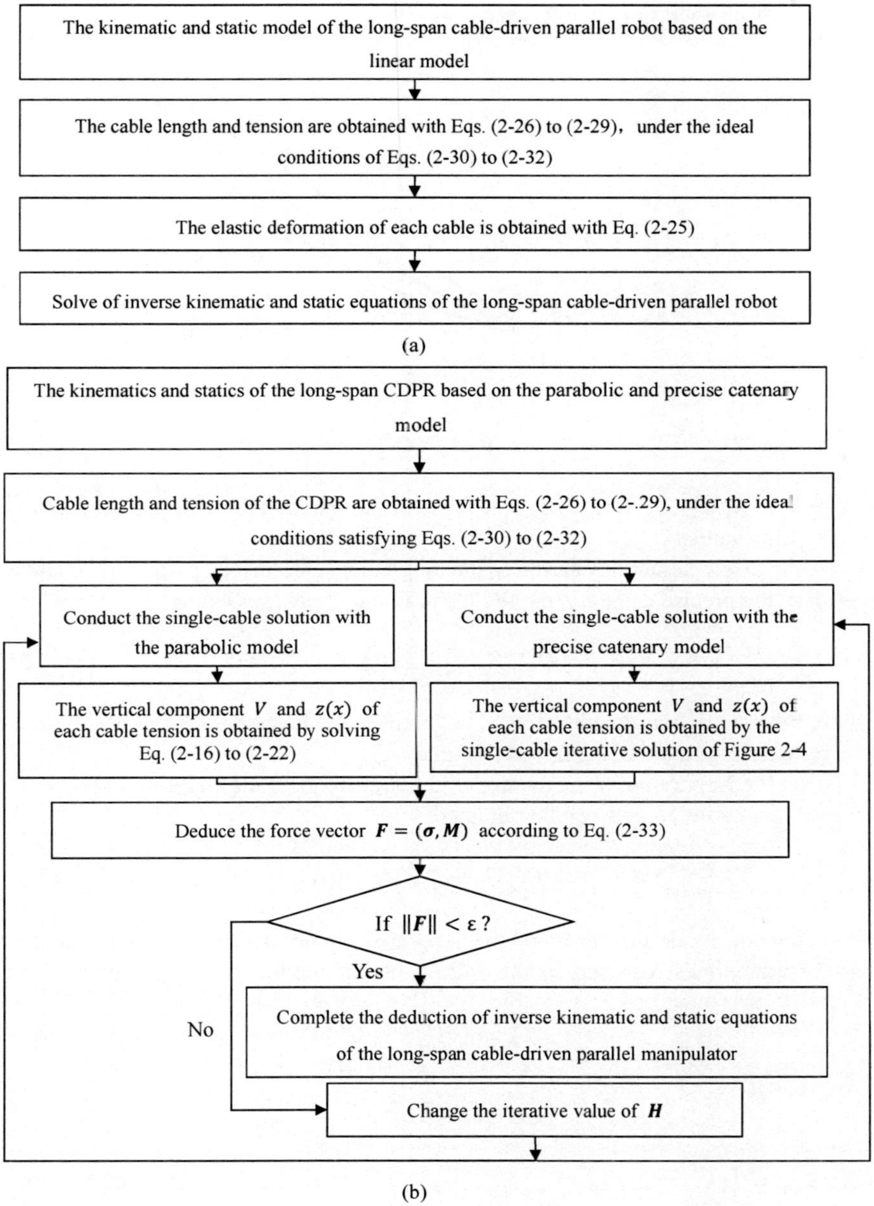

Fig. 2.8 Modeling method of the long-span cable-driven parallel mechanism. **a** Modeling of the long-span CDPR based on the linear model; **b** Modeling of the long-span CDPR based on the parabolic and catenary model

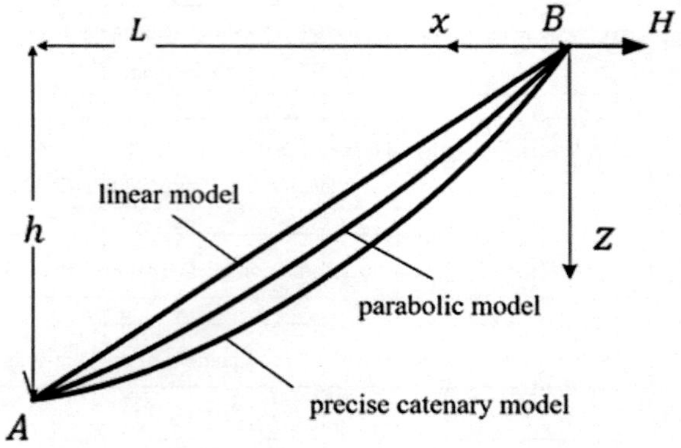

Fig. 2.9 3 single-cable models

$$\boldsymbol{l} = \mathcal{F}(\boldsymbol{O}') \tag{2.35}$$

where $\boldsymbol{l} = (l_1, l_2, \ldots, l_i)(i = 1, 2, \ldots, m)$ is the length vector of the cable-driven parallel mechanism.

The inverse kinematic equation of the long-span cable-driven parallel mechanism based on the precise catenary, parabola and linear are expressed as:

$$\boldsymbol{l}_c = \mathcal{F}_c(\boldsymbol{O}'); \boldsymbol{l}_p = \mathcal{F}_p(\boldsymbol{O}'); \boldsymbol{l}_l = \mathcal{F}_l(\boldsymbol{O}') \tag{2.36}$$

where the equation with subscripts c, p, l denote the inverse equation based on the precise catenary, parabola and linear respectively.

Using the Newton iteration method, the forward kinematic equation based on the three single-cable models can be obtained:

$$\mathcal{G}_c(\boldsymbol{l}_c) = \boldsymbol{O}', \mathcal{G}_p(\boldsymbol{l}_p) = \boldsymbol{O}', \mathcal{G}_l(\boldsymbol{l}_l) = \boldsymbol{O}' \tag{2.37}$$

Taking the modeling result of the long-span cable-driven parallel mechanism based on the precise catenary as the truth value, the model errors are expressed as:

$$\boldsymbol{\varepsilon_p} = \mathcal{G}_c(\boldsymbol{l}_p) - \boldsymbol{O}' \tag{2.38}$$

$$\boldsymbol{\varepsilon_l} = \mathcal{G}_c(\boldsymbol{l}_l) - \boldsymbol{O}' \tag{2.39}$$

where $\boldsymbol{\varepsilon_p} = \left[\varepsilon x_p, \varepsilon y_p, \varepsilon z_p, \varepsilon \alpha_p, \varepsilon \beta_p, \varepsilon \gamma_p\right]^{\mathrm{T}}$ and $\boldsymbol{\varepsilon_l} = \left[\varepsilon x_l, \varepsilon y_l, \varepsilon z_l, \varepsilon \alpha_l, \varepsilon \beta_l, \varepsilon \gamma_l\right]^{\mathrm{T}}$ are the terminal errors of the cable-driven parallel mechanism caused by the modeling with the parabolic and linear models respectively.

It can be found that the cable length is only related to its tension, position and self-weight. Therefore, this book adopts the high order polynomial as the error compensation equation, which can be expressed in the global workspace as:

$$\varepsilon l_{pi} = f_1(L, H, h, \rho, A_0); \varepsilon l_{li} = f_2(L, H, h, \rho, A_0) \quad i = 1, 2 \ldots, m \tag{2.40}$$

The degree number of the high order polynomial $f_*(L, H, h, \rho, A_0)$ depends on the modeling precision requirement.

For a long-span CDPR, its cable density and cable cross-sectional area are constant, and Eq. (2.40) can be simplified as:

$$\varepsilon l_{pi} = f_1(L, H, h); \varepsilon l_{li} = f_2(L, H, h) i = 1, 2 \ldots, m \tag{2.41}$$

The error compensation polynomial can be expressed as:

$$f_*(L, H, h) = a_1 L^n + b_1 h^n + c_1 H^n + \ldots + a_n L + b_n h + c_n H + d n = 1, 2, \ldots \tag{2.42}$$

To keep the terminal error of the simplified model within the maximum allowable error $[\varepsilon_0, \varepsilon_{angle}]^{\mathrm{T}}$, the error compensation objective for the long-span cable-driven parallel mechanism is presented as:

$$\begin{cases} \text{parabola: } min\left(\sqrt{\frac{\sum_1^{t_0}(\varepsilon x_{pa}^2 + \varepsilon y_{pa}^2 + \varepsilon z_{pa}^2)}{t}}\right) \\ \text{linear: } min\left(\sqrt{\frac{\sum_1^{t_0}(\varepsilon x_{line}^2 + \varepsilon y_{line}^2 + \varepsilon z_{line}^2)}{t}}\right) \end{cases} \quad t = 1, 2, \ldots \tag{2.43}$$

where t is the sample point in the required workspace of the long-span cable-driven parallel mechanism; t_0 is the sum of sample points.

The error compensation conditions are:

$$\begin{cases} \sqrt{\varepsilon x_{pa}^2 + \varepsilon y_{pa}^2 + \varepsilon z_{pa}^2} < \varepsilon_0; \\ \sqrt{\varepsilon x_{line}^2 + \varepsilon y_{line}^2 + \varepsilon z_{line}^2} < \varepsilon_0; \end{cases} \tag{2.44}$$

$$\begin{cases} \varepsilon \alpha_{pa} < \varepsilon_{angle}; \varepsilon \beta_{pa} < \varepsilon_{angle}; \varepsilon \gamma_{pa} < \varepsilon_{angle} \\ \varepsilon \alpha_{line} < \varepsilon_{angle}; \varepsilon \beta_{line} < \varepsilon_{angle}; \varepsilon \gamma_{line} < \varepsilon_{angle} \end{cases} \tag{2.45}$$

where,

$$\boldsymbol{\varepsilon pa} = \mathcal{G}_c(\boldsymbol{l}_p + \varepsilon \boldsymbol{l}_p) - \boldsymbol{O}'; \boldsymbol{l}_p = \mathcal{F}_p(\boldsymbol{O}') \tag{2.46}$$

$$\boldsymbol{\varepsilon line} = \mathcal{G}_c(\boldsymbol{l}_l + \varepsilon \boldsymbol{l}_l) - \boldsymbol{O}'; \boldsymbol{l}_l = \mathcal{F}_l(\boldsymbol{O}') \tag{2.47}$$

$$\boldsymbol{\varepsilon pa} = \left(\varepsilon x_{pa}, \varepsilon y_{pa}, \varepsilon z_{pa}, \varepsilon \alpha_{pa}, \varepsilon \beta_{pa}, \varepsilon \gamma_{pa}\right)^{\mathrm{T}} \tag{2.48}$$

$$\boldsymbol{line} = (\varepsilon x_{line}, \varepsilon y_{line}, \varepsilon z_{line}, \varepsilon \alpha_{line}, \varepsilon \beta_{line}, \varepsilon \gamma_{line})^{\mathrm{T}} \tag{2.49}$$

The error compensation equation can be optimized with the genetic algorithm and the coefficients can be deduced, considering above optimization objectives and conditions. However, in the process, the complex and time-consuming forward equation of the long-span CDPR based on the precise catenary makes it more difficult to deduce the optimized coefficients of the error compensation equation. An alternative method that employs the fast deducing speed of the forward algorithm based on the linear model, is adopted to deduce the optimization coefficients for the high order compensation polynomial.

Firstly, the modeling errors of the cable-driven parallel mechanism based on the parabola and linear are:

$$\boldsymbol{\varepsilon_p} = \mathcal{G}_l(\boldsymbol{l}_p) - \mathcal{G}_l(\boldsymbol{l}_c) \tag{2.50}$$

$$\boldsymbol{\varepsilon_l} = \mathcal{G}_l(\boldsymbol{l}_l) - \mathcal{G}_l(\boldsymbol{l}_c) \tag{2.51}$$

Compared with Eqs. (2.38) and (2.39), the precise terminal position of the cable-driven parallel mechanism is not $\boldsymbol{O}^{'}$, but replaced with $\mathcal{G}_l(\boldsymbol{l}_c)$. In the optimization algorithm above, cable length of the compensated parabola and linear are used to approximate the precise terminal position $\boldsymbol{O}^{'}$ for a faster deduction. The cable length of the compensated parabola and linear is used to approximate the precise terminal position $\mathcal{G}_l(\boldsymbol{l}_c)$.

It can be found that Eqs. (2.46) to (2.47) in the optimization algorithm of the compensation polynomial can be expressed as:

$$\boldsymbol{\varepsilon pa} = \mathcal{G}_l(\boldsymbol{l}_p + \varepsilon \boldsymbol{l}_p) - \mathcal{G}_l(\boldsymbol{l}_c); \boldsymbol{l}_p = \mathcal{F}_p(\boldsymbol{O}') \tag{2.52}$$

$$\boldsymbol{\varepsilon line} = \mathcal{G}_l(\boldsymbol{l}_l + \varepsilon \boldsymbol{l}_l) - \mathcal{G}_l(\boldsymbol{l}_c); \boldsymbol{l}_l = \mathcal{F}_l(\boldsymbol{O}') \tag{2.53}$$

The same optimization algorithm as above can be adopted to deduce the coefficients of the compensation polynomial, which is used to compensate for the terminal error of the parabolic and linear models.

2.4 Modeling of the FAST First-Level Feed Support System

In this section, the first-level feed support system of the FAST, i.e., the six-cable CDPR, is modeled. Figure 2.10 shows the six-cable CDPR of the FAST. Firstly, the coordinate systems are established: one is the inertial coordinate system $\Re$: $O - XYZ$, the origin of which is located at the center of the base platform with the Z-axis upward; the other is the moving coordinate system $\Re'$: $O' - X'Y'Z'$, the origin of which is located at the center of the moving platform with the Z'-axis upward along the normal of the moving platform. $B_i (i = 1, 2, \ldots, m)$ is the cable

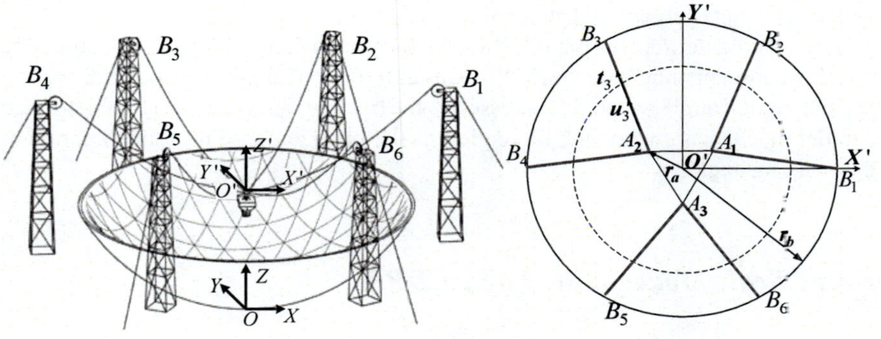

Fig. 2.10 Model of the six-cable CDPR in FAST

exit point of the base platform and $A_i (i = 1, 2, \ldots, m)$ is the cable anchor point of the moving platform.

To model the six-cable CDPR, the symbols are defined as follows: $\boldsymbol{O}^{\prime\Re}$ is the position vector of the origin of the moving coordinate system O' in the inertial coordinate system; $\boldsymbol{B}_i^{\Re}$ is the position vector of B_i in the inertial coordinate system; $\boldsymbol{A}_j^{\Re}$ is the position vector of A_j in the inertial coordinate system; $\boldsymbol{A}_j^{\Re'}$ is the position vector of A_j in the moving coordinate system; r_b is the radius of the base platform, i.e. the radius of the distribution circle of the cable exit points; r_a is the radius of the moving platform; h is the height of the cable exit points.

According to Fig. 2.10, the cable vector is deduced as:

$$\boldsymbol{B}_i^{\Re} = \left[r_b \cos(i-1)60°, r_b \sin(i-1)60°, h\right]^{\mathrm{T}} (i = 1,2, \ldots,6) \tag{2.54}$$

$$\boldsymbol{A}_j^{\Re'} = \left[r_a \cos(4j-3)30°, r_a \sin(4j-3)30°, 0\right]^{\mathrm{T}} (j = 1, 2, 3) \tag{2.55}$$

$$\boldsymbol{A}_j^{\Re} = \boldsymbol{R}\boldsymbol{A}_j^{\Re'} + \boldsymbol{O}^{\prime\Re} \tag{2.56}$$

where $\boldsymbol{R}$ is the rotation matrix from moving to inertial coordinate system.

Gosselin C M proposed a coordinate transformation method based on T&T (Tilt-and-Torsion) angle. As for this method, there are also three turning angles: the azimuth angle φ, the tilt angle θ, and the torsion angle ψ.

$$\begin{aligned}\boldsymbol{R} &= \begin{bmatrix} c\phi & -s\phi & 0 \\ s\phi & c\phi & 0 \\ 0 & 0 & 1 \end{bmatrix} \begin{bmatrix} c\theta & 0 & s\theta \\ 0 & 1 & 0 \\ -s\theta & 0 & c\theta \end{bmatrix} \begin{bmatrix} c\phi & s\phi & 0 \\ -s\phi & c\phi & 0 \\ 0 & 0 & 1 \end{bmatrix} \begin{bmatrix} c\psi & -s\psi & 0 \\ s\psi & c\psi & 0 \\ 0 & 0 & 1 \end{bmatrix} \\ &= \begin{bmatrix} c\phi c\theta c(\psi-\phi) - s\phi s(\psi-\phi) & -c\phi c\theta s(\psi-\phi) - s\phi c(\psi-\phi) & c\phi s\theta \\ s\phi c\theta c(\psi-\phi) + c\phi s(\psi-\phi) & -s\phi c\theta s(\psi-\phi) + c\phi c(\psi-\phi) & s\phi s\theta \\ -s\theta c(\psi-\phi) & s\theta c(\psi-\phi) & c\theta \end{bmatrix}\end{aligned} \tag{2.57}$$

where cϕ denotes cosϕ, sϕ denotes sinϕ, and so on.

The coordinate transfer matrix $\boldsymbol{R}$ of the long-span CDPR adopts the coordinate transformation method based on T&T (Tilt-and-Torsion) angle. The azimuth angle φ and the torsion angle ψ can be expressed as the tracking direction of the cable-driven parallel mechanism, while the tilt angle θ is the pitch angle of it in the direction of the tracking angle.

2.4.1 Scale Model of the FAST CDPR

To investigate the modeling and error compensation method of the long-span CDPR with different spans, the modeling of the 40 m six-cable CDPR is studied first.

Table 2.1 gives the dimensions of the 40 m six-cable CDPR of the kinematic similar model of FAST established by the Miyun Observing Station of NAOC in Beijing.

Figure 2.11 shows the workspace of the 40 m six-cable CDPR, which is a spherical dome with a radius of 9.6 m, and the maximum observation angle is 0–40°. The observation pitch angle of the cable-driven parallel mechanism is 0–16°. Both angles are linear distributed.

As shown in Fig. 2.12, according to the model error analysis of the long-span cable-driven parallel mechanism in Sect. 2.2, the model error of the six-cable CDPR based on the parabolic and linear model in the global workspace can be obtained from Eqs. (2.48) and (2.49).

The analysis of Fig. 2.12a shows that the modeling end error of the 40 m similar model CDPR based on the parabolic model is less than 1 mm. The parabola can approximate the precise catenary model in this case, and the CDPR can be controlled directly by the modeling method based on the parabolic cable model.

Since the parabolic deducing process requires one iteration, the modeling based on the linear model needs to be further analyzed, to speed up the deduction. From Fig. 2.12b, it can be found that when the linear model is used for the modeling of

Table 2.1 Dimensions of the kinematic similar model of the FAST CDPR

Symbols	Parameters	Value (unit)
r_a	Radius of the moving platform	0.5 (m)
r_b	Radius of the distribution circle of the cable exit points	20 (m)
h_{tower}	Height of the cable exit points	18 (m)
d	Diameter of the cable	8 (mm)
ρ	Density of the cable	0.259 (Kg/m)
E	Elastic modulus of the cable	1.6×10^{11}(Pa)
m_0	Mass of the moving platform	300 (Kg)
$[\varepsilon_0, \varepsilon_{angle}]^{\mathrm{T}}$	The maximum allowable error	[1 mm, 0.5°]

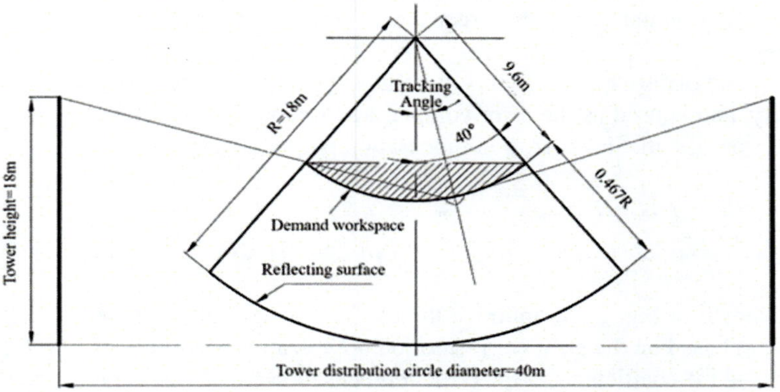

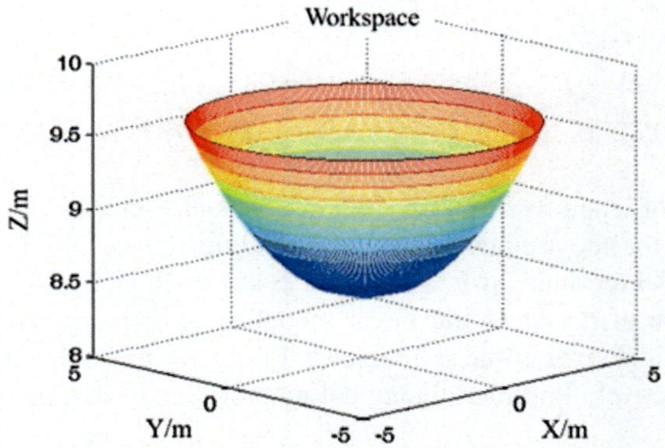

Fig. 2.11 Workspace of the 40 m six-cable CDPR

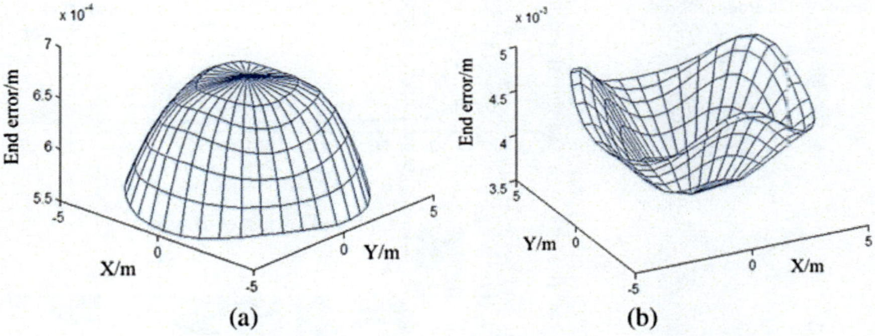

Fig. 2.12 Terminal error of the CDPR based on the simplified model. **a** Terminal error cf based on the parabolic model; **b** Terminal error based on the linear model

CDPR, the terminal errors are larger than 1 mm, which does not meet the design requirement.

The parabola is a quadratic curve, satisfies the requirement of approximating the catenary, and is used as the error compensation polynomial for linear modeling in this section. Let the linear error compensation polynomial be:

$$\begin{aligned} \mathrm{E}l_l &= f_2(L, H, h_{tower}) \\ &= a_1 L^2 + a_2 L + b_1 h_{tower}^2 + b_2 h_{tower} + c_1 H^2 + c_2 H + d \end{aligned} \tag{2.58}$$

As shown in Fig. 2.13, optimize the coefficients of Eq. (2.58) with the genetic algorithm based on the error compensation optimization algorithm of Eqs. (2.43) to (2.53), and the coefficients of the error compensation equation can be obtained. The error compensation polynomial based on the linear model is:

$$\begin{aligned} \varepsilon l_l &= f_2(L, H, h_{tower}) \\ &= (8.3721 \times L^2 - 3.5328 \times L - 9.8747 \times h_{tower}^2 - 7.1498 \times h_{tower} \\ &\quad - 5.0527 \times 10^{-4} \times H^2 + 0.0894 \times 10^{-2} \times H - 3.2483) \times 10^{-6} \end{aligned}$$

where the unit of length is m and the unit of cable tension is N.

Substituting the above equation into the modeling of the CDPR based on the linear model, the model terminal error is obtained as shown in Fig. 2.14.

The adoption of the quadratic linear modeling error compensation polynomial makes the terminal error of the six-cable CDPR based on the linear model less than 1 mm, and effectively improves the modeling precision of the linear model.

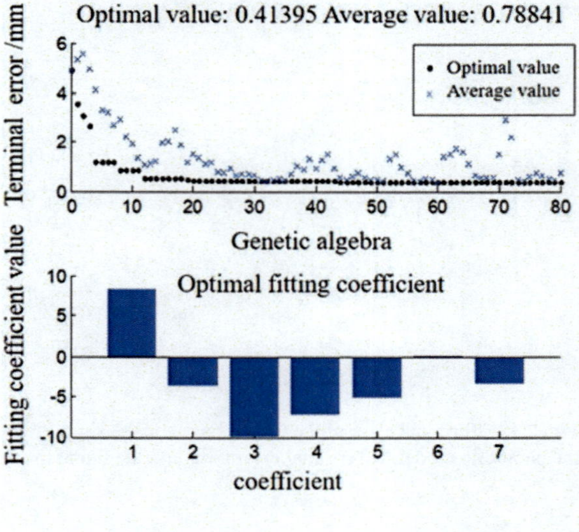

Fig. 2.13 Coefficients of the compensation polynomial based on the linear model

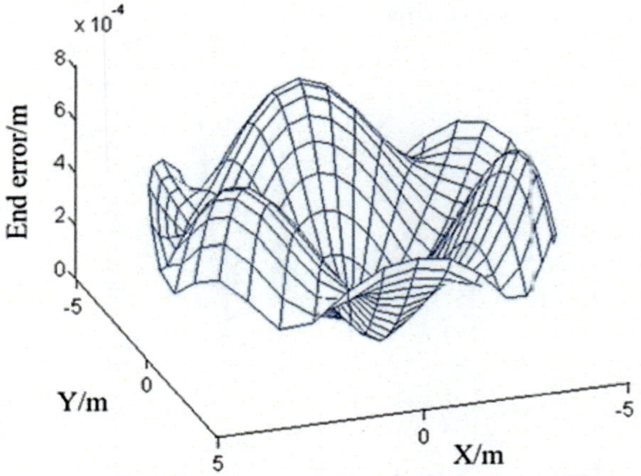

Fig. 2.14 Terminal error of the six-cable CDPR based on the linear model after compensation

Through the study of the modeling analysis and error compensation of the 40 m six-cable CDPR, the modeling and model error compensation method of the cable-driven parallel mechanism based on the linear model can be applied to the practical control. Next, a linear trajectory will be implemented to study the terminal error of the linear modeling and compensation method with the 40 m CDPR.

A linear experimental trajectory is given as shown in Fig. 2.15. The coordinates of the two ends of the trajectory are G1 = (0, 0, 8.4 m) and G2 = (0,0,10 m), the practical control precision of the cable-driven parallel mechanism is required to be RMS10 mm, and the experimental speed is v = 15.3 mm/s.

The terminal errors of the six-cable driven parallel mechanism in the three translational directions are obtained as shown in Fig. 2.16.

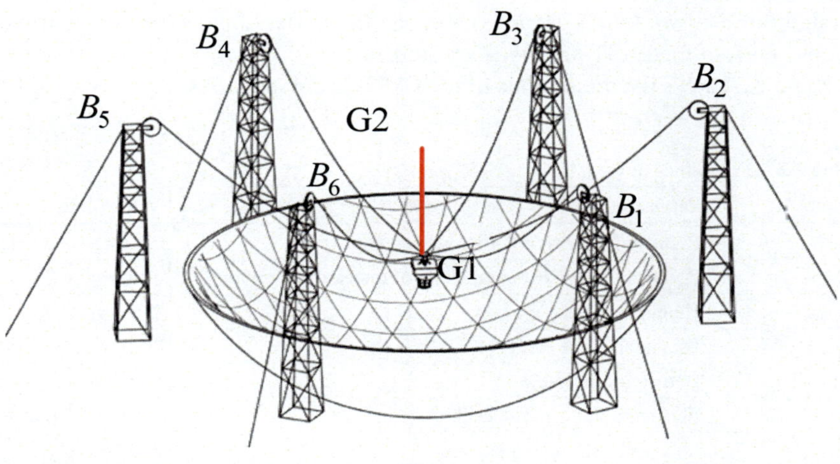

Fig. 2.15 A linear trajectory in the six-cable CDPR

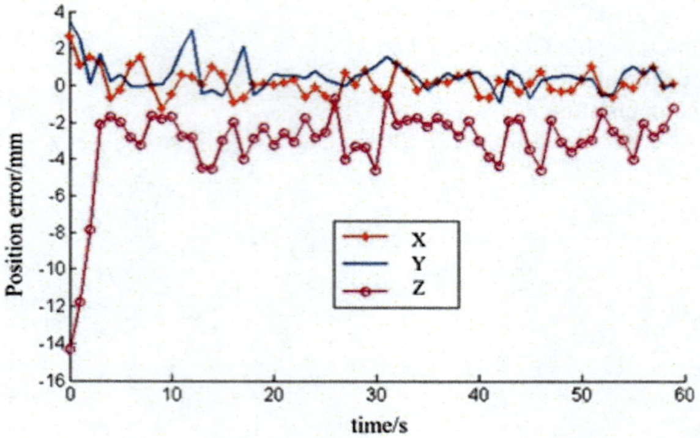

Fig. 2.16 Terminal error of the six-cable CDPR

From Fig. 2.16, it can be found that the position error of the six-cable driven parallel mechanism in the motion trajectory is relatively balanced, and the error in a single direction generally does not exceed 3 mm, except for the larger error at the beginning of motion,. The error analysis of under this trajectory reveals that the terminal error is RMS3.69 mm, which meets the practical control precision requirement. It also proves that the simplified model and the model error compensation method in this chapter can be applied to practical motion control to meet the given precision requirement.

2.4.2 *The CDPR of the FAST Prototype*

The modeling and compensation polynomial will change when parameters of the span and end precision requirement are different, even for the same cable-driven parallel mechanism. In this section, the modeling analysis and error compensation of the CDPR of the FAST prototype are studied.

Table 2.2 shows the dimensions of the FAST prototype [25].

Table 2.2 Dimensions of the CDPR in FAST prototype

Symbols	Parameters	Value (unit)
r_a	Radius of the moving platform	7.5 (m)
r_b	Radius of the distribution circle of the cable exit points	300 (m)
h_{tower}	Height of the cable tower	275 (m)
d	Diameter of the cable	40 (mm)
ρ	Density of the cable	576 (Kg/m)
E	Elastic modulus of the cable	1.6×10^{11} (Pa)
m_0	Mass of the moving platform	300,000 (Kg)
$[\varepsilon_0, \varepsilon_{angle}]^T$	The maximum allowable error	[10 mm, 0.5°]

Figure 2.17 shows the workspace of the CDPR, which is a spherical dome with a radius of 160 m, and the observation pitch angle of the CDPR is ±16°.

Figure 2.18 illustrates the terminal error of the six-cable CDPR of the FAST prototype based on the parabolic and linear modeling in the global workspace. It can be found that both the parabolic model and the linear model of the long-span CDPR cannot meet the modeling requirement, while the modeling precision of the parabolic model is closer to the required precision. In this case, the parabolic-based modeling error compensation can constrain the modeling error to the required range faster.

Establish the quadratic compensation polynomial of the modeling error based on the parabolic modeling as:

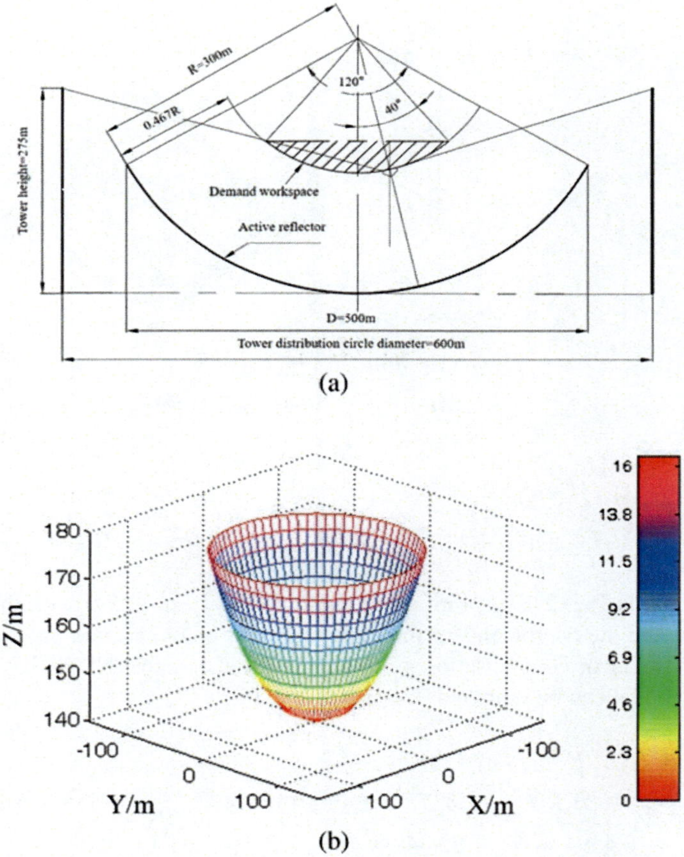

Fig. 2.17 Workspace of the six-cable CDPR of the FAST prototype

Fig. 2.18 Terminal error of the cable-driven parallel mechanism

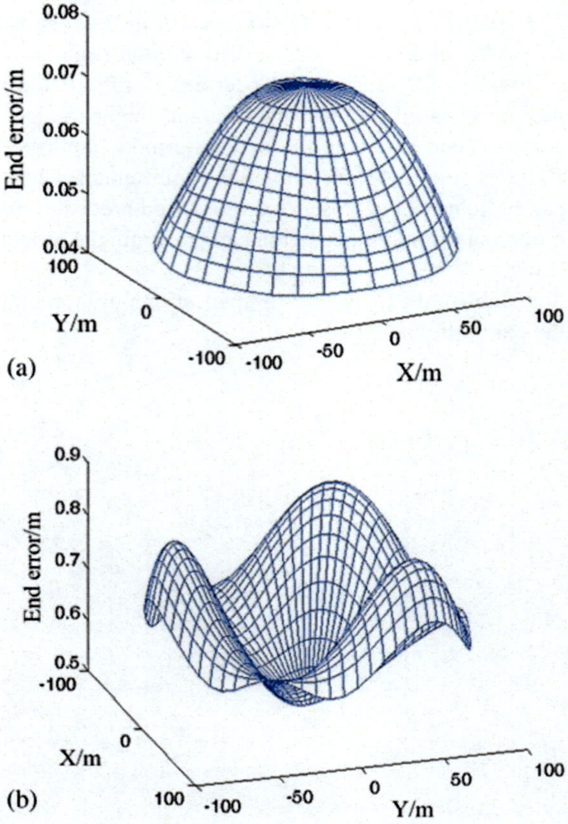

$$
\begin{aligned}
El_p &= f_1(L, H, h_{tower}) \\
&= a_1 L^2 + a_2 L + b_1 h_{tower}^2 + b_2 h_{tower} + c_1 H^2 + c_2 H + d
\end{aligned} \tag{2.59}
$$

As shown in Fig. 2.19, optimize the coefficients of Eq. (2.59) with the genetic algorithm, and the coefficients of the modeling error compensation polynomial are obtained. The error compensation polynomial of the six-cable CDPR based on the parabolic model can be written as:

$$
\begin{aligned}
\varepsilon l_p &= f_1(L, H, h_{tower}) \\
&= (0.1683 \times L^2 + 1.8275 \times L + 1.4596 \times h_{tower}^2 - 3.558 \times h_{tower} \\
&\quad - 0.4594 \times H^2 - 8.9449 \times H - 0.9931 \times 10^2) \times 10^{-6}
\end{aligned}
$$

where the unit of length is m and the unit of cable tension is kN.

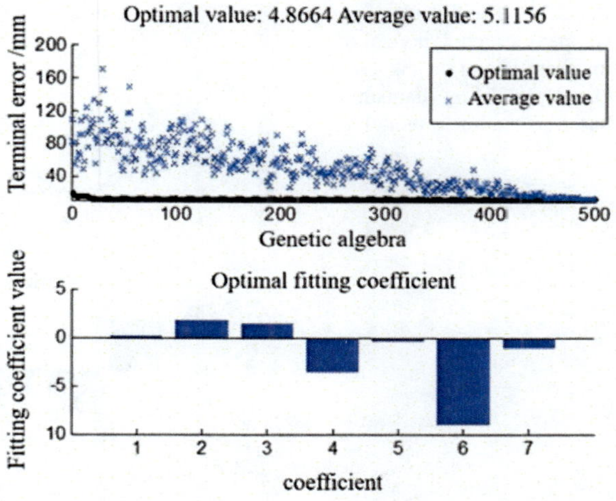

Fig. 2.19 Coefficients of the compensation polynomial based on the parabolic model

Substituting the above equation into the modeling of the cable-driven parallel mechanism based on the parabolic model, the terminal error of the model is obtained as shown in Fig. 2.20.

Since parabolic modeling requires a quadratic curve for error compensation, it showcases that the error compensation by a quadratic polynomial from linear modeling alone cannot meet the precision requirement and can only increase the modeling precision to the level of parabolic modeling. This conclusion is further verified as follows:

Based on the analysis of the parametric similarity ratio of the six-cable driven parallel mechanism and its similar model, the quadratic compensation polynomial for the model error based on the linear modeling is given as:

$$El_l = f_2(L, H, h_{tower})$$

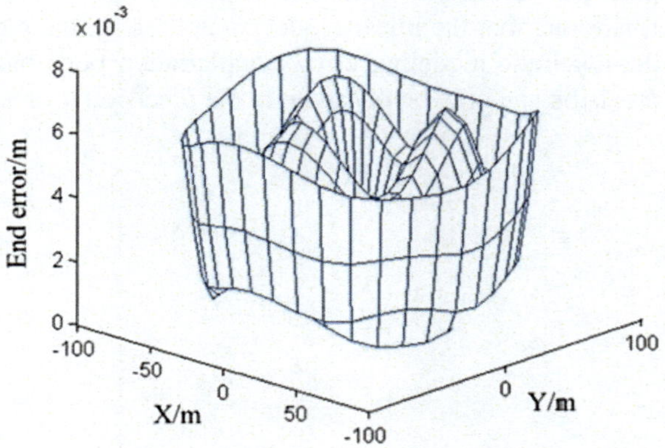

Fig. 2.20 Terminal error of the six-cable CDPR of the FAST prototype based on the parabolic model after error compensation

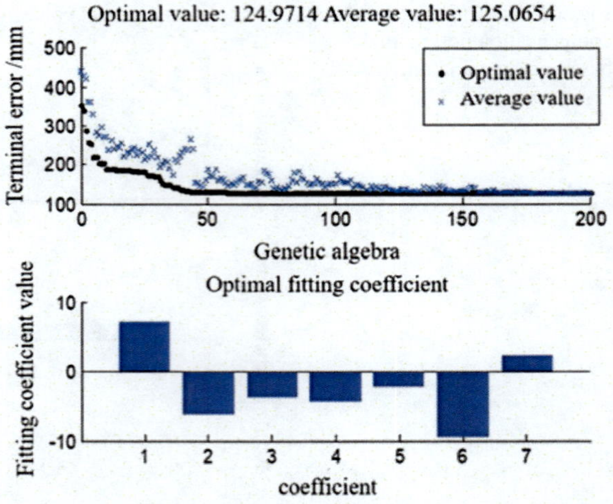

Fig. 2.21 Coefficients of the compensation polynomial of the prototype of the six-cable driven parallel mechanism based on the linear model

$$= a_1 L^2 + a_2 L + b_1 h_{tower}^2 + b_2 h_{tower} + c_1 H^2 + c_2 H + d \tag{2.60}$$

Figure 2.21 shows the optimization algorithm of the error compensation of Eqs. (2.43) to (2.53). Optimize the coefficients of Eq. (2.60) with the genetic algorithm, the coefficients of the modeling error compensation polynomial are obtained. The error compensation polynomial based on the linear model can be written as:

$$\begin{aligned} \varepsilon l_l &= f_2(L, H, h_{tower}) \\ &= (7.1718 \times L^2 - 6.1115 \times 10^2 \times L - 3.5668 \times h_{tower}^2 - 4.3521 \times 10^2 \\ &\quad \times h_{tower} - 2.0202 \times H^2 - 9.3496 \times 10^2 \times H \end{aligned}$$

where the unit of length is m and the unit of cable tension is kN.

Substituting the above equation into the cable-driven parallel mechanism based on the linear model, the terminal error of the modeling is obtained as shown in Fig. 2.22. It is found that the linear model cannot realize the required accuracy if only adopt the quadratic modeling error compensation polynomial, and the linear modeling precision can only be raised up to the precision level of the parabolic modeling.

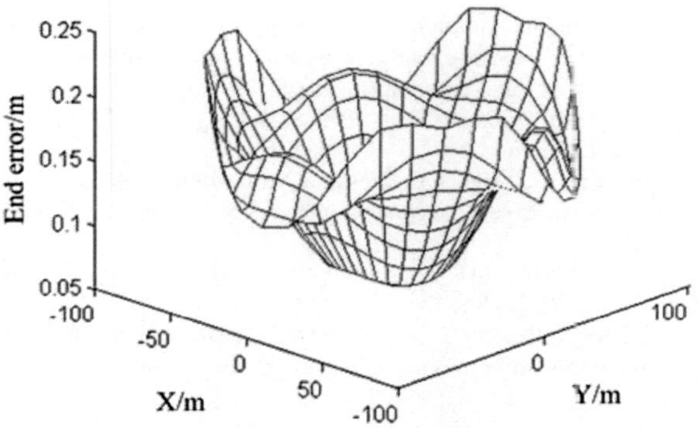

Fig. 2.22 Terminal error of the six-cable driven parallel mechanism in the FAST prototype based on the linear model after error compensation

2.5 Conclusion

(1) The CDPR can realize a large range of motion based on low cost, and the long-span cable-driven parallel mechanism can give full play to the advantages of the large workspace, which has been applied to the lifting and positioning conditions of large workspace.
(2) As for long-span working conditions, typical single-cable models include the linear model, the parabolic model and the catenary model. The modeling of the long-span CDPR can first adopt the deduction result of the ideal model (ignore the elasticity and mass of the cable) as the initial iterative values, and complete the iterative solution with the mechanism and the single-cable equivalent equations.
(3) To achieve high efficiency and precision in deducing the long-span CDPR, an error compensation polynomial is proposed, and the coefficients of the polynomial are optimized by the genetic algorithm to achieve error compensation in the global workspace.
(4) The modeling of the first-level feed support system of FAST—the six-cable CDPR is completed, and a control algorithm with the linear modeling and error compensation polynomial is presented. The algorithm can effectively reduce the amount of computation while ensuring that the precision meets the requirement and can be coordinated with the closed-loop control to achieve the precision control requirement of FAST.

References

1. Gouttefarde M, Collard JF, Riehl N, et al. Simplified static analysis of large-dimension parallel cable-driven robots. In: 2012 IEEE international conference on robotics and automation. IEEE; 2012.

2. Nguyen DQ, Gouttefarde M, Company O, et al. On the analysis of large-dimension reconfigurable suspended cable-driven parallel robots. In: Proceedings—IEEE international conference on robotics and automation, 2014.
3. Izard J-B, Dubor A, Hervé P-E, et al. Large-scale 3D printing with cable-driven parallel robots. Constr Robot. 2017.
4. Vasko FJ, Landquist E, Kresge G, Tal A, Jiang Y, Papademetris X. A simple and efficient strategy for solving very large-scale generalized cable-trench problems. Networks. 2016;67(3):199–208.
5. Kärcher HJ, Li H, Sun J, et al. Proposed design concepts of the FAST focus cabin suspension. Proc SPIE Int Soc Opt Eng.; 2008.
6. Zheng Y. Research on key theoretical issues of wire-driven parallel kinematic manipulators and the application to wind tunnel support systems. Ph.D. thesis. Quanzhou: Huaqiao University; 2004.
7. Krishna P. Cable-suspended roofs. New York: McGraw-Hill; 1978. p. 27–65.
8. Irivne H. Cable structures. MA: MIT Press; 1981. p. 16–20.
9. Kozak K, Zhou Q, Wang J. Static analysis of cable-driven manipulators with non-negligible cable mass. IEEE Trans Rob. 2006;22(3):425–33.
10. Lambert C, Nahon M, Chalmers D. Implementation of an aerostat positioning system with cable control. IEEE/ASME Trans Mech. 2007;12(1):32–40.
11. Nahon M, Gilardi G, Lambert C. Dynamics/control of a radio telescope receiver supported by a tethered aerostat. J Guidance Control Dyn. 2002;25(6):1107–15.
12. Taghirad HD, Nahon MA. Kinematic analysis of a macro–micro redundantly actuated parallel manipulator. Adv Robot. 2008;22:657–87.
13. Taghirad HD, Nahon MA. Dynamic analysis of a macro–micro redundantly actuated parallel manipulator. Adv Robot. 2008;22:949–81.
14. Su YX, Duan BY, Nan RD, et al. Development of a large parallel-cable manipulator for the feed-supporting system of a next-generation large radio telescope. J Robot Syst. 2001;18(11):633–43.
15. Qiang W, Qiu YY, Duan BY. Locus program of cabin movement for the large radio telescope with eight cables. China Mech Eng. 2002;12(23):2036–9.
16. Qu L, Tang X, Yao R, et al. Error analysis and compensation of cable driven parallel manipulators for the forty-meter aperture radio telescope. Chin High Tech Lett.
17. Li H, Zhu WB, Pan GF. Equilibrium analysis of fast rope-drive parallel manipulator based on rope force optimization. Eng Mech. 2011;28(4):185–93.
18. Lu Y, Zhu W, Ren G. Feedback control of a cable-driven Gough-Stewart platform. IEEE Trans Robot. 2006;22(1):198–202.
19. Lu YJ, Zhu WB, Zhu LC. Study of a real time filter based on frequency analysis. Sci Technol Eng. 2007;7(11):2476–83.
20. Qiu Y, Duan B, Sheng Y, et al. Similarity between the distorted model and the prototype of the large radio telescope. J Xidian Univ (Natural Science). 2004;31(4):493–500.
21. Wei Q, Qiu Y, Duan B. Wind induced vibration analysis of a 50 meter scaled model of large radio telescope. Chin J Appl Mech. 2003;20(1):59–63.
22. Duan BY, Du JL, Bao H. Dynamic modeling and active control of a cable-suspended parallel robot. Mechatronics. 2008;18:1–12.
23. Su YX, Duan BY, Nan RD. Mechatronics design of stiffness enhancement of the feed supporting system for the square-kilometer array. IEEE/ASME Trans Mechatron. 2003;8(4):425–30.
24. Duan X, Qiu Y, Duan B, et al. Adaptive interactive PID supervisory control of the macro-micro parallel manipulator. J Mech Eng. 2010;46(1):10–7.
25. Nan R et al. Preliminary design of the 500-meter spherical radio telescope (FAST) project (feeder support fascicle). National Astronomical Observatory of the Chinese Academy of Sciences; 2008.

Chapter 3
The Static Characteristic Analysis of the Long-Span Cable-Driven Parallel Robot

The cable in the cable-driven parallel mechanism can only bear tension, not pressure, and the unidirectional force-bearing characteristic of the cable will affect the motion and stability of the cable-driven mechanism. Therefore, it is necessary to analyze the cable tension of the cable-driven parallel mechanism, and the static characteristic is an important aspect in the optimization design of the long-span cable-driven parallel mechanism. This chapter completes the integrated static modeling and deduction of the long-span cable-driven parallel mechanism based on the self-weight and elastic deformation of the cable. Based on it, this chapter analyzes the static characteristic of the long-span cable-driven parallel mechanism according to the deduction results of the precise static model.

To ensure the controllability of the long-span cable-driven parallel mechanism, the cable tension is required to satisfy certain conditions, but it is still necessary to explore the indices describing the cable tension characteristics. In this chapter, the indices of the local and global cable tension characteristics are proposed to describe the cable tension characteristics of the long-span cable-driven parallel mechanism in the workspace.

The long-span cable-driven parallel mechanism itself has a large scale, which is difficult to build and control, and the end errors of the mechanism are unavoidable. This chapter will further analyze the effect of the end errors on the cable tension of the long-span cable-driven parallel mechanism based on Chap. 2, and establish corresponding indices to quantify the effect of the allowable end errors on the cable tension of the long-span cable-driven parallel mechanism.

This chapter firstly introduces the static characteristic of the long-span cable-driven parallel mechanism and the cable tension indices; in Sect. 3.2, the error model of the long-span cable-driven parallel mechanism is established, the cable tension characteristics in the end error space of the long-span cable-driven parallel mechanism are analyzed, and the cable tension characteristic indices in the allowable error space are put forward; Sect. 3.3 takes the six-cable driven parallel mechanism in the first-level feed support system of the FAST as an example to study its global

© Tsinghua University Press 2024

X. Tang et al., *Research and Application of Cable-Driven and Rigid Parallel Robots*,
https://doi.org/10.1007/978-981-99-7452-8_3

cable tension characteristics and the cable tension characteristics in the allowable error space, and verify the feasibility of the force characteristic indices described in this chapter, which can provide a theoretical foundation for the force control of the long-span cable-driven parallel mechanism.

The main contents of this chapter are:

(1) Cable tension characteristic indices of the long-span cable-driven parallel mechanism;
(2) End errors and the cable tension characteristics;
(3) Force characteristics analysis of the similar model of the first-level feed support system of FAST.

3.1 Cable Tension Characteristic Indices of the Long-Span CDPR

When the tension of a cable in the cable-driven parallel mechanism is close to zero or the difference with that of other cables is great, the cable-driven parallel mechanism will have poor controllability or unstable motion, which is called the instability or pseudo-drag phenomenon of the long-span cable-driven parallel mechanism. To prevent this phenomenon, a condition on tension is defined in the design of the mechanism, and effective cable tension characteristic indices are established to avoid instability or pseudo-drag with the attitude analysis in the workspace. Therefore, the cable tension characteristic is an important index to measure the performance of the cable-driven parallel mechanism.

In terms of the cable tension indices, early studies mainly focused on the minimum value of the cable tension and the ratio of the maximum and minimum values of it to reflect the its equilibrium, thus these indices are mainly applied in the design of the cable-driven parallel robots [1]. For example, Takeda [4] analyzed the force characteristics of the cable-driven mechanism with n degrees of freedom by using a simplified equivalent telescopic link, and introduced a force transfer coefficient TI_{w} obtained from the Jacobian matrix of the mechanism. Depending on the geometric conditions of the mechanism, this index is generally used to measure the force transfer performance of the cable to the moving platform; Takeda and Funabashi [5] proposed a seven-cable driven parallel mechanism with six degrees of freedom, and firstly conducted analysis and optimization design of the force transfer performance of the push–pull cable-driven parallel mechanism. Zheng Yaqing [6], based on the coefficients of the force transfer performance proposed by Takeda, analyzed the deduction process of the index in detail and adopted it to analyze the workspace of a new cable-driven parallel mechanism with six degrees of freedom.

In the motion control of the long-span cable-driven parallel mechanism, to ensure that each cable remains tensioned during the moving process and to make the mechanism operate stably, the relationship between the maximum and minimum cable tension is used to measure and evaluate the force characteristics of the mechanism at a certain attitude in the workspace. Two types of indices are adopted to indicate the cable tension characteristics of the long-span cable-driven parallel mechanism: the index of the difference between the maximum and minimum cable tension; the ratio between the maximum and minimum cable tension. And the equation of the local and global force characteristic indices is established to quantify the mechanical property of the mechanism.

3.1.1 Performance Indices of the Local Cable Tension of the Long-Span CDPR

When the end of the long-span cable-driven parallel mechanism is located at a certain attitude in the workspace, two cable tension characteristic indices TCI, namely the difference between the maximum and minimum cable tension as well as the ratio between the maximum and minimum cable tension, are defined to measure the cable tension characteristics.

The difference $\mathrm{TCI}_{\max}$ between the maximum and minimum cable tension:

$$\mathrm{TCI}_{\max} = \max\left(\sigma_i - \sigma_j\right) i = 1, \ldots, m;\ j = 1, \ldots, m; i \neq j \tag{3.1}$$

To make the mechanism operate stably, the smaller $\mathrm{TCI}_{\max}$ is, the smaller the variation of the cable tension of the mechanism and the more smoothly it moves. Therefore, for a certain attitude in the same mechanism, the smaller $\mathrm{TCI}_{\max}$ is, the better.

At some attitudes, each cable tension itself isn't great, but the cable tension may differ by a multiple, thus another measure index of the cable tension—the ratio $\mathrm{TCI}_{\mathrm{rmax}}$ between the maximum and minimum cable tension is required:

$$\mathrm{TCI}_{\mathrm{rmax}} = \max\left(\frac{\sigma_i}{\sigma_j}\right) i = 1, \ldots, m;\ j = 1, \ldots, m; i \neq j \tag{3.2}$$

$\mathrm{TCI}_{\mathrm{rmax}} \geq 1$, the smaller it is, the better.

The tension constraint:

$$\begin{cases} \boldsymbol{\sigma} \geq (\sigma_{\min}, \ldots, \sigma_{\min})^{\mathrm{T}} \\ \boldsymbol{\sigma} \leq (\sigma_{\max}, \ldots, \sigma_{\max})^{\mathrm{T}} \end{cases} \tag{3.3}$$

where $\sigma_{\min}$ and $\sigma_{\max}$ denote the allowable minimum and maximum values of the cable tension in a cable-driven parallel mechanism respectively.

The above two cable tension indices are used to measure the tension characteristics of the same mechanism at a certain attitude, and can also be considered as an ability of the mechanism to avoid pseudo-drag at that attitude.

3.1.2 *Performance Indices of the Global Cable Tension in the Workspace of the Long-Span CDPR*

The following global indices are established based on the two force characteristic indices proposed in Sect. 3.1.1. The global tension characteristic index GTCI is described in the form of RMS.

The difference $\mathrm{GTCI}_{\max}$ between the global maximum and minimum cable tension:

$$\mathrm{GTCI}_{\max} = \sqrt{\frac{\sum_{t=0}^{t_0}\left(\max(\sigma_i(t) - \sigma_j(t))\right)^2}{t_0}} i = 1, \ldots, m;\ j = 1, \ldots, m;\ i \neq j \tag{3.4}$$

The global ratio $\mathrm{GTCI}_{\mathrm{rmax}}$ between the maximum and minimum cable tension:

$$\mathrm{GTCI}_{\mathrm{rmax}} = \sqrt{\frac{\sum_{t=0}^{t_0}\left(\max\left(\frac{\sigma_i(t)}{\sigma_j(t)}\right)\right)^2}{t_0}} i = 1, \ldots, m;\ j = 1, \ldots, m;\ i \neq j \tag{3.5}$$

where t is the sample point in the workspace of the long-span cable-driven parallel mechanism; t_0 is the sum of sample points.

The above indices are not applicable when there is a particularly poor tension position in the mechanism, thus the tension constraint is given as:

$$\begin{cases} \boldsymbol{\sigma} \geq (\sigma_{\min}, \ldots, \sigma_{\min})^{\mathrm{T}} \\ \boldsymbol{\sigma} \leq (\sigma_{\max}, \ldots, \sigma_{\max})^{\mathrm{T}} \end{cases}$$

The two global tension characteristic indices in this section aim to analyze the force characteristics of the same type of long-span cable-driven parallel mechanism to obtain the global mechanical property in the global workspace, which can also be considered as an evaluation of the ability of the cable-driven mechanism to avoid pseudo-drag and ensure reliable motion in the global workspace.

3.2 Analysis of End Errors and the Cable Tension Characteristics

The long-span cable-driven parallel mechanism is large in structural size with flexible cables, therefore, its end errors are greater compared with the rigid mechanism. In the previous studies on the statics or dynamics of the long-span cable-driven parallel mechanism, only the characteristics of the cable tension in the workspace of the mechanism are analyzed, while the influence of the end errors on the characteristics of the cable tension is neglected.

When the required workspace of the mechanism is at the boundary of the controlled workspace, although the characteristics of the cable tension at this attitude can satisfy the design requirements, it is difficult to guarantee the mechanical characteristics of the mechanism in its allowable end error space because of the inevitable control errors of the mechanism. If the end of the mechanism is at the attitude point with poor cable tension characteristics in the allowable error space, the mechanical property of the mechanism will be poor, which is detrimental to the accuracy and stability control of it. Therefore, it is an important proposition to evaluate and ensure that the cable-driven parallel mechanism meets the design requirements of the force characteristics even in the allowable end error space of a certain attitude.

In this section, the error model of the long-span cable-driven parallel mechanism will be established based on the modeling error compensation method, and the evaluation cable tension characteristics indices in the allowable end error space of the mechanism will be proposed.

3.2.1 *Error Analysis of the Long-Span CDPR*

The accuracy of the long-span cable-driven parallel mechanism is an important performance index, and the mechanical property of the mechanism affect each other. The end control errors of the cable-driven parallel mechanism are affected by many factors, such as driving unit errors, installation errors, manufacturing tolerances and clearances, structural design, etc. As for the parallel mechanism, the number of error sources is more than 100. To simplify the deduction, only the errors of joints and the length of branch chains are considered as error sources in literature for the estimation of end errors.

As to a fully constrained long-span cable-driven parallel mechanism, the errors of the single branch chain of the mechanism are shown in Fig. 3.1. The error model is described using a vector closed-loop, with $\Re : O - XYZ$ being the coordinate system on the static platform and $\Re' : O' - X'Y'Z'$ being the coordinate system of the end-effector on the moving platform. In the mechanism, $B_i (i = 1, 2, \ldots, m)$ is the connection point between the cable platform and the static platform, and $A_i (i = 1, 2, \ldots, m)$ is the connection point between the cable platform and the moving platform.

The errors of the cable-driven parallel mechanism include the error of the length of the branch chain $\delta l_{ci} (i = 1, 2, \ldots, m)$ and the position errors $\delta \boldsymbol{O'A}_i$ and $\delta \boldsymbol{OB}_i$ of the connection point of the upper and lower platforms. The output error of the cable-driven parallel mechanism, i.e., the end error of the moving platform is $\delta \boldsymbol{X} = \left[\delta \boldsymbol{OO'}^{\mathrm{T}}, \delta \boldsymbol{\Omega}^{\mathrm{T}}\right]^{\mathrm{T}}$, where $\delta \boldsymbol{OO'} = [\delta x, \delta y, \delta z]^{\mathrm{T}}$ is the position error and $\delta \boldsymbol{\Omega} = \left[\delta \omega_x, \delta \omega_y, \delta \omega_z\right]^{\mathrm{T}}$ is the attitude error.

According to the error model of the single branch chain of the long-span cable-driven parallel mechanism shown in Fig. 3.1, the kinematic equation of the branch chain can be deducted from Chap. 2 as:

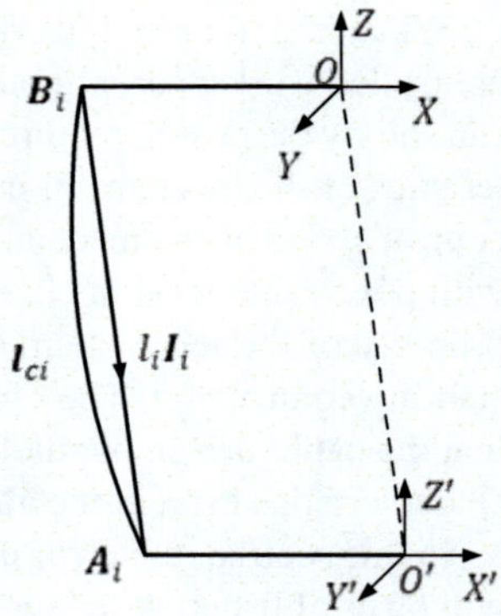

Fig. 3.1 Error model of the single branch chain of the long-span cable-driven parallel mechanism

$$\boldsymbol{l}_i = \boldsymbol{RO'A}_i + \boldsymbol{OO'} - \boldsymbol{OB}_i (i = 1, 2, \ldots, m) \quad (3.6)$$

The above equation can be expressed as:

$$l_i \boldsymbol{I}_i = \boldsymbol{RO'A}_i + \boldsymbol{OO'} - \boldsymbol{OB}_i \quad (3.7)$$

The above equation is the inverse kinematic equation in the cable linear model. To obtain the model of the single branch chain of the cable-driven parallel mechanism in accurate modeling, the kinematic modeling can be conducted according to the error modeling compensation method proposed in Chap. 2, and the above equation can be written as:

$$(l_{ci} - \varepsilon_{linei})\boldsymbol{I}_i = \boldsymbol{RO'A}_i + \boldsymbol{OO'} - \boldsymbol{OB}_i (i = 1, 2, \ldots, m) \quad (3.8)$$

where l_{ci} is the length of the i-th cable of the long-span cable-driven parallel mechanism based on precise catenary modeling, ε_{linei} is the compensation term of the length error of the i-th cable of the long-span cable-driven parallel mechanism based on linear modeling, and $\boldsymbol{R}$ is the coordinate rotation matrix.

Differentiating Eq. (3.8) obtains:

$$\delta(l_{ci} - \varepsilon_{linei})\boldsymbol{I}_i + (l_{ci} - \varepsilon_{linei})\delta\boldsymbol{I}_i = \delta\boldsymbol{RO'}^{\boldsymbol{A}_i} + \boldsymbol{R}\delta\boldsymbol{O'A}_i + \delta\boldsymbol{OO'} - \delta\boldsymbol{OB}_i \quad (3.9)$$

where,

$$\delta\boldsymbol{R} = \boldsymbol{R}_\Omega\boldsymbol{R} = \begin{bmatrix} 0 & -\delta\omega_z & \delta\omega_y \\ \delta\omega_z & 0 & -\delta\omega_x \\ -\delta\omega_y & -\delta\omega_x & 0 \end{bmatrix}\boldsymbol{R} \quad (3.10)$$

Substituting Eq. (3.10) into Eq. (3.9) obtains:

$$\delta(l_{ci} - \varepsilon_{linei})\boldsymbol{I}_i + (l_{ci} - \varepsilon_{linei})\delta\boldsymbol{I}_i = \delta\boldsymbol{OO'} + \delta\boldsymbol{RO'}^{\boldsymbol{A}_i} + \boldsymbol{R}\delta\boldsymbol{O'A}_i - \delta\boldsymbol{OB}_i$$

$$= \delta\boldsymbol{O}\boldsymbol{O}' + \boldsymbol{R}_{\Omega}\boldsymbol{R}\boldsymbol{O}'^{A_i} + \boldsymbol{R}\delta\boldsymbol{O}'\boldsymbol{A}_i - \delta\boldsymbol{O}\boldsymbol{B}_i$$
$$= \delta\boldsymbol{O}\boldsymbol{O}' + \delta\boldsymbol{\Omega} \times \left(\boldsymbol{R}\boldsymbol{O}'^{A_i}\right) + \boldsymbol{R}\delta\boldsymbol{O}'\boldsymbol{A}_i - \delta\boldsymbol{O}\boldsymbol{B}_i \tag{3.11}$$

where $\delta\boldsymbol{\Omega} = \left[\delta\omega_x, \delta\omega_y, \delta\omega_z\right]^{\mathrm{T}}$.

Since $\boldsymbol{I}_i^{\mathrm{T}}$ is the unit vector of the branch chain, it can be found from $\boldsymbol{I}_i^{\mathrm{T}}\boldsymbol{I}_i = 1, \quad \boldsymbol{I}_i^{\mathrm{T}}\delta\boldsymbol{I}_i = 0$ that Eq. (3.11) contains the variation of the direction of the vector of the branch chain with time.

The variation of the length of the branch chain can be separated from this equation by doing dot product on both sides of the equation with the unit vector $\boldsymbol{I}_i^{\mathrm{T}}$ synchronously:

$$\delta(l_{ci} - \varepsilon_{linei}) = \boldsymbol{I}_i^{\mathrm{T}}\delta\boldsymbol{O}\boldsymbol{O}' + \boldsymbol{I}_i^{\mathrm{T}}\delta\boldsymbol{\Omega} \times \left(\boldsymbol{R}\boldsymbol{O}'\boldsymbol{A}_i\right)$$
$$+ \boldsymbol{I}_i^{\mathrm{T}}\boldsymbol{R}\delta\boldsymbol{O}'\boldsymbol{A}_i - \boldsymbol{I}_i^{\mathrm{T}}\delta\boldsymbol{O}\boldsymbol{B}_i \tag{3.12}$$

After analyzing the above error model of the single branch chain, the kinematic error model of the long-span cable-driven parallel mechanism can be expressed as:

$$\boldsymbol{J}\delta\boldsymbol{X} = \delta\boldsymbol{L} + \delta\boldsymbol{\varepsilon} + \boldsymbol{D}_1\delta\boldsymbol{O}'\boldsymbol{A}_i + \boldsymbol{D}_2\delta\boldsymbol{O}\boldsymbol{B}_i \tag{3.13}$$

where,

$$\delta\boldsymbol{X} = \left[\delta\boldsymbol{O}\boldsymbol{O}'^{\mathrm{T}}, \delta\boldsymbol{\Omega}^{\mathrm{T}}\right]^{\mathrm{T}}$$

$$\delta\boldsymbol{L} = \begin{bmatrix} \delta l_{c1} \\ \vdots \\ \delta l_{cm} \end{bmatrix}$$

$$\delta\boldsymbol{\varepsilon} = \begin{bmatrix} -\delta\varepsilon_{line1} \\ \vdots \\ -\delta\varepsilon_{linem} \end{bmatrix}$$

$$\boldsymbol{J} = \begin{bmatrix} \boldsymbol{I}_1^{\mathrm{T}} & \boldsymbol{I}_1^{\mathrm{T}} \times (\boldsymbol{R} \cdot \boldsymbol{O}'\boldsymbol{A}_1) \\ \vdots & \vdots \\ \boldsymbol{I}_m^{\mathrm{T}} & \boldsymbol{I}_m^{\mathrm{T}} \times (\boldsymbol{R} \cdot \boldsymbol{O}'\boldsymbol{A}_m) \end{bmatrix}$$

$$\boldsymbol{D}_1 = \begin{bmatrix} -\boldsymbol{I}_1^{\mathrm{T}} \cdot \boldsymbol{R} & \cdots & 0 \\ \vdots & \ddots & \vdots \\ 0 & \cdots & -\boldsymbol{I}_m^{\mathrm{T}} \cdot \boldsymbol{R} \end{bmatrix}$$

$$\boldsymbol{D}_2 = \begin{bmatrix} \boldsymbol{I}_1^{\mathrm{T}} & \cdots & 0 \\ \vdots & \ddots & \vdots \\ 0 & \ldots & \boldsymbol{I}_m^{\mathrm{T}} \end{bmatrix}$$

where $\delta \boldsymbol{X}$ is the position error and attitude error at the end of the cable-driven parallel mechanism; $\delta \boldsymbol{L}$ is the cable length error; $\delta \boldsymbol{\varepsilon}$ is the modeling error of the cable length after adopting the linear model and its modeling error compensation; $\delta \boldsymbol{O}'\boldsymbol{A}_i$ is the error vector of the cable connection points of the moving platform; $\delta \boldsymbol{O}\boldsymbol{B}_i$ is the error vector of the cable connection points of the static platform.

If the Jacobi matrix $\boldsymbol{J}$ is invertible, i.e., $\boldsymbol{J}^{-1}$ exists, Eq. (3.13) can be expressed as:

$$\delta \boldsymbol{X} = \boldsymbol{J}^{-1}\delta \boldsymbol{L} + \boldsymbol{J}^{-1}\delta \boldsymbol{\varepsilon} + \boldsymbol{J}^{-1}\boldsymbol{D}_1\delta \boldsymbol{O}'^{\boldsymbol{A}_i} + \boldsymbol{J}^{-1}\boldsymbol{D}_2\delta \boldsymbol{O}\boldsymbol{B}_i \tag{3.14}$$

Ultimately, the kinematic error equation of the long-span cable-driven parallel mechanism can be expressed as:

$$\delta \boldsymbol{X} = \boldsymbol{E}\delta \boldsymbol{\eta} \tag{3.15}$$

where,

$$\boldsymbol{E} = \left[\boldsymbol{J}^{-1}, \boldsymbol{J}^{-1}, \boldsymbol{J}^{-1}\boldsymbol{D}_1, \boldsymbol{J}^{-1}\boldsymbol{D}_2\right]$$
$$\delta \boldsymbol{\eta} = \left[\delta \boldsymbol{L}^{\mathrm{T}}, \delta \boldsymbol{\varepsilon}^{\mathrm{T}}, \delta \boldsymbol{O}'\boldsymbol{A}_i^{\mathrm{T}}, \delta \boldsymbol{O}\boldsymbol{B}_i^{\mathrm{T}}\right]^{\mathrm{T}}$$

$\boldsymbol{E}$ is the error transfer matrix of the long-span cable-driven parallel mechanism, where several main geometric configuration factors are considered, therefore, through the matrix $\boldsymbol{E}$, all error sources can be transferred to the end of the cable-driven parallel mechanism.

3.2.2 *Study of Mechanical Characteristics in Error Space*

To ensure the stability of the cable-driven parallel mechanism, it is required that the cable tension in the workspace meets certain conditions. However, due to the influence of processing, installation and control, there are errors in practical work of the mechanism. When the cable-driven parallel mechanism moves to the boundary of the workspace, the end errors of the mechanism may make its mechanical property get worse dramatically.

In the previous design of the cable-driven parallel mechanism, only the mechanism with good force characteristics in the workspace is considered, while the influence of the mechanism's allowable end error on the performance of the cable tension is ignored, which may lead to pseudo-drag in practical applications and affect the stability of the mechanism. Therefore, to measure the force characteristics of the long-span cable-driven parallel mechanism at a certain attitude in the error space, a controllable volume index TCGVI is defined to quantify the force characteristics of the mechanism at that attitude.

Assuming that the end attitude of the long-span cable-driven parallel mechanism is $\left[x_0, y_0, z_0, \alpha_0, \beta_0, \gamma_0\right]^{\mathrm{T}}$, and setting its allowable error at this attitude as $\left[\varepsilon x, \varepsilon y, \varepsilon z, \varepsilon\alpha, \varepsilon\beta, \varepsilon\gamma\right]^{\mathrm{T}}$:

$$\varepsilon x \in (-x_r, x_r);\ \varepsilon y \in (-y_r, y_r);\ \varepsilon z \in (-z_r, z_r);$$
$$\varepsilon\alpha \in (-\alpha_r, \alpha_r);\ \varepsilon\beta \in (-\beta_r, \beta_r);\ \varepsilon\gamma \in (-\gamma_r, \gamma_r);$$

Then the allowable error space at this attitude is $\forall_{\mathrm{err}}$:

$$\forall_{\mathrm{err}} = \left\{ x, y, z, \alpha, \beta, \gamma \,\middle|\, \begin{array}{c} x \in (-x_r + x_0, x_r + x_0) \\ y \in (-y_r + y_0, y_r + y_0) \\ z \in (-z_r + z_0, z_r + z_0) \\ \alpha \in (-\alpha_r + \alpha_0, \alpha_r + \alpha_0) \\ \beta \in (-\beta_r + \beta_0, \beta_r + \beta_0) \\ \gamma \in (-\gamma_r + \gamma_0, \gamma_r + \gamma_0) \end{array} \right\} \tag{3.16}$$

The generalized volume $\mathrm{V}_{\forall_{\mathrm{err}}}$ of the error space $\forall_{\mathrm{err}}$ can be expressed as:

$$\mathrm{V}_{\forall_{\mathrm{err}}} = \int\int\int\int\int\underset{\forall_{\mathrm{err}}}{\int} dxdydzd\alpha d\beta d\gamma \tag{3.17}$$

To ensure the stability of the cable-driven parallel mechanism, the cable tension is required to satisfy certain conditions, i.e., $\sigma_{\max} \geq \sigma_{\mathrm{i}} \geq \sigma_{\min}$, where $\sigma_{\min}$ is the allowable minimum value of the cable tension and $\sigma_{\max}$ is the allowable maximum value of the cable tension. To describe the force characteristics in the allowable error space of the long-span cable-driven parallel mechanism at a certain attitude, the controllable space $\forall_{\mathrm{control}}$ is defined as the set of the attitude of the mechanism that satisfies the tension conditions at a certain attitude in the allowable error space, i.e.:

$$\begin{aligned} &\forall_{\mathrm{control}} = \{x_c, y_c, z_c, \alpha_c, \beta_c, \gamma_c | \sigma_{\max} \geq \sigma_i \geq \sigma_{\min}, i = 1, \ldots, m\} \\ &\forall_{\mathrm{control}} \in \forall_{\mathrm{err}} \end{aligned} \tag{3.18}$$

Therefore, the generalized volume $\mathrm{V}_{\forall_{\mathrm{control}}}$ of the controllable space in the error space at a particular attitude of the long-span cable-driven parallel mechanism can be obtained as:

$$\mathrm{V}_{\forall_{\mathrm{control}}} = \int\int\int\int\int\underset{\forall_{\mathrm{control}}}{\int} dxdydzd\alpha d\beta d\gamma \tag{3.19}$$

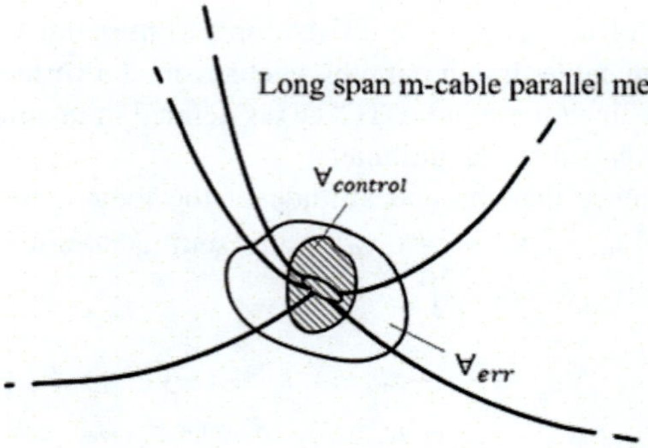

Fig. 3.2 Allowable error space and controllable space of the long-span cable-driven parallel mechanism at a certain attitude

The relationship between the allowable error space and the controllable space at a certain attitude of the long-span cable-driven parallel mechanism is shown in Fig. 3.2.

To quantify the force characteristics of the long-span cable-driven parallel mechanism at a certain attitude in the error space, a Tension Controllability Generalized Volume Index ($\Theta_\forall$, TCGVI) is used to express:

$$\Theta_\forall = \frac{V_{\forall_{control}}}{V_{\forall_{err}}} \quad (\Theta_\forall \in (0, 1]) \tag{3.20}$$

where $\Theta_\forall$ is the ratio of the generalized volume $V_{\forall_{control}}$ of the controllable space $\forall_{control}$ in the error space to the generalized volume $V_{\forall_{err}}$ of the allowable error space $\forall_{err}$.

$\Theta_\forall$ can quantify the controllability of the mechanism at this attitude. When $\Theta_\forall$ tends to 1, it means that the force characteristics of the mechanism at that attitude are fine and there is a low probability of pseudo-drag. Similarly, when $\Theta_\forall$ tends to 0, it means that the force characteristics of the mechanism at that attitude are poor and there is a high probability of pseudo-drag. In the optimization design of the mechanism based on force characteristics, $\Theta_\forall$ should be 1 as much as possible.

To analyze the cable tension characteristics at each attitude within the error space in the global workspace, RMS is proposed to indicate the global force characteristic GTCGVI (Tension Controllability Generalized Volume Index), which is defined as:

$$\mathrm{GTCGVI} = \sqrt{\frac{\sum_{t=0}^{t_0} (\Theta_\forall(t))^2}{t_0}} t = 1, 2, \ldots, t_0 \tag{3.21}$$

where t is the sample point in the workspace of the long-span cable-driven parallel mechanism; t_0 is the sum of sample points.

This section discusses the force characteristic indices in the error space at a specific attitude of the long-span cable-driven parallel mechanism and the global force characteristic indices in the global workspace, aiming to analyze the mechanical property of the mechanism in the allowable error space at each attitude in the global workspace, which can also be considered as an evaluation of the ability of the mechanism to avoid pseudo-drag in the global workspace.

3.3 Force Characteristics Analysis of the Similar Model of the First-Level Feed Support System of FAST

The basic modeling method of the six-cable driven parallel mechanism and its similar models in FAST has been studied in Chap. 2. In this section, the force characteristics are studied using the two moving trajectories given by the similar models in Chap. 2, including force characteristics and attitude relationship analysis, error analysis and its force characteristics analysis in the allowable error space.

3.3.1 Study of Force Characteristics and Attitude Under Two Specific Trajectories

Figure 3.3 shows the workspace of the 40 m six-cable driven parallel mechanism. The workspace of its moving platform is a spherical dome with a radius of 9 6 m and a maximum observation angle ±40°.

Two boundary trajectories of the workspace are selected as the study objects, as shown in Fig. 3.4. Figure 3.4a shows trajectory 1, which is a circle formed at the highest point of the workspace of the six-cable driven parallel mechanism with a radius of 6.17 m and a height of 10.6 m from the ground, requiring the attitude angle of the cable-driven parallel mechanism to be the pitch angle of 40°; Fig. 3.4b shows

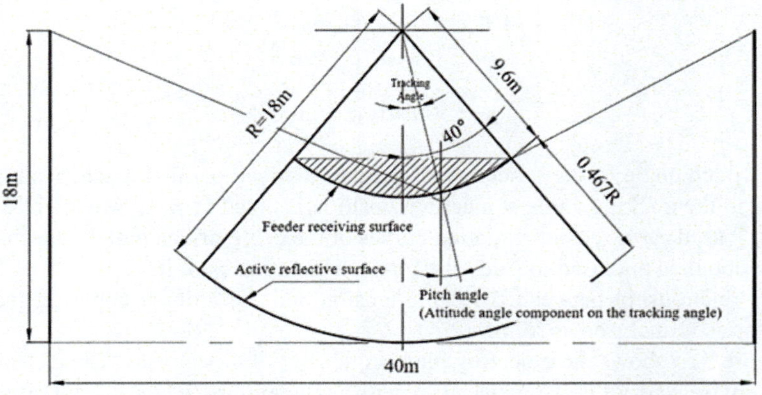

Fig. 3.3 The workspace of the 40 m six-cable driven parallel mechanism

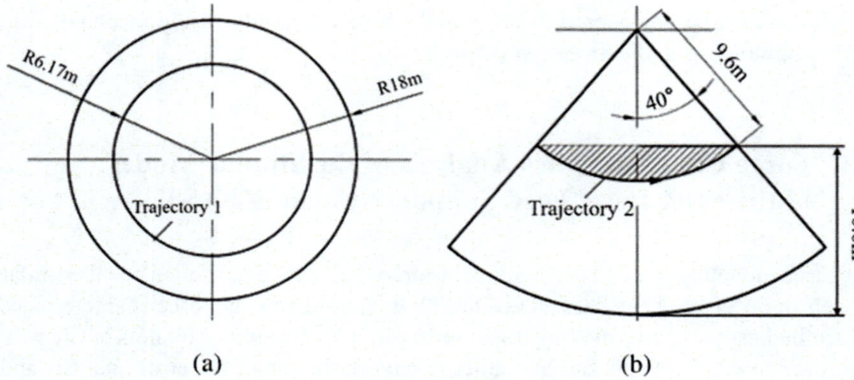

Fig. 3.4 Two moving trajectories in the workspace of the six-cable driven parallel mechanism. **a** Trajectory 1; **b** trajectory 2

trajectory 2, which is a contour line on the surface of the workspace of the six-cable driven parallel mechanism.

To prevent the pseudo-drag phenomenon and the uncontrollability of the mechanism, the minimum pretightening force $\sigma_{\min}$ and the maximum tension $\sigma_{\max}$ are given for the cable-driven parallel mechanism. Let the feed at a certain position requires the trajectory tracking pitch angle α, and the cable-driven parallel mechanism can achieve the trajectory pitch angle β. The equation of the trajectory attitude angle of the cable-driven parallel mechanism based on the tension constraint can be expressed as:

Objective function:

Approaching the tracking angle as closely as possible $\min|\alpha - \beta|$

Constraint function:

$$\begin{cases} \boldsymbol{\sigma} \geq [\sigma_{\min}, \ldots, \sigma_{\min}]^{\mathrm{T}} \\ \boldsymbol{\sigma} \leq [\sigma_{\max}, \ldots, \sigma_{\max}]^{\mathrm{T}} \end{cases}$$

$$\sigma_{\min} = 500\mathrm{N}\sigma_{\max} = 3000\mathrm{N}$$

The pitch angle of the trajectory of the cable-driven parallel mechanism that is closest to the tracking angle at a certain position is noted as $\beta_{\to\alpha}$. Meanwhile, a set of $\beta_{\to\alpha}$-based cable tension variation curves of the cable-driven parallel mechanism can be obtained and used to study the force characteristics of it.

The relationship between the cable tension and the attitude angle on the two trajectories is analyzed as follows.

Figure 3.5a shows the trajectory pitch angle $\beta_{\to\alpha}$ of trajectory 1. The distribution of $\beta_{\to\alpha}$ of trajectory 1 is symmetrical but with angular jumps, which cannot satisfy the maximum observation angle requirement. The variation curves of the cable tension of trajectory 1 is shown in Fig. 3.5b, and it meets the cable tension requirement with the tension constraint function, but the variation of cable tension curves is not smooth due to the jump of the attitude angle.

To ensure the continuity of the cable tension, it is necessary to take the variation curve of β^{opt} as a continuous curve, and trajectory 1 is at the boundary of the workspace, which requires a large attitude angle, thus taking β^{opt} as the inner tangent of $\beta_{\rightarrow\alpha}$, as shown in Fig. 3.5c, then the maximum pitch angle 16.9° of the six-cable driven parallel mechanism at this position can be obtained, and the optimization curves of the cable tension are shown in Fig. 3.5d.

Similarly, the variation curve $\beta_{\rightarrow\alpha}$ of trajectory 2 and the comparison figure of the demanding pitch angle can be obtained. As shown in Fig. 3.6a, the pitch angle of the six-cable driven parallel mechanism is at $x = [-6.17\,\text{m}, -1.85\,\text{m}] \cup [1.85\,\text{m}, 6.17\,\text{m}]$, which cannot satisfy the requirement.

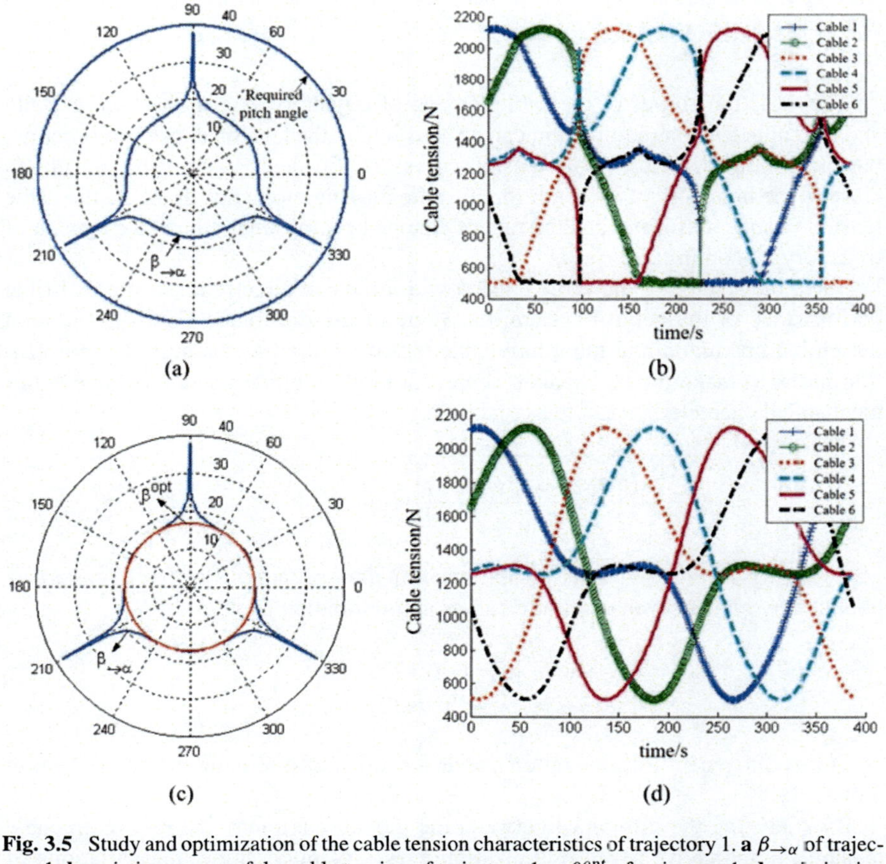

Fig. 3.5 Study and optimization of the cable tension characteristics of trajectory 1. **a** $\beta_{\rightarrow\alpha}$ of trajectory 1; **b** variation curves of the cable tension of trajectory 1; **c** β^{opt} of trajectory 1; **d** optimization curves of the cable tension of trajectory 1

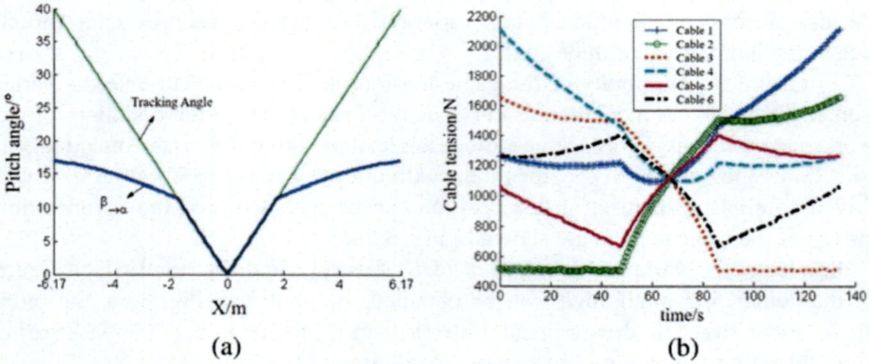

Fig. 3.6 Study of the cable tension characteristics of trajectory 2. **a** The demand tracking angle and $\beta_{\to\alpha}$ angle of the six-cable driven parallel mechanism; **b** variation curves of the cable tension of trajectory 2

The variation curves of the cable tension of trajectory 2 is shown in Fig. 3.6b, and the cable tension requirement can be satisfied by the tension constraint function, but the continuity of the cable tension curves is not ideal, as the cable tension is close to the boundary of its condition in some moving moments. To make the cable tension change uniformly and the rate of change become small, the cable tension of trajectory 2 is optimized.

Since the change of the cable tension with time can directly affect the real-time performance of the control system, the value of the difference $\mathrm{GTCI}_{\max}$ between the global maximum and minimum cable tension is used as the attitude optimization index to make the mechanism achieve a larger attitude angle and have better mechanical characteristics at the same time.

$$\mathrm{GTCI}_{\max} = \sqrt{\frac{\sum_{t=0}^{t_0}\left(\max\left(\sigma_i(t) - \sigma_j(t)\right)\right)^2}{t_0}} i = 1, \ldots, m;\ j = 1, \ldots, m;\ i \neq j$$

As shown in Fig. 3.7, a set of optimization curve functions of pitch angle β is obtained by giving seven quadratic curves approximating $\beta_{\to\alpha}$:

$$\beta_k^{opt} = \begin{cases} a_k x + b_k & x = [-6.17, 0] \\ a_k x - b_k & x = [0, 6.17] \end{cases} \quad k = 1, 2, \ldots, 7$$

The coefficients of optimization curve functions of pitch angle β are shown in Table 3.1.

By analyzing the difference between the global maximum and minimum cable tension for the above seven optimization curves of pitch angle, the variations of $\mathrm{GTCI}_{\max}$ with the change of pitch angle curves are obtained as shown in Fig. 3.8.

Through the above analysis, it can be found that with the optimization curves of the attitude angle getting farther away from $\beta_{\to\alpha}$, the curve $\mathrm{GTCI}_{\max}$ shows a trend

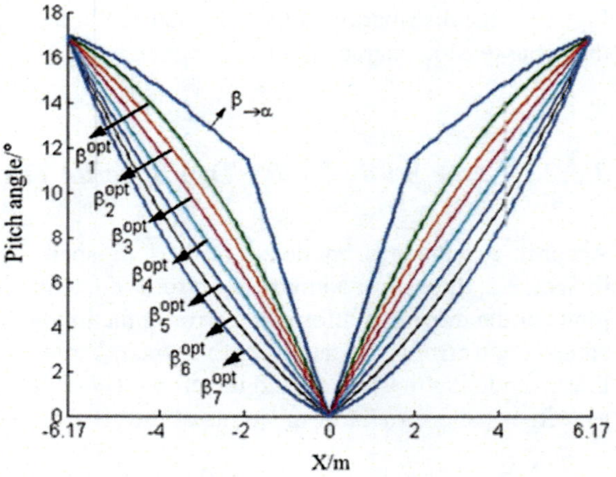

Fig. 3.7 Seven quadratic curves approximating $\beta_{\to\alpha}$

Table 3.1 The coefficients of optimization curve functions of pitch angle $\boldsymbol{\beta}$

Coefficients	$k = 1$	$k = 2$	$k = 3$	$k = 4$	$k = 5$	$k = 6$	$k = 7$
a_k	−0.3	−0.2	−0.1	−2.74	0.1	0.2	0.3
b_k	−4.59	−3.97	−3.36	0	−2.12	−1.51	−0.89

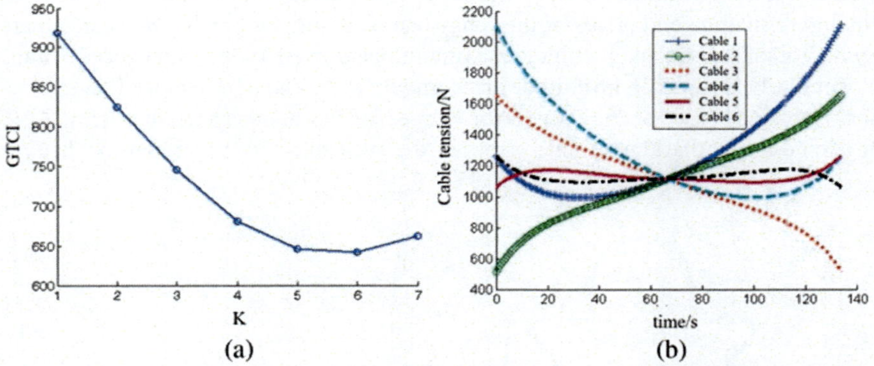

Fig. 3.8 Optimization curves of trajectory 2. **a** Curve $GTCI_{max}$; **b** optimization variation curve soft the cable tension cf trajectory 2

of first decreasing and then increasing. It can be seen that the optimization curves of the pitch angle β_k^{opt} (k= 4, 5, 6) have better mechanical characteristics to ensure the uniform variation of the tension of the cable-driven parallel mechanism. To get a larger trajectory attitude angle, β_5^{opt} is chosen here as the variation curve of the pitch angle on this trajectory for further study. Figure 3.8b shows the variation of the cable tension on the β_5^{opt}-based pitch angle curve on this trajectory, and compared with

Fig. 3.6b, the distribution of the cable tension of the mechanism is more uniform and the cable tension characteristics are better.

3.3.2 Error Study Under Two Specific Trajectories

According to the error modeling of the long-span cable-driven parallel mechanism in Sect. 3.2, three error sources are introduced, that is, cable length error, error of the joints of the moving platform and error of the joints of the static platform, where the cable length error is the cable length error εl_{ine} after the error compensation based on linear modeling in Chap. 2, and the error $\varepsilon \boldsymbol{OB}$ of the joints of the moving platform as well as the error $\varepsilon' \boldsymbol{A}$ of the joints of the static platform are defined as:

$$|\varepsilon l_{ine}|_{\max} = 2\,\text{mm}$$
$$\|\varepsilon \boldsymbol{OB}\|_{\max} = 0.8\,\text{mm}$$
$$\left\|\varepsilon \boldsymbol{O'A}\right\|_{\max} = 0.8\,\text{mm}$$

The two moving trajectories in Fig. 3.4 are used to study the maximum position error and attitude error of the six-cable driven parallel mechanism under them.

Figure 3.9 shows the maximum position error amount at the end of trajectory 1 and trajectory 2 obtained based on the above error sources, and it can be found from the figure that the end errors of the long-span cable-driven parallel mechanism are symmetrically distributed with the maximum position error does not exceed 7 mm.

Similarly, the maximum attitude error amount at the end of trajectory 1 and trajectory 2 obtained based on the above error sources can be obtained as shown in Fig. 3.10. It can be found that the attitude errors of the six-cable driven parallel mechanism

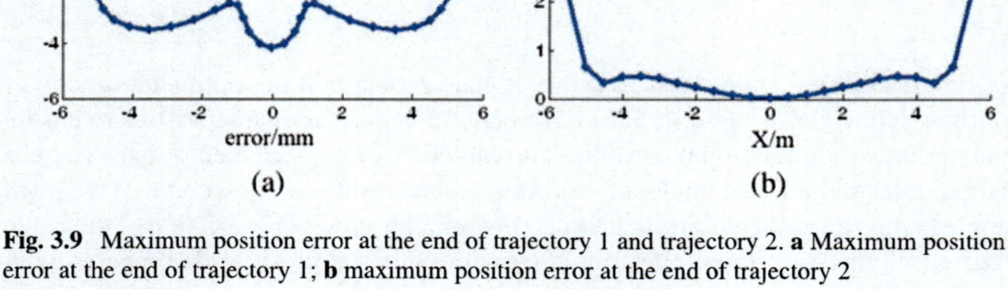

Fig. 3.9 Maximum position error at the end of trajectory 1 and trajectory 2. **a** Maximum position error at the end of trajectory 1; **b** maximum position error at the end of trajectory 2

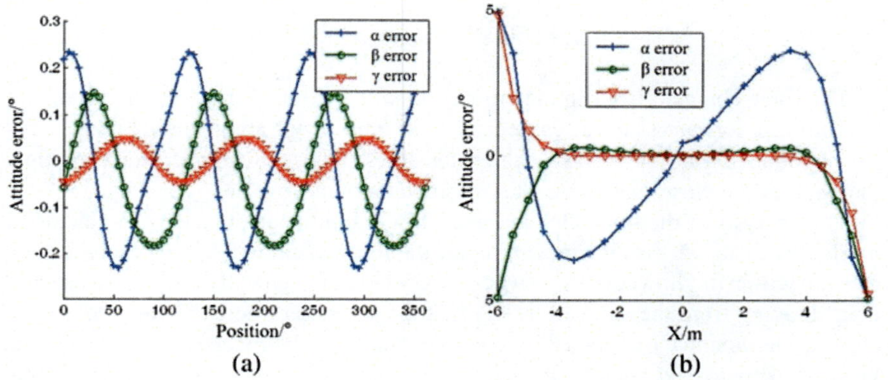

Fig. 3.10 Maximum attitude error at the end of trajectory 1 and trajectory 2. **a** Maximum attitude error at the end of trajectory 1; **b** maximum attitude error at the end of trajectory 2

are also symmetrically distributed with the maximum attitude error does not exceed 0.5°.

3.3.3 Study of Force Characteristics in Error Space Under Two Specific Trajectories

To verify the validity of the cable tension characteristics of the long-span cable-driven parallel mechanism at a certain attitude in error space quantified by the Tension Controllability Generalized Volume Index Θ_v, i.e., TCGVI proposed in this chapter, two moving trajectories of the similar model of the six-cable driven parallel mechanism in FAST are simulated and analyzed.

The relationship between the attitude angle and force characteristics of the two trajectories in the six-cable driven parallel mechanism is given in Sect. 3.3.1, and the variation curves of the six-cable tension on the two trajectories are also obtained. To study the force characteristics of the two moving trajectories in their error spaces, the maximum error is first given as:

$$\sqrt{x_r^2 + y_r^2 + z_r^2} = 6\text{mm};$$

$$\alpha_r = \pm 0.5^\circ;\ \beta_r = \pm 0.5^\circ;\ \gamma_r = \pm 0.5^\circ;$$

Similarly, its cable tension is required to satisfy:

$$\begin{cases} \boldsymbol{\sigma} \geq (\sigma_{\min}, \ldots, \sigma_{\min})^{\mathrm{T}} \\ \boldsymbol{\sigma} \leq (\sigma_{\max}, \ldots, \sigma_{\max})^{\mathrm{T}} \end{cases}$$

$$\sigma_{\min} = 500\,\text{N} \quad \sigma_{\max} = 3000\,\text{N}$$

The force characteristics in the error space of trajectory 1 are first analyzed.

The variation curves of the cable tension of trajectory 1 are given in Fig. 3.11a. It can be found that although all six cables' tension satisfies the cable tension condition, their values are close to the boundary at some moments.

The analysis of the force characteristic TCGVI in the error space of trajectory 1 in Fig. 3.11b shows that at some attitudes, the mechanism will have a cable tension overrun within its allowable error (when TCGVI < 1), which will easily cause pseudo-drag. Judging from the value of TCGVI, at the positions where the cable tension is close to the boundary of the cable tension condition, the mechanism is more than twice as likely to have pseudo-drag in the allowable error space than at the uniformly distributed positions of the cable tension, and the end errors of the mechanism at these positions will seriously affect the stability of the mechanism's motion. The comparison with Fig. 3.11a shows that the variation of the curves of the force characteristic indices is consistent with the variation of the uniformity of the cable tension, indicating that the force characteristic index TCGVI can quantify the cable tension characteristics of the mechanism.

Similarly, the mechanical characteristics in the error space of trajectory 2 can be studied.

The variation curves of the cable tension of trajectory 2 is given in Fig. 3.12a. It can be seen that the cable tension is distributed uniformly at all times except for the cable tension at the ends of the trajectory closes to the boundary of the cable tension condition.

Therefore, it can be seen from Fig. 3.12b that the value of TCGVI of the force characteristics in the error space of the six-cable driven parallel mechanism at all attitudes is 1 except for the ends of the moving trajectory are susceptible to the end

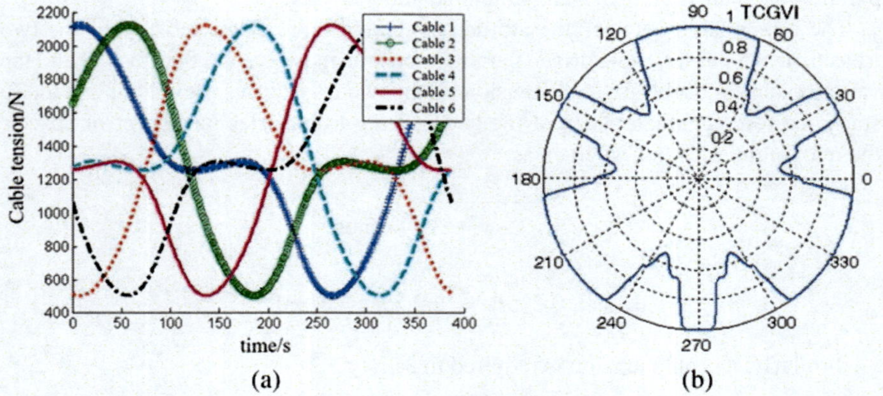

Fig. 3.11 Force characteristics of trajectory 1 in its error space. **a** Optimization variation curves of the cable tension of trajectory 1; **b** curve TCGVI

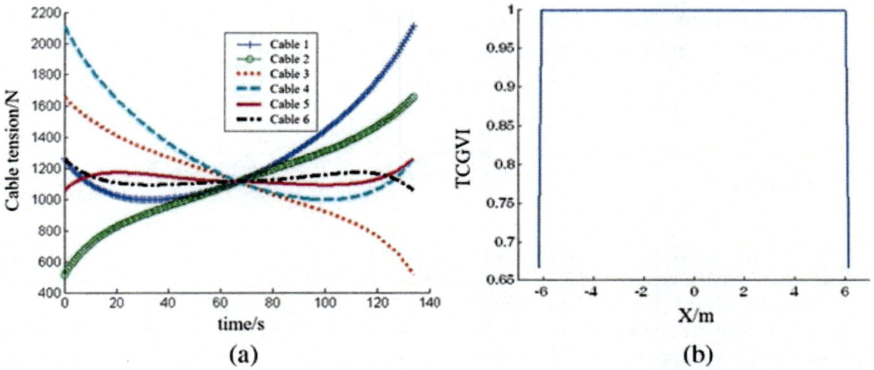

Fig. 3.12 Force characteristics of trajectory 2 in its error space. **a** Optimization variation curves of the cable tension of trajectory 2; **b** curve TCGVI

position errors. The comparison with Fig. 3.12a reveals that the variation of the curves of the force characteristic indices is consistent with the variation of the uniformity of the cable tension, indicating that the force characteristic index TCGVI can quantify the cable tension characteristics of the mechanism.

Based on the simulation study in this section, it can be seen that the value of TCGVI, that is, the force characteristic index in the error space of the long-span cable-driven parallel mechanism can effectively quantify the force characteristics of the mechanism in its workspace, thus can be used for the optimization design and control strategy study of the long-span cable-driven parallel mechanism.

3.4 Conclusion

(1) In the analysis of the mechanical characteristics of the cable-driven parallel mechanism, two cable tension indices are commonly adopted: the difference between the maximum and minimum cable tension, as well as the ratio between the maximum and minimum cable tension. Based on the local indices, the global force characteristic indices can be established in the form of RMS.
(2) Based on the error transfer relationship in the workspace of the long-span cable-driven parallel robot and the force characteristics in the error space, the Tension Controllability Generalized Volume Index is proposed, which can also be established in the form of RMS.
(3) In this chapter, the force characteristics and error characteristics of the first-level feed support system of the similar 40 m model of FAST are simulated. The results show that the workspace of the first-level feed support system needs to consider the cable tension characteristics as well as that in the allowable error space. At the same time, it is difficult for the first-level feed support system to directly realize the pitch angle required by FAST, thus an angular mechanism is

added to reduce the angular requirement of the cable-driven parallel mechanism and improve its accuracy and reliability.

References

1. Landsberger S E, Sheridan T B. A minimal linkage: the tension-compression parallel link manipulator. In: Toshi Takampri and K. Tsuchiya, editors. Robotics, Mechatronics and Manufacturing System, Elseriver, Newyork, USA; 1993, p. 81–8.
2. Sun X. Nonlinear static, kinematic and dynamical theories and methods for the suspension feed support system of large radio telescopes [PhD thesis]. Xi'an University of Science and Technology; 2001.
3. Su YX, Duan BY. The mechanical design and kinematics accuracy analysis of a fine tuning stable platform for the large spherical radio telescope. Mechatronics. 2000;10:819–34.
4. Takeda Y, Funabashi H. A transmission index for in-parallel wire-driven mechanism. Trans Japanese Soc Mech Eng. 1999;65:2521–7.
5. Takeda Y, Funabashi H. Kinematic synthesis of spatial in-parallel wire-driven mechanism with six-degrees of freedom with high force transmissibility. In: Proceedings of the ASME-DETC2000; 2000.
6. Yaqing, Z. Research on key theoretical issues of wire—driven parallel kinematic manipulators and the application to wind tunnel support systems [PhD thesis]. Quanzhou: Huaqiao University; 2004.

Chapter 4
Stiffness Analysis of the Long-Span Cable-Driven Parallel Robot

Since most of the long-span cable-driven parallel robots are installed in the outdoor environment and will be disturbed by wind, they need to be stiff enough to ensure accuracy and stability under the influence of external forces. Among them, the stiffness characteristic is one of the important indices to evaluate the performance of the long-span CDPR.

The stiffness of the long-span CDPR is related to the cable force, cable length and other factors. At present, two methods are mainly used to study the stiffness of the long-span CDPR: the first one is theoretical research for the establishment of the stiffness matrix. As it is difficult to determine the yield behavior of the mechanism's actuator, the friction between the cable and the cable-sheave, as well as the friction between the cable and the moving platform, they are generally ignored, and only the overall stiffness performance of the mechanism is approximated; the second one is the prototype experiment, that is, to obtain accurate stiffness values of the mechanism through experimental studies, but it is difficult to realize because of the expensive building cost of the prototype of the long-span CDPR.

To address the problem, this chapter organically combines theoretical analysis and experimental studies. Firstly, the stiffness matrix is established based on the cable force and cable length obtained from the kinematic and static modeling of the long-span CDPR in Chap. 2, and the stiffness matrix is simplified by employing the modeling error compensation method to obtain the theoretical stiffness values of the mechanism; then, the similarity theory is adopted to establish the physical stiffness similarity model of the long-span cable-driven parallel robot; finally, the experimental results on the stiffness similarity model is used to verify the theoretical results and analyze the stiffness of the prototype.

In this chapter, Sect. 4.1 discusses the theoretical stiffness analysis method of the long-span cable-driven parallel mechanism, establishes the stiffness matrix, presents the stiffness characteristic indices in the global workspace, and proposes the theoretical foundation for the comprehensive design of the long-span cable-driven parallel mechanism; Sect. 4.2 studies the stiffness similarity method of the long-span CDPR

© Tsinghua University Press 2024

X. Tang et al., *Research and Application of Cable-Driven and Rigid Parallel Robots*,
https://doi.org/10.1007/978-981-99-7452-8_4

based on the similarity theory; Sect. 4.3, for the scheme of the first-level support system for FAST (the long-span CDPR), to complete the modeling and experiments of the stiffness similarity model, deduce the stiffness of the prototype, and verify the feasibility of combining theoretical analysis and similarity model in the optimization design of the long-span CDPR.

The main contents of this chapter are:

(1) Simplified stiffness analysis of the long-span CDPR;
(2) Similarity stiffness method for the long-span CDPR;
(3) Modeling and experiments of the stiffness similarity model for the four-cables scheme of the first-level support system for FAST.

4.1 Simplified Stiffness Analysis of the Long-Span CDPR

The cable-driven parallel robot consists of the frame (the stationary platform), the cable and the moving platform. As the stiffness of the frame is generally two orders of magnitude higher than that of the cable, and the friction between the cable and the cable-sheave is difficult to determine, in the theoretical studies of the stiffness of the long-span cable-driven parallel mechanism, the stiffness of it can only be deduced approximately. To obtain more accurate values of the stiffness, a combination of physical experiments is required.

In the theoretical studies, Kino [1] divided the stiffness of the cable-driven parallel mechanism into the stiffness due to internal forces along the direction of cable tension and the stiffness due to internal forces along the direction of vertical cable tension and confirmed that redundant cables in the redundant mechanism can increase the stiffness of the mechanism. Verhoeven [2] studied the variation of the stiffness of the moving platform of the mechanism due to the pretightening force of the cable and derived the workspace conditions with stiffness. According to Verhoeven, Yaqing [3], together with Lafourcade, proposed a research method for the stiffness of the moving platform, including the homogeneous stiffness matrix, its condition number and its geometric mean stiffness. Saeed [4] neglected the effect of friction at the connection point between the cable and the cable-sheave and simplified the solution to obtain the values of stiffness of the cable-driven parallel mechanism by taking the planar cable-driven parallel mechanism as an example.

This chapter aims to adopt the theoretical analysis of stiffness as the theoretical foundation for stiffness analysis of similarity models, thus the theoretical studies are carried out by using the simplified stiffness method. In the theoretical studies of the simplified stiffness of the long-span cable-driven parallel mechanism, it is first assumed that the cable-driven parallel mechanism satisfies the following conditions:

(1) The self-weight and elastic deformation of the cable cannot be neglected, and the cable is within its elastic deformation zone;
(2) The cable can only bear tension, not pressure;

(3) The friction between the cable and the cable-sheave, and the friction between the cable and the frame, as well as the connection point of the moving platform are neglected.

Then, define the stiffness of the cable as follows:

k_0 is the unit stiffness of the cable in the long-span cable-driven parallel mechanism:

$$k_0 = EA \tag{4.1}$$

where E is the elastic modulus of the cable and A is the nominal cross-sectional area. Thus the stiffness of a single cable in the cable-driven parallel mechanism can be obtained as:

$$k_i = \frac{k_0}{l_{ci}}(i = 1, 2, \ldots, m) \tag{4.2}$$

where l_{ci} is the length of the i-th cable derived from the exact model of the cable-driven parallel mechanism based on the self-weight and elastic deformation of the cable.

The same type of cable is generally used in a cable-driven parallel mechanism, thus the stiffness matrix $\boldsymbol{K}$ of the long-span cable-driven parallel mechanism can be obtained as:

$$\boldsymbol{K} = k_0 \boldsymbol{J}_c^{\mathrm{T}} \mathrm{diag}\left(\frac{1}{l_1}, \ldots, \frac{1}{l_m}\right) \boldsymbol{J}_c \tag{4.3}$$

where $\boldsymbol{J}_c^{\mathrm{T}}$ is the force transfer matrix of the end of the long-span cable-driven parallel mechanism based on the exact catenary modeling, and l_i is the actual length of the cable under tension.

$$l_i = l_{ci} - \Delta l_{ci} \tag{4.4}$$

where Δl_{ci} is the elastic deformation amount of the cable.

Previous stiffness modeling studies generally focus on the short-span cable-driven parallel mechanism, and the effect of the elastic deformation of the cable is usually ignored. In this section, the stiffness matrix will be established based on the modeling of the elastic deformation of the cable in Sect. 2.2, and the actual length of the cable can be expressed as:

$$l_i = \int_l \frac{1}{\frac{\sigma_i}{EA} + 1} dl \tag{4.5}$$

where σ_i is the tension of the i-th cable.

The establishing method of the stiffness matrix based on the exact catenary model is more complicated to deduce the stiffness values of the mechanism, thus the simplified stiffness matrix $\tilde{\boldsymbol{K}}$ can be established by adopting the modeling error compensation method in Chap. 2.

Firstly, the actual length of the cable is written as:

$$\tilde{l}_i = l_{li} + \varepsilon l_{li} - \Delta l_{li} \tag{4.6}$$

where l_{li} is the cable length deduced based on the linear model; εl_{li} is the modeling error compensation value from linear modeling; Δl_{li} is the elastic deformation amount of the cable, which can be expressed as:

$$\Delta l_{li} = \frac{\sigma_i (l_{li} + \varepsilon l_{li})}{EA} \tag{4.7}$$

The simplified stiffness matrix $\tilde{\boldsymbol{K}}$ can be obtained as:

$$\tilde{\boldsymbol{K}} = k_0 \boldsymbol{J}^T diag\left(\frac{1}{\tilde{l}_1}, \ldots, \frac{1}{\tilde{l}_m}\right) \boldsymbol{J} = \boldsymbol{J}^{\mathrm{T}} \begin{bmatrix} \frac{k_0}{\tilde{l}_1} & 0 & 0 \\ 0 & \ddots & 0 \\ 0 & 0 & \frac{k_0}{\tilde{l}_m} \end{bmatrix} \boldsymbol{J} \tag{4.8}$$

where $\boldsymbol{J}^T$ is the force transfer matrix of the cable-driven parallel mechanism for linear modeling, which can be referred to Eq. (2.32).

After establishing the stiffness matrix $\tilde{\boldsymbol{K}}$ of the long-span cable-driven parallel mechanism, the stiffness equation of it can be obtained as:

$$\boldsymbol{F} = \tilde{\boldsymbol{K}} \Delta \tag{4.9}$$

where Δ is the elastic displacement of the stiffness test point of the mechanism; $\boldsymbol{F}$ is the force excitation of that.

According to the obtained simplified stiffness matrix $\tilde{\boldsymbol{K}}$, the stiffness in three translational directions and three rotational directions of the mechanism at any attitude can be obtained by conducting experiments on the stiffness test points of the moving platform of the mechanism, which can be expressed as:

Translational stiffness: k_x, k_y, k_z.

Rotational stiffness: $k_\alpha, k_\beta, k_\gamma$.

Compared with the rigid mechanism, the stiffness of the cable-driven parallel mechanism is smaller, which will affect its stability and control accuracy. To obtain higher control accuracy and stability, the stiffness characteristic of the mechanism needs to be considered in the comprehensive design.

To effectively evaluate the stiffness characteristic of the mechanism, two stiffness indices are given as follows:

1. **The stiffness index S_{G} in a single direction in the global workspace**

For a single attitude,

$$\boldsymbol{S} = \left[S_x, S_y, S_z, S_\alpha, S_\beta, S_\gamma\right] \tag{4.10}$$

$$S_x = k_x \tag{4.11}$$

$S_y, S_z, S_\alpha, S_\beta, S_\gamma$ can be expressed in the same way.

The unidirectional stiffness index in the global workspace can be expressed by the RMS of the stiffness:

$$\boldsymbol{S}_{\mathrm{G}} = \left[S_{\mathrm{G}x}, S_{\mathrm{G}y}, S_{\mathrm{G}z}, S_{\mathrm{G}\alpha}, S_{\mathrm{G}\beta}, S_{\mathrm{G}\gamma}\right] \tag{4.12}$$

$$S_{\mathrm{G}x} = \sqrt{\frac{\sum_{t=0}^{t_0}(k_x(t))^2}{t_0}} t = 1, 2, \ldots, t_0 \tag{4.13}$$

$S_{\mathrm{G}y}, S_{\mathrm{G}z}, S_{\mathrm{G}\alpha}, S_{\mathrm{G}\beta}, S_{\mathrm{G}\gamma}$ can be expressed in the same way. t is the sample point in the workspace of the long-span cable-driven parallel mechanism and t_0 is the sum of sample points.

The global stiffness index $\boldsymbol{S}_{\mathrm{G}}$ cannot express the possible minimum of stiffness in the workspace of the mechanism, thus it is necessary to give the minimum condition of the stiffness:

$$\begin{cases} k_x \geq k_{x\min}, k_y \geq k_{y\min}, k_z \geq k_{z\min} \\ k_\alpha \geq k_{\alpha\min}, k_\beta \geq k_{\beta\min}, k_\gamma \geq k_{\gamma\min} \end{cases} \tag{4.14}$$

The stiffness index in a single direction in the global workspace aims to analyze the stiffness characteristic of the same type of the long-span cable-driven parallel mechanism and obtain the global stiffness performance in the global workspace.

2. **The variation rate of stiffness in a single direction in the global workspace**

For a single attitude,

$$\boldsymbol{S}_\delta = \left[S_{\delta x}, S_{\delta y}, S_{\delta z}, S_{\delta\alpha}, S_{\delta\beta}, S_{\delta\gamma}\right] \tag{4.15}$$

$$S_{\delta x}(t-1) = \left|\frac{k_x(t) - k_x(t-1)}{k_x(t)}\right| \times 100\% t = 2, \ldots, t_0 \tag{4.16}$$

$S_{\delta y}, S_{\delta z}, S_{\delta\alpha}, S_{\delta\beta}, S_{\delta\gamma}$ can be expressed in the same way.

The index of the variation rate of stiffness in the global workspace can be expressed by RMS:

$$\boldsymbol{S}_{\mathrm{G}\delta} = \left[S_{\mathrm{G}\delta x}, S_{\mathrm{G}\delta y}, S_{\mathrm{G}\delta z}, S_{\mathrm{G}\delta\alpha}, S_{\mathrm{G}\delta\beta}, S_{\mathrm{G}\delta\gamma}\right] \tag{4.17}$$

$$S_{G\delta x} = \sqrt{\frac{\sum_{t=0}^{t_0}(S_{\delta x}(t))^2}{t_0}} t = 1, 2, \ldots, t_0 \tag{4.18}$$

$S_{G\delta y}$, $S_{G\delta z}$, $S_{G\delta\alpha}$, $S_{G\delta\beta}$, $S_{G\delta\gamma}$ can be expressed in the same way. t is the sample point in the demanded workspace of the long-span cable-driven parallel mechanism and t_0 is the sum of sample points.

The stiffness condition is presented as:

$$\begin{cases} k_x \geq k_{x\min}, k_y \geq k_{y\min}, k_z \geq k_{z\min} \\ k_\alpha \geq k_{\alpha\min}, k_\beta \geq k_{\beta\min}, k_\gamma \geq k_{\gamma\min} \end{cases} \tag{4.19}$$

The index of the variation rate of stiffness in a single direction in the global workspace aims to analyze the stiffness characteristic of the same type of the long-span cable-driven parallel mechanism and obtain the global variation of stiffness in the global workspace.

4.2 The Stiffness Similarity Method Based on the Similarity Theory

Stiffness theoretical studies of the long-span cable-driven parallel mechanism can only yield approximate numerical results of the stiffness of the mechanism. Although the accurate value of the stiffness of the mechanism can be obtained experimentally, it is very expensive to build the experimental prototype of the long-span cable-driven parallel mechanism. Therefore, this chapter proposes to study the stiffness characteristic of the mechanism by establishing a physical stiffness similarity model. There are two research objectives:

Firstly, to verify the feasibility of the stiffness similarity model, which can reflect the magnitude and change law of the actual stiffness through experiments. Based on this result, the relationship between design parameters and stiffness performance can be studied by adopting the modeling method on the stiffness theory, which provides a theoretical foundation for the dimensional design of the mechanism;

Secondly, to propose an experimental method on exact stiffness that applies to the long-span CDPR, study the application of the similarity theory in the CDPR, and establish a low-cost experimental stiffness method for the long-span CDPR.

Although the similarity theory has been widely used in the fields of geology and architecture, it is still challenging to apply it in the complicated mechanism [5]. The cable-driven parallel mechanism has a simple structure and few dimensional parameters compared to the rigid parallel mechanism, thus can be studied on the stiffness characteristic by adopting the similarity theory.

4.2.1 *Description of the Basic Similarity Method*

The currently used similarity theory is generally based on two types: the Buckingham Π theorem [6, 7] and the Differential Equations [8–10].

The Buckingham Π theorem adopts the dimensionless quantity. It associates all physical quantities of the mechanism dimensionlessly to form a set of dimensionless similarity criteria and obtain all parameter values of the similarity model, which can describe any system, but it needs to examine and analyze all characteristic quantities of the system, otherwise it will directly affect the accuracy of the results and lead to that the established similarity model may not be physically meaningful.

The Differential Equations Method deduces similarity indices and criteria by using known physical equations that govern the characteristic laws and describe the features of the system. It is required to specify the relationship between all physical quantities in the mechanism, and then conduct similarity analysis on each physical equation to obtain the dimensions of the similarity model. This method yields a more realistic correspondence, but it is not easy to build an complete equation for the complicated system.

In this chapter, the stiffness similarity model of the long-span CDPR will be established by combining these two methods. The similarity ratio of the mechanism is obtained by combining the relationship between physical quantities through dimensional analysis. The specific process is:

(1) Determining the physical parameters affecting the system model;
(2) Determining the basic dimensions of the model and selecting the same number of basic physical quantities;
(3) Nondimensionalize other physical quantities with the basic physical quantities to obtain the corresponding dimensionless equations;
(4) According to the similarity criterion that dimensionless physical quantities must be 1 and other institutional qualifications, the similarity relationship between physical quantities, as well as the similarity ratio between the similarity model and the relevant physical quantities in the prototype are obtained.

In general, if the system has a physical quantity m and a basic physical quantity n, the similarity criterion that can be established is $m - n$ [11, 12].

4.2.2 *The Stiffness Similarity Modeling Method of the Long-Span CDPR*

For a long-span cable-driven parallel mechanism, based on the theoretical studies of its stiffness in Sect. 4.1, the basic dimensional parameters can be known. The three basic dimensions are selected: the length dimension L, the mass dimension M and the time dimension T. The dimensions of other parameters can be expressed according to these basic dimensions.

Table 4.1 Dimensional parameters of the cable-driven parallel mechanism and their dimensions

Symbols	Parameters	Dimensions
l	Length	L
m	Mass	M
ρ	Linear density	ML^{-1}
ρ_v	Bulk density	ML^{-3}
σ	Force	MLT^{-2}
E	Elastic modulus	$ML^{-1}T^{-2}$
g	Gravitational acceleration	LT^{-2}

The relevant parameters and their dimensions are presented in Table 4.1 [13].

Based on the dimensions of the parameters in the table, the relationship can be written in the form of a dimensional matrix:

	m	l	ρ	ρ_v	σ	g	E
M	1	0	1	1	1	0	1
L	0	1	−1	−3	1	1	−1
T	0	0	0	0	−2	−2	−2

When transforming the dimensional matrix, we would like to have one or several parameters that can be used as basic parameters to nondimensionalize other parameters. By analyzing the above dimensional matrix, it is found that other parameters cannot be nondimensionalized by m and l only, thus a new basic parameter $\sqrt{ml/\sigma}$ is given to obtain the dimensionless expression of other parameters:

	m	l	$\sqrt{ml/\sigma}$	$\rho l/m$	$\rho_v l^3/m$	mg/σ	El^2/σ
M	1	0	0	0	0	0	0
L	0	1	0	0	0	0	0
T	0	0	1	0	0	0	0

Based on the above dimensional matrix, the similarity criterion for the long-span cable-driven parallel mechanism is obtained as:

$$\pi_1 = \frac{\rho l}{m} = \frac{ML^{-1}L}{M} = 1 \tag{4.20}$$

$$\pi_2 = \frac{\rho_v l^3}{m} = \frac{ML^{-3}L^3}{M} = 1 \tag{4.21}$$

$$\pi_3 = \frac{mg}{\sigma} = \frac{MLT^{-2}}{MLT^{-2}} = 1 \tag{4.22}$$

$$\pi_4 = \frac{EA}{\sigma} = \frac{El^2}{\sigma} = \frac{ML^{-1}T^{-2}L^2}{MLT^{-2}} = 1 \tag{4.23}$$

where A in π_4 is the cross-sectional area of the cable.

Based on the above similarity criterion for the long-span cable-driven parallel mechanism, the similarity ratio λ between the similarity model of the mechanism and the prototype can be obtained (the subscripts indicate the physical quantities of the similarity ratio).

Let λ_g be the similarity ratio of gravitational acceleration; λ_l be the similarity ratio of length; λ_d be the similarity ratio of diameter; λ_A be the similarity ratio of area; λ_m be the similarity ratio of mass; λ_ρ be the similarity ratio of linear density; λ_{ρ_v} be the similarity ratio of bulk density; λ_σ be the similarity ratio of force; λ_E be the similarity ratio of elastic modulus; λ_K be the stiffness similarity ratio of the cable-driven parallel mechanism. The parameters containing the subscript μ are that of the similarity model of the mechanism, and those containing the subscript p are that of the prototype of the mechanism. Therefore, the similarity ratio of the mechanism can be obtained as:

$$\lambda_g = \frac{g_\mu}{g_p} = 1 \tag{4.24}$$

$$\lambda_l = \frac{l_\mu}{l_p}; \lambda_d = \frac{d_\mu}{d_p} \tag{4.25}$$

$$\lambda_A = \frac{A_\mu}{A_p} = \lambda_d^2 \tag{4.26}$$

$$\lambda_m = \frac{m_\mu}{m_p} = \lambda_\rho \lambda_l \tag{4.27}$$

$$\lambda_\rho = \frac{\rho_\mu}{\rho_p} = \lambda_{\rho_v} \lambda_A \tag{4.28}$$

$$\lambda_{\rho_v} = \frac{\rho_{v\mu}}{\rho_{vp}} \tag{4.29}$$

$$\lambda_\sigma = \frac{\sigma_\mu}{\sigma_p} = \lambda_E \lambda_A = \lambda_\rho \lambda_l \tag{4.30}$$

$$\lambda_E = \frac{E_\mu}{E_p} \tag{4.31}$$

$$\lambda_K = \frac{K_\mu}{K_p} = \frac{\lambda_E \lambda_A}{\lambda_l} = \frac{\lambda_E \lambda_d^2}{\lambda_l} \tag{4.32}$$

In the analysis of the above similarity ratios, there is a problem: the elastic modulus and the density of the cable are correlated quantities, thus it is difficult to find similarity cables that satisfy both the similarity ratio of elastic modulus and that of density when building the similarity model [14, 15]. To solve this problem, similarity modeling can be conducted by adopting the counterweight cable scheme:

Firstly, the same cable as the prototype is used to ensure the similarity ratio $\lambda_E = 1$ of elastic modulus of the cable; thus, the similarity ratio of cable density can be obtained;

Secondly, a suitable counterweight cable is selected to meet the similarity ratio of cable density together with a similarity cable.

In the counterweight cable scheme, there are two types of cables: the similarity cable and the counterweight cable. The former is a cable of the similarity model obtained from the similarity ratio of diameter; the latter is used only to meet the similarity ratio of linear density and does not bear the tension.

To define the diameter of the counterweight cable as d_c. The similarity model adopts the same cable as the prototype, therefore, the similarity ratio of the elastic modulus of the cable to its density can be obtained:

$$\lambda_E = 1, \lambda_{\rho_v} = 1 \tag{4.33}$$

The similarity ratio of the similarity cable is:

$$\lambda_d = \sqrt{\lambda_\rho \lambda_l} \tag{4.34}$$

The similarity ratio of the counterweight cable is:

$$\lambda_{d_c} = \sqrt{\lambda_\rho (1 - \lambda_l)} \tag{4.35}$$

The similarity ratio of cable density is:

$$\lambda_\rho = \lambda_{\rho_v} \left(\lambda_d^2 + \lambda_{d_c}^2\right) = \lambda_d^2 + \lambda_{d_c}^2 \tag{4.36}$$

Therefore, the stiffness similarity ratio of the long-span cable-driven parallel mechanism can be expressed as:

$$\lambda_K = \frac{\lambda_E \lambda_d^2}{\lambda_l} = \frac{\lambda_d^2}{\lambda_l} \tag{4.37}$$

The above stiffness similarity ratio is the stiffness similarity ratio of the test points on the moving platform of the mechanism obtained according to Eq. (4.8).

4.3 Experiment of the Stiffness Similarity Model for the Four-Cables Scheme of the First-Level Support System for FAST

To verify the stiffness modeling of the long-span cable-driven parallel mechanism and its experimental similarity method, this section will conduct arithmetic validation and experimental studies by adopting a CDPR with three translational degrees of freedom.

The four-cable driven parallel mechanism in Fig. 4.1 is an early optional scheme for the first-level support system for FAST with three translational degrees of freedom. It is modeled as:

Firstly, to establish two coordinate systems: one is the inertial coordinate system $\Re : O - XYZ$, the origin of which is located at the center of the stationary platform of the cable-driven parallel mechanism with the Z-axis upward; the other is the moving coordinate system $\Re' : ' - X'Y'Z'$, the origin of which is located at the center of the moving platform of the cable-driven parallel mechanism with the Z'-axis upward along the normal of the moving platform. In the mechanism, $B_i\,(i = 1, 2, \ldots, 4)$ is the cable connection point of the stationary platform, and A is the cable connection point of the moving platform.

To model the four-cable driven parallel mechanism, the symbols are defined as: $\boldsymbol{O}^{\prime\Re}$ is the vector representation of the origin O' of the moving coordinate system in the inertial coordinate system; $\boldsymbol{B}_i^{\Re}$ is the vector representation of B_i in the inertial coordinate system; $\boldsymbol{A}^{\Re'}$ is the vector representation of A in the inertial coordinate system; $\boldsymbol{A}^{\Re'}$ is the vector representation of A in the moving coordinate system; r_b is the radius of the stationary platform of the mechanism, i.e., the radius of the distribution circle of the cable tower; h is the height of the cable tower.

According to Fig. 4.1, the vector representation of the cable is:

$$B_i^{\Re} = \left[r_b \cos(2i - 1)45^\circ, r_b \sin(2i - 1)45^\circ, h\right]^T (i = 1, 2, \ldots, 4) \tag{4.38}$$

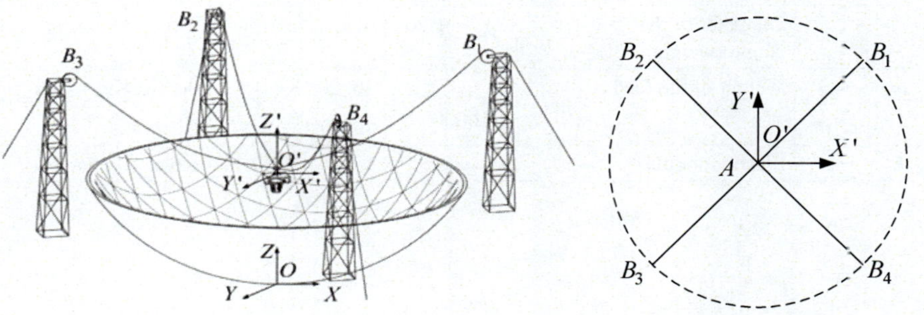

Fig. 4.1 The scheme of the four-cable driven parallel mechanism for FAST

$$A^{\Re'} = [0, 0, 0]^T \tag{4.39}$$

$$\boldsymbol{A}^{\Re} = \boldsymbol{R}\boldsymbol{A}^{\Re'} + \boldsymbol{O}'^{\Re} \tag{4.40}$$

where $\boldsymbol{R}$ is the transfer matrix from moving to inertial coordinate system.

4.3.1 The Stiffness Similarity Model of the Long-Span CDPR

According to the modeling method of the stiffness similarity model of the long-span cable-driven parallel mechanism, the similarity ratio of the parameters of the four-cable driven parallel mechanism with similar dimensions of 1:400 is obtained, as shown in Table 4.2.

According to the dimensional parameters of the similarity model obtained from Table 4.2, the experimental similarity model is established as shown in Fig. 4.2.

The experimental model adopts a double-wheel mechanism and a counterweight mechanism to ensure the tension of the cable, and bearings are designed under the cable puller to ensure that the direction of the cable puller is consistent with that of the cable and to reduce the error of the mechanism's cable puller. The above counterweight mechanism and double-wheel mechanism are subsequently applied in the establishment of the 40 m similarity model of FAST.

Table 4.2 Dimensional parameters of the four-cable driven parallel mechanism and its similarity model

Symbols	Parameters	Similarity Ratio	Dimensions of the prototype	Dimensions of the similarity model
r_b	Radius of the distribution circle of the cable tower	$\lambda_l = 1{:}400$	600 m	1.5 m
h	Height of the cable tower	$\lambda_l = 1{:}400$	265 m	0.6625 m
ρ	Cable density	$\lambda_\rho = 1{:}400$	7.716 Kg/m	0.024723 Kg/m
d	Cable diameter	$\lambda_d = 1{:}400$	40 mm	0.1 mm
d_c	Diameter of the counterweight cable	$\lambda_{d_c} = 1{:}20$	0 m	2 m
m_0	Weight of the feed	$\lambda_m =$ 1:160000	30, 000 Kg	0.1875 Kg
E	Elastic modulus	$\lambda_E = 1$	1.6×10^{11} Pa	1.6×10^{11} Pa

Fig. 4.2 The similarity model

4.3.2 *The Stiffness Similarity Experiment of the Long-Span CDPR*

After the mechanism passed the basic calibration, the stiffness experiment is conducted. Firstly, a trajectory is selected as the experimental trajectory: a linear trajectory from 140 to 220 m in the Z-direction of the prototype is shown in Fig. 4.3.

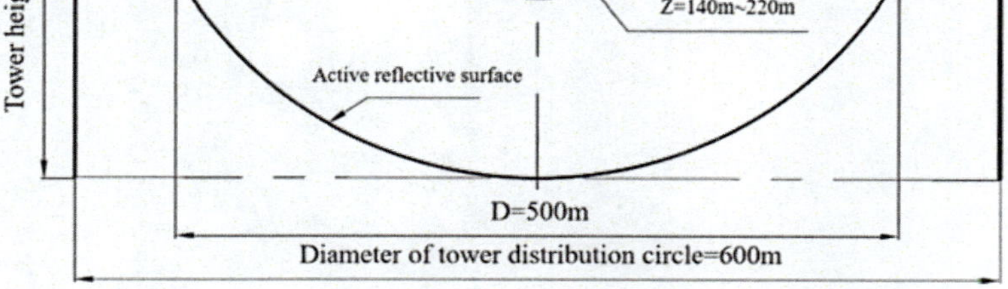

Fig. 4.3 The experimental trajectory

Firstly, the stiffness values of the prototype on this trajectory can be obtained according to the stiffness modeling method in Sect. 4.1, and then according to the stiffness similarity method in Sect. 4.2, the stiffness values of the similarity model can be obtained. The Z-directional stiffness on this experimental trajectory is simulated by selecting 9 test points (*A* to *I*), and the theoretical stiffness values of its prototype and similarity model are given in Table 4.3.

As shown in Fig. 4.4, the Z-directional stiffness of the four-cable driven parallel mechanism is experimented with a micrometer, and each test point is tested twice, then taking the average value as the experimental stiffness value for that test point.

Taking point C as an example, the stiffness of the mechanism is tested on the moving platform, and its load–deflection curve is obtained as shown in Fig. 4.5.

The experimentally measured stiffness value at point C is:

$$K_c = \frac{709.46 + 674.8}{2} = 692.13\,\text{N/m}$$

Table 4.3 Theoretical and similarity values of Z-directional stiffness of the four-cable driven parallel mechanism

	A	B	C	D	E
Z-directional height values/m	140	150	160	170	180
Theoretical stiffness values/(10^5 N/m)	4.7872	4.1929	3.6172	3.0529	2.5134
Stiffness similarity values/(N/m)	1196.6	1049.3	904.2	763.2	628.4
	F	G	H	I	
Z-directional height values/m	190	200	210	220	
Theoretical stiffness values/(10^5 N/m)	2.0067	1.5414	1.1252	0.7656	
Stiffness similarity values/(N/m)	501.7	385.3	281.3	191.4	

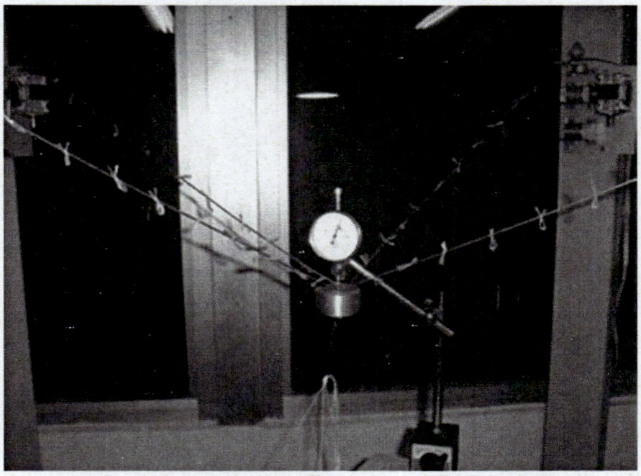

Fig. 4.4 The Z-directional stiffness experiment of the four-cable driven parallel mechanism

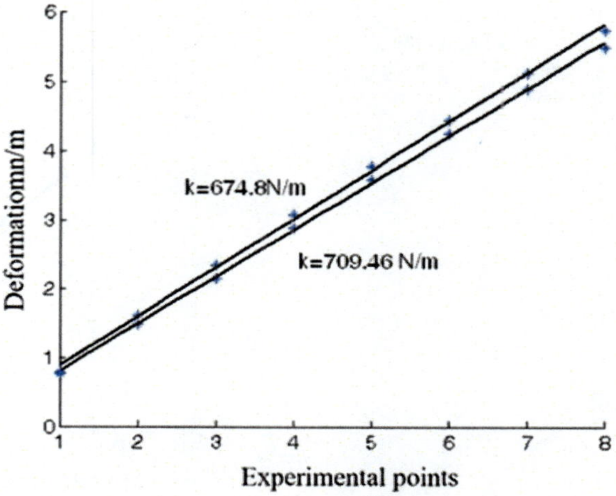

Fig. 4.5 Experimental results of Z-directional stiffness of point C

The obtained Z-directional experimental stiffness of the mechanism at the nine test points and the comparison with the stiffness similarity values of the similarity model obtained from Table 4.3 are shown in Fig. 4.6. As can be seen from the figure, the experimental stiffness values are smaller than the theoretical stiffness values at most of the experimental points, but the variation rate of the experimental stiffness values is smoother than that of the theoretical values. It is analyzed below.

Firstly, it can be found from Eq. (4.8) that the stiffness of the cable-driven parallel mechanism is inversely proportional to the cable length, thus when the cable-driven parallel mechanism runs from high to low, the cable length increases and the stiffness decreases;

Secondly, in terms of the theoretical calculation, the cable length refers to the length from the cable puller to the center point of the moving platform, but in practice,

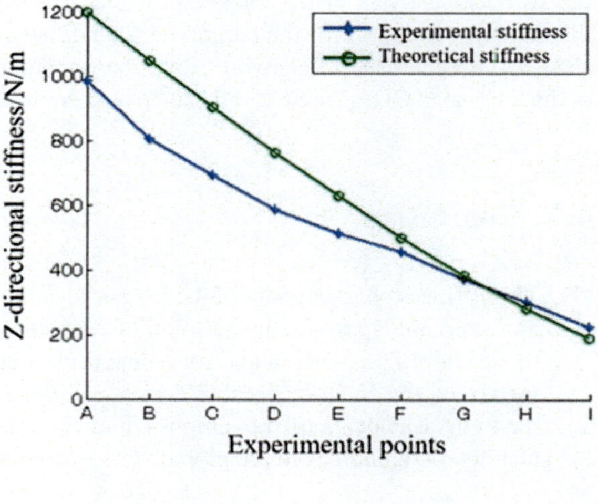

Fig. 4.6 Experimental results of Z-directional stiffness at the experimental points

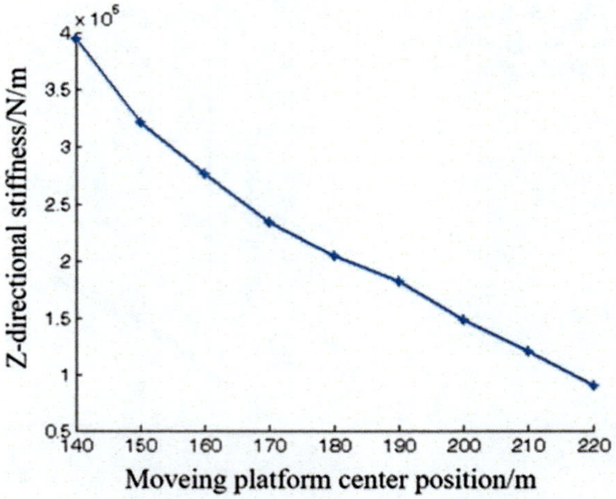

Fig. 4.7 Z-directional stiffness values on the experimental trajectory of the prototype

the cable passes through the cable control mechanism such as the double-wheel structure, which makes the actual length of the cable larger than the theoretical value, therefore, the experimental stiffness of the CDPR is smaller;

Finally, since the stiffness of the long-span cable-driven parallel mechanism is nonlinear, the tension of the cable will affect the stiffness values of the mechanism, which causes the difference between the experimental stiffness and the calculated stiffness.

Meanwhile, the variation curve of the Z-directional stiffness values of the prototype of the four-cable driven parallel mechanism on this trajectory can be obtained according to the stiffness similarity ratio, as shown in Fig. 4.7.

Overall, the experimental results show that the stiffness model of the long-span CDPR established in Sect. 4.2 can approximately represent the stiffness characteristic and the trend of the mechanism, and the stiffness modeling method can be used for the stiffness analysis of the long-span cable-driven parallel mechanism and lay a theoretical foundation for the comprehensive dimensional design of it based on the stiffness characteristic. Meanwhile, the experimental method of the similarity model of the long-span CDPR in Sect. 4.3 can be further applied to its stiffness experiments.

4.4 Conclusion

(1) The stiffness characteristic of the long-span cable-driven parallel mechanism is analyzed, the approximate solution of each cable length is deduced according to the simplified model and error compensation method in Chap. 2, the stiffness matrix of the long-span cable-driven parallel mechanism is established, the theoretical analysis of the simplified stiffness is completed, and the stiffness characteristic indices in the global workspace are given;

(2) This chapter quantitatively establishes the physical stiffness similarity model of the long-span cable-driven parallel mechanism based on Buckingham Π theorem, and adopts the scale model to solve the problem that the equivalent scale model for the long-span cable-driven parallel mechanism is difficult to be established;
(3) Taking the scheme of the four-cable CDPR for FAST as an example, the stiffness modeling and experimental studies are conducted by combining theoretical analysis and similarity model experiments, thus the stiffness characteristic of the prototype are determined. At the same time, this experiment proves the effectiveness of the proposed method of stiffness analysis of the long-span cable-driven parallel mechanism based on the experiments of the stiffness similarity model.

References

1. Kino H, Yabe S, Shimamoto T. Stiffness increase method of wire driven systems using interference of wire tension with mechanical constraint directions. In: Proceeding of international conference on machine automation, Osaka; 2000, p. 63–8.
2. Verhoeven R. Analysis of the workspace of tendon-based Stewart platforms. Duisburg: Gerhard Mercator University; 2004.
3. Lafourcade P, Zheng YQ, Liu XW. Stiffness analysis of wire-driven parallel kinematic manipulators. In Proceedings of the 11th world congress on theory of machines and mechanisms, Tianjin, China; 2004, p. 1878–82.
4. Behzadipour S, Khajepour A. Stiffness of cable-based parallel manipulators with application to stability analysis. J Mech Des. 2006;128:303–10.
5. Yuanying Q, Baoyan D, Ying S, et al. Similarity between the distorted model and the prototype of the large radio telescope. Similarity Between Distort Model Prototype Large Radio Telescope. 2004;31(4):493–500.
6. Baker WE, Westine PS, Dodge FT. Similarity method in engineering dynamics: theory and practice of scale modeling. Amsterdam: Elsevier; 1991.
7. Zhou ML. Similarity engineering. Beijing: Machinery Industry Press; 1998.
8. Xiaoqiang T, Zhu W, et al. Similarity model of feed support system for FAST. Experiment Astron;2011,29(3):177–87.
9. Lu Y, Zhu W, Ren G. Feedback control of a cable-driven Gough-Stewart platform. IEEE Trans Robot;2006,22(1):198–202.
10. Bluman GW, Cole JD. Similarity methods for differential equations. New York: Springer-Verlag; 1974.
11. Zhou ML. Similarity science. Beijing: China Science and Technology Press; 1993.
12. Tiehua W. Phenomenological similarity, similarity theorem and composition of similar indicators. Technol Product Modern. 1997;4:3–14.
13. Yao R, Tang XQ, Wang JS. Dimensional optimization design of the four-cable driven parallel manipulator in FAST. IEEE/ASME Trans Mechatron. 2010;15(6):932–41.
14. Ren GX, et al. Coupling of mobile cart-feeder stabilization system for the large radio telescope FAST. Tsinghua University; 2003.
15. Lu YJ. Positioning and orientating control study on the feed support system of a large radio telescope [PhD thesis]. Beijing: Tsinghua University; 2007.

Chapter 5
Dimensional Synthetic Optimization Design of the Cable-Driven Parallel Robot

Based on the theoretical analysis of the performance of the mechanism, the optimization of structural parameters is an important element of mechanism design and prototype development. In this chapter, the optimization design of the dimension parameters of the long-span cable-driven parallel robot is carried out, including the dimensions of the moving and stationary platform and the diameters of the cable. The purpose is to enable the mechanism to have better stability and controllability while meeting the requirements of the workspace. Therefore, the optimization is mainly based on the requirements of the workspace, taking the force and stiffness characteristics of the mechanism into account.

In this chapter, Sect. 5.1 adopts the sensitivity of dimension parameters to performance as the optimization objective, and establishes the performance index system and optimization design method. Section 5.2 uses the force characteristics indices of the long-span CDPR proposed in Chap. 3 to establish the optimization function of dimension parameters; Sect. 5.3 investigates the dimensional optimization based on the stiffness characteristics indices. In Sect. 5.4, for the optimization theory and method proposed in this chapter, the dimensional optimization function based on the force characteristics and stiffness characteristics is used for the dimensional optimization design of the six-cable CDPR of the feed first-level support system for FAST, and the dimensional optimization solution is obtained.

This chapter describes the synthetic design method of dimension parameters for the long-span CDPR and the establishment method for the optimization function based on force characteristics and stiffness characteristics, which provides the theoretical foundation and effective method for the dimensional optimization design of the long-span CDPR.

The Main Contents of This Chapter Are

(1) Performance index system and optimization method;
(2) Dimensional synthetic design of the long-span CDPR based on force characteristics;

© Tsinghua University Press 2024

X. Tang et al., *Research and Application of Cable-Driven and Rigid Parallel Robots*,
https://doi.org/10.1007/978-981-99-7452-8_5

(3) Dimensional synthetic design of the long-span CDPR based on stiffness characteristics;
(4) Optimization analysis of the six-cable CDPR of the feed first-level support system for FAST.

5.1 Performance Index System and Optimization Method

The dimensional optimization design of the cable-driven parallel robot is based on meeting the requirements of the workspace, and the optimization parameters can be divided into: the cable tension (force transfer performance), the mass of the workspace, the dynamic characteristics, stiffness and the kinematic characteristics, etc. [1–6]. To realize the dimensional optimization design of the long-span CDPR and make it achieve the expected performance objectives, it is necessary to establish the evaluation indices of the effects of the design parameters on the performance of the mechanism.

Assuming that W is one of the performance indices of the long-span CDPR, whose equation is a function of multiple design variables:

$$W = \Gamma(b_1, b_2, \ldots, b_i) \quad i = 1, 2, \ldots \tag{5.1}$$

where $b_1, b_2, \ldots, b_i$ is the dimensional design parameters of the mechanism, and Eq. (5.1) is expressed as the sensitivity S_Γ of the effects of the parameter b_i on the mechanism performance W by taking the partial derivative of the design parameter b_i:

$$S_\Gamma = \frac{\partial \Gamma}{\partial b_i} \approx \frac{\Gamma(b_i + \Delta b_i) - \Gamma(b_i)}{\Delta b_i} \quad i = 1, 2, \ldots \tag{5.2}$$

S_Γ indicates the magnitude of the rate-of-change of the performance of the long-span CDPR with the design parameter b_i, which is also the sensitivity of the performance with the changing parameter. $S_\Gamma > 0$ indicates that as b_i increases, the value of the performance index W increases; similarly, $S_\Gamma < 0$ indicates that as b_i increases, the value of the performance index W decreases. When the absolute value of S_Γ is larger, it indicates that the design parameter b_i has a greater impact on the performance of the mechanism; similarly, when the absolute value of S_Γ is smaller, it indicates that the design parameter b_i has a smaller impact on the performance of the mechanism.

Figure 5.1 shows a long-span m-cable driven parallel mechanism with n degrees of freedom, according to the performance-parameter variable function equation of Eq. (5.1), which can be written as:

$$W = \Gamma(\boldsymbol{B}_1, \boldsymbol{B}_2, \ldots, \boldsymbol{B}_i; \boldsymbol{A}_1, \boldsymbol{A}_2, \ldots, \boldsymbol{A}_i; d) \quad i = 1, 2, \ldots, m \tag{5.3}$$

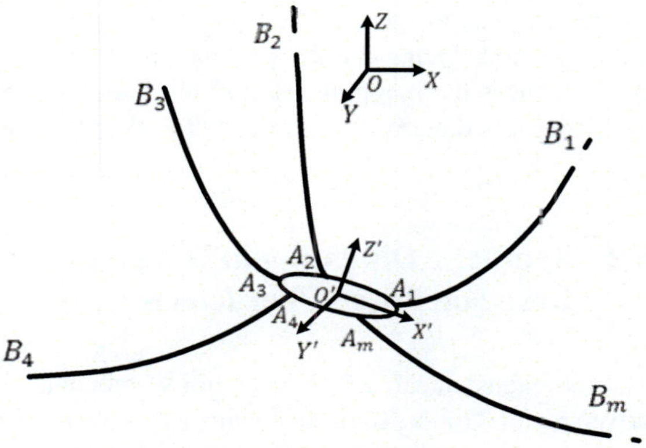

Fig. 5.1 The long-span *m*-cable driven parallel mechanism with *n* degrees of freedom

where $\boldsymbol{B}_i$ is the vector of the cable exit point of the *i*-th cable in the inertial coordinate system; $\boldsymbol{A}_i$ is the coordinate vector of the anchor point of the *i*-th cable to the moving platform; d is the diameter of the cable.

The location of the joints on the stationary and moving platform of the long-span cable-driven parallel mechanism is distributed in a circle. The radius of the distribution circle of the stationary platform is r_b, the radius of the distribution circle of the moving platform is r_a, the distribution angle of the joints on the stationary platform is $\boldsymbol{\theta}_b$, the distribution angle of the joints on the moving platform is $\boldsymbol{\theta}_a$, the height of the cable exit point in the inertial coordinate system is $h_1, h_2, \ldots, h_i$, $i = 1, 2, \ldots, m$, and the mass of the moving platform is m_0. Then Eq. (5.3) can be written as:

$$W = \Gamma(h_1, h_2, \ldots, h_i; r_b; r_a; \boldsymbol{\theta}_b; \boldsymbol{\theta}_a; d; m_0) \quad i = 1, 2, \ldots, m \tag{5.4}$$

To reduce the effects of dimension parameters on the optimization design of the long-span cable-driven parallel mechanism, the dimensional design parameters of the mechanism are dimensionlessly expressed. When nondimensionalize the dimension parameters in the performance function w, one of the preferred design parameters of length and angle can be chosen. After determining the radius of the stationary platform r_b, the distribution angle θ_{a1} of one joint on the moving platform and the mass of the moving platform, the dimensionless analysis of Eq. (5.4) can be conducted:

$$W = \Gamma\left(\frac{h_1}{r_b}, \frac{h_2}{r_b}, \ldots, \frac{h_i}{r_b}; \frac{r_a}{r_b}; \frac{\boldsymbol{\theta}_b}{\theta_{a1}}; \frac{d}{r_b}; m_0\right) \quad i = 1, 2, \ldots, m \tag{5.5}$$

For the long-span cable-driven parallel mechanism, the performance function (.) is nonlinear and complex, thus it is difficult to calculate its partial derivatives of the design parameters directly. Therefore, this chapter, adopting a numerical calculation method, giving the design interval and using dimensionless design parameters, deduces the mapping relationship between the mechanism performance and dimensionless dimension parameters for its optimization design.

5.2 Synthetic Dimensional Design of the Long-Span CDPR Based on Force Characteristics

The cable tension can affect the stability and controllability of the long-span cable-driven parallel robot. To make the mechanism meet the end accuracy requirements in the demanded workspace and have better force characteristics, the long-span CDPR needs to be designed based on the force characteristics for dimensional optimization.

According to the static modeling of the long-span CDPR in Chap. 2, the cable tension of the mechanism at a certain attitude can be obtained. On the basis of the force characteristics indices of the long-span CDPR proposed in Chap. 3, the design function W_σ for dimensional optimization of the CDPR based on the force characteristics will be established in this section.

Based on the difference between the global maximum and minimum cable tension $\mathrm{GTCI}_{\max}$, the design function $W_{\sigma 1}$ for dimensional optimization can be established as:

$$W_{\sigma 1} = \Gamma\left(\frac{h_1}{r_a}, \frac{h_2}{r_a}, \ldots, \frac{h_i}{r_a}; \frac{r_b}{r_a}; \frac{\theta_b}{\theta_{a1}}; \frac{d}{r_a}; m_0\right) \quad i = 1, 2, \ldots, m \tag{5.6}$$

$$W_{\sigma 1} = \mathrm{GTCI}_{\max} = \sqrt{\frac{\sum_{t=0}^{t_0} (\mathrm{TCI}_{\max}(t))^2}{t_0}} \tag{5.7}$$

$$\mathrm{TCI}_{\max}(t) = \max\bigl(\sigma_i(t) - \sigma_j(t)\bigr)\begin{cases} i = 1, 2, \ldots, m \\ j = 1, 2, \ldots, m \end{cases} i \neq j \quad t = 1, 2, \ldots t_0 \tag{5.8}$$

where t is the sample point; t_0 is the sum of sample points; $\mathrm{TCI}_{\max}(t)$ is the difference between the maximum and minimum cable tension at an attitude; $\sigma_i(t)(i = 1, 2, \ldots, m)$ is the cable tension.

To make the long-span CDPR have better force characteristics, the $\mathrm{GTCI}_{\max}$-based design function for dimensional optimization and the constraint are given.

The objective function for optimization:

$$\min(W_{\sigma 1}) \tag{5.9}$$

The optimization constraint:

$$\begin{cases} \boldsymbol{\sigma} \geq [\sigma_{\min}, \ldots, \sigma_{\min}]^{\mathrm{T}} \\ \boldsymbol{\sigma} \leq [\sigma_{\max}, \ldots, \sigma_{\max}]^{\mathrm{T}} \\ (x, y, z, \alpha, \beta, \gamma) \in \text{the demanded workspace} \end{cases} \tag{5.10}$$

where $\boldsymbol{\sigma}$ is the cable tension vector of the long-span CDPR; $\sigma_{\min}$ is the allowable minimum value of the cable tension; $\sigma_{\max}$ is the allowable maximum value of the cable tension; $(x, y, z, \alpha, \beta, \gamma)$ is the reachable working attitude of the center of the moving platform.

Based on the ratio $\mathrm{GTCI}_{\mathrm{r\,max}}$ between the global maximum and minimum cable tension of the long-span CDPR, a dimensional optimization function $W_{\sigma 2}$ can be established:

$$W_{\sigma 2} = \Gamma\left(\frac{h_1}{r_a}, \frac{h_2}{r_a}, \ldots, \frac{h_i}{r_a}; \frac{r_b}{r_a}; \frac{\boldsymbol{\theta}_b}{\theta_{a1}}; \frac{d}{r_a}; m_0\right) \quad i = 1, 2, \ldots, m \tag{5.11}$$

$$W_{\sigma 2} = \mathrm{GTCI}_{\mathrm{rmax}} = \sqrt{\frac{\sum_{t=0}^{t_0} (\mathrm{TCI}_{\mathrm{rmax}})^2}{t_0}} \tag{5.12}$$

$$\mathrm{TCI}_{\mathrm{r\,max}} = \max\left(\frac{\sigma_i}{\sigma_j}\right)\begin{cases} i = 1, 2, \ldots, m \\ j = 1, 2, \ldots, m \end{cases} i \neq j \quad t = 1, 2, \ldots t_0 \tag{5.13}$$

where t is the sample point; t_0 is the sum of sample points; $\mathrm{TCI}_{\mathrm{r\,max}}$ is the difference between the maximum and minimum cable tension at an attitude; $\sigma_i(t)(i = 1, 2, \ldots, m)$ is the cable tension.

To make the long-span CDPR have stable force characteristics, the $\mathrm{GTCI}_{\mathrm{max}}$-based design function for dimensional optimization and the constraint are given.

The objective function for optimization:

$$\min(W_{\sigma 2}) \tag{5.14}$$

The optimization constraint:

$$\begin{cases} \boldsymbol{\sigma} \geq [\sigma_{\min}, \ldots, \sigma_{\min}]^{\mathrm{T}} \\ \boldsymbol{\sigma} \leq [\sigma_{\max}, \ldots, \sigma_{\max}]^{\mathrm{T}} \\ (x, y, z, \alpha, \beta, \gamma) \in \text{the demanded workspace} \end{cases} \tag{5.15}$$

where $\boldsymbol{\sigma}$ is the cable tension vector of the long-span CDPR; $\sigma_{\min}$ is the allowable minimum value of the cable tension; $\sigma_{\max}$ is the allowable maximum value of the cable tension; $(x, y, z, \alpha, \beta, \gamma)$ is the reachable working attitude of the center of the moving platform.

Based on the controllability generalized volume index GTCGVI of the global cable tension of the long-span CDPR, a dimensional optimization function $W_{\sigma 3}$ can be established:

$$W_{\sigma 3} = \Gamma\left(\frac{h_1}{r_a}, \frac{h_2}{r_a}, \ldots, \frac{h_i}{r_a}; \frac{r_b}{r_a}; \frac{\boldsymbol{\theta}_b}{\theta_{a1}}; \frac{d}{r_a}; m_0\right) \quad i = 1, 2, \ldots, m \tag{5.16}$$

$$W_{\sigma 3} = \text{GTCGVI} = \sqrt{\frac{\sum_{t=0}^{t_0}(\Theta_{\forall}(t))^2}{t_0}} \tag{5.17}$$

$$\Theta_{\forall} = \frac{V_{\forall\text{control}}}{V_{\forall\text{err}}}(\Theta_{\forall} \in [0, 1]) \quad t = 1, 2, \ldots, t_0 \tag{5.18}$$

where t is the sample point; t_0 is the sum of sample points; $\Theta_{\forall}$is the ratio between the generalized volume $V_{\forall\text{control}}$ of the controllable space $\forall_{\text{control}}$ in the error space and the generalized volume $V_{\forall\text{err}}$ of the allowable error space $\forall_{\text{err}}$.

To make the long-span CDPR have the same good force characteristics in its allowable error space, the GTCGVI -based design function for dimensional optimization and the constraint are given.

The objective function for optimization:

$$\max(W_{\sigma 3}) \tag{5.19}$$

The optimization constraint:

$$\begin{cases} \boldsymbol{\sigma} \geq [\sigma_{\min}, \ldots, \sigma_{\min}]^{\mathrm{T}} \\ \boldsymbol{\sigma} \leq [\sigma_{\max}, \ldots, \sigma_{\max}]^{\mathrm{T}} \\ (x, y, z, \alpha, \beta, \gamma) \in \text{the allowable error space} \end{cases} \tag{5.20}$$

where $\boldsymbol{\sigma}$ is the cable tension vector of the long-span CDPR; $\sigma_{\min}$ is the allowable minimum value of the cable tension; $\sigma_{\max}$ is the allowable maximum value of the cable tension; $(x, y, z, \alpha, \beta, \gamma)$ is the reachable working attitude of the center of the moving platform.

5.3 Dimensional Synthetic Design of the Long-Span CDPR Based on Stiffness Characteristics

The stiffness of the CDPR is related to the cable material, cable tension and cable length, which is an important performance index of it. To ensure that the mechanism has sufficient stiffness that changes smoothly and the motion stability, the long-span CDPR needs dimensional optimization design based on stiffness characteristics.

According to the establishment method of the stiffness matrix of the long-span CDPR in Chap. 4, the stiffness values of the mechanism at each degree of freedom can be obtained, and based on the given stiffness indices, the design function W_K for dimensional synthetic optimization based on the stiffness characteristics will be established in this section.

Based on the stiffness index $\boldsymbol{S}_{\mathbf{G}}$ in a single direction in the global workspace, the dimensional optimization function W_{K1} can be established:

$$W_{K1} = \Gamma\left(\frac{h_1}{r_a}, \frac{h_2}{r_a}, \ldots, \frac{h_i}{r_a}; \frac{r_b}{r_a}; \frac{\boldsymbol{\theta}_b}{\theta_{a1}}; \frac{d}{r_a}; m_0\right) \quad i = 1, 2, \ldots, m \tag{5.21}$$

$$\boldsymbol{S}_{\mathbf{G}} = \left[S_{\mathrm{G}x}, S_{\mathrm{G}y}, S_{\mathrm{G}z}, S_{\mathrm{G}\alpha}, S_{\mathrm{G}\beta}, S_{\mathrm{G}\gamma}\right] \tag{5.22}$$

$$W_{Kx1} = S_{\mathrm{G}x} = \sqrt{\frac{\sum_{t=0}^{t_0}(k_x(t))^2}{t_0}} \quad t = 1, 2, \ldots, t_0 \tag{5.23}$$

where $k_x(t)$ is the stiffness value in the X-direction at a certain attitude $(x, y, z, \alpha, \beta, \gamma)$ of the long-span CDPR, and similarly, $W_{Ky1}, W_{Kz1}, W_{K\alpha1}, W_{K\beta1}, W_{K\gamma1}$ can be expressed; t is the sample point in the workspace of the long-span CDPR t_0 is the sum of sample points.

To definite:

$$W_{K1}xyz = \frac{W_{Kx1} + W_{Ky1} + W_{Kz1}}{3}$$

$$W_{K1}\alpha\beta\gamma = \frac{W_{K\alpha1} + W_{K\beta1} + W_{K\gamma1}}{3}$$

To make the long-span CDPR have better stiffness, the $\boldsymbol{S}_{\mathbf{G}}$-based design function for dimensional optimization and the constraint are given.

The objective function for optimization:

$$\max(W_{K1}xyz); \max(W_{K1}\alpha\beta\gamma) \tag{5.24}$$

The optimization constraint:

$$\begin{cases} k_x \geq k_{x\min}, k_y \geq k_{y\min}, k_z \geq k_{z\min} \\ k_\alpha \geq k_{\alpha\min}, k_\beta \geq k_{\beta\min}, k_\gamma \geq k_{\gamma\min} \end{cases} \tag{5.25}$$

$\left[k_{x\min}, k_{y\min}, k_{z\min}, k_{\alpha\min}, k_{\beta\min}, k_{\gamma\min}\right]$ is the allowable minimum stiffness of the long-span CDPR.

Based on the rate-of-change index $\boldsymbol{S}_{\delta}$ of stiffness in a single direction in the global workspace, the dimensional optimization function W_{K2} can be established:

$$W_{K2} = \Gamma\left(\frac{h_1}{r_a}, \frac{h_2}{r_a}, \ldots, \frac{h_i}{r_a}; \frac{r_b}{r_a}; \frac{\boldsymbol{\theta}_b}{\theta_{a1}}; \frac{d}{r_a}; m_0\right) \quad i = 1, 2, \ldots, m \tag{5.26}$$

$$\boldsymbol{S}_{\delta} = \left[S_{\delta x}, S_{\delta y}, S_{\delta z}, S_{\delta\alpha}, S_{\delta\beta}, S_{\delta\gamma}\right] \tag{5.27}$$

$$W_{Kx2} = S_{\delta x} = \sqrt{\frac{\sum_{t=0}^{t_0}(S_{\delta x}(t))^2}{t_0}} t = 1, 2, \ldots, t_0 \tag{5.28}$$

where,

$$S_{\delta x}(t-1) = \left|\frac{k_x(t) - k_x(t-1)}{k_x(t-1)}\right| \times 100\% \quad t = 2, \ldots, t_0 \tag{5.29}$$

where $k_x(t)$ is the stiffness value in the X-direction at a certain attitude $(x, y, z, \alpha, \beta, \gamma)$ of the long-span CDPR and similarly, $W_{Ky2}, W_{Kz2}, W_{K\alpha2}, W_{K\beta2}, W_{K\gamma2}$ can be expressed; t is the sample point in the workspace of the long-span CDPR t_0 is the sum of sample points.

To definite:

$$W_{K2}xyz = \frac{W_{Kx2} + W_{Ky2} + W_{Kz2}}{3}$$

$$W_{K2}\alpha\beta\gamma = \frac{W_{K\alpha2} + W_{K\beta2} + W_{K\gamma2}}{3}$$

To make the long-span CDPR have better stiffness, the $\boldsymbol{S}_\delta$-based design function for dimensional optimization and the constraint are given.

The objective function for optimization:

$$\min(W_{K2}xyz); \min(W_{K2}\alpha\beta\gamma) \tag{5.30}$$

The optimization constraint:

$$\begin{cases} k_x \geq k_{x\min}, k_y \geq k_{y\min}, k_z \geq k_{z\min} \\ k_\alpha \geq k_{\alpha\min}, k_\beta \geq k_{\beta\min}, k_\gamma \geq k_{\gamma\min} \end{cases} \tag{5.31}$$

5.4 Optimization Analysis of the Six-Cable CDPR of the Feed First-Level Support System for FAST

Figure 5.2 shows the basic dimensions and workspace of the six-cable driven parallel mechanism for FAST. According to the data of FAST provided by NAOC [7], it is known that the aperture of the reflecting surface of FAST is 500 m, the radius of the stationary platform of the six-cable driven parallel mechanism is $r_b = 300$ m, and the weight of the feed, namely, the load-bearing at the center of the moving platform is about 30,000 kg. To ensure observation, the height of the cable tower should not be higher than 290 m. The demanded workspace of the six-cable driven parallel mechanism for FAST is a spherical dome with a radius of 160 m, and the maximum

(a)

(b)

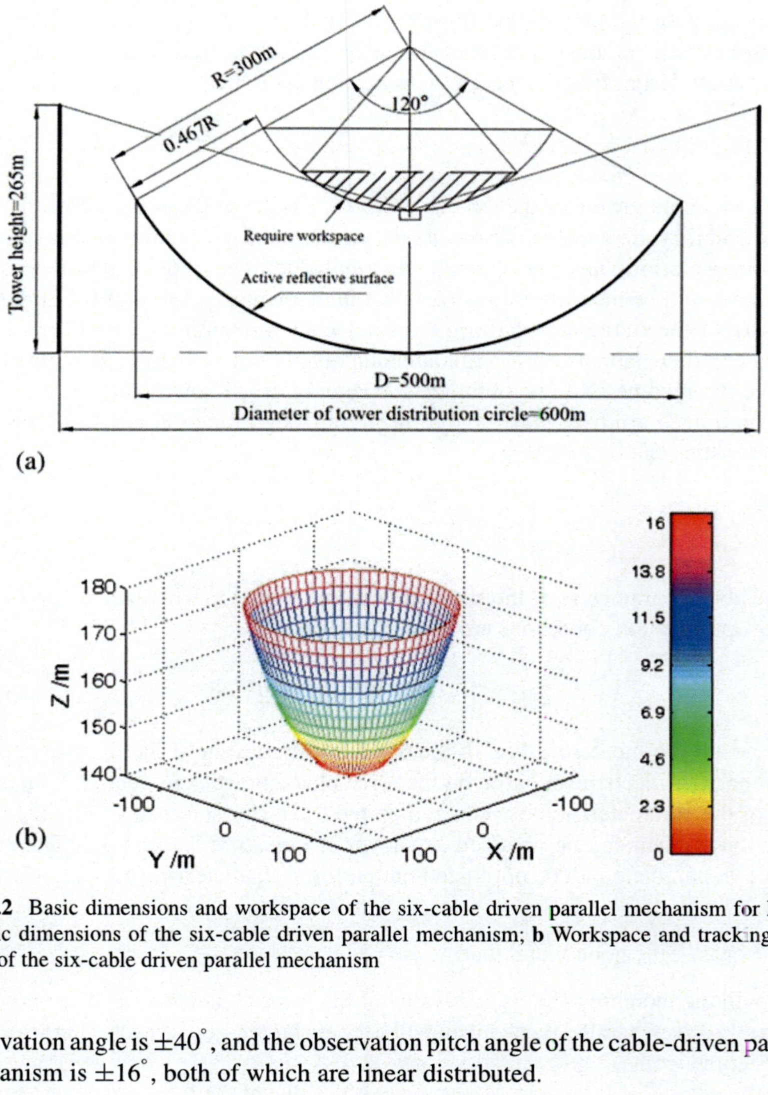

Fig. 5.2 Basic dimensions and workspace of the six-cable driven parallel mechanism for FAST. **a** Basic dimensions of the six-cable driven parallel mechanism; **b** Workspace and tracking pitch angle of the six-cable driven parallel mechanism

observation angle is $\pm 40^{\circ}$, and the observation pitch angle of the cable-driven parallel mechanism is $\pm 16^{\circ}$, both of which are linear distributed.

5.4.1 Dimensional Optimization Based on Force Characteristics

There are seven basic parameters for the optimization of the six-cable driven parallel mechanism for FAST: the height of the cable tower $h_i (i = 1, 2, \ldots, 6)$, the distribution circle radius of the cable tower r_b, the distribution angle of the cable tower

θ_b, the radius of the moving platform r_a, the distribution angle of the joints of the moving platform θ_a, the diameter of the cable d, and the weight of the feed m_0. Its optimization design function based on force characteristics is:

$$W_\sigma = \Gamma(h_1, h_2, \ldots, h_i; r_b; r_a; \theta_b; \theta_a; d; m_0) \quad (i = 1, 2, \ldots, 6) \tag{5.32}$$

The six cable towers of the six-cable driven parallel mechanism are of the same height, and they are evenly distributed on a circumference of radius r_b. The joints of the moving platform are symmetrically distributed, thus $\theta_b = 60^\circ$ can be determined first, and it will be nondimensionalized as a fundamental quantity of the angle. The radius r_b of the stationary platform is also a known quantity, $r_b = 300\,\text{m}$, which will be nondimensionalized as a fundamental quantity of the length. In addition, the mass of the feed needs to be optimized separately. It is known that the parameters to be optimized are five, thus the optimization design function based on the force characteristics can be written as:

$$W_\sigma = \Gamma\left(\frac{h}{r_b}; \frac{r_a}{r_b}; \frac{\theta_a}{\theta_b}; \frac{d}{r_b}; m_0\right) \tag{5.33}$$

The above parameters of the six-cable driven parallel mechanism are optimized, and its optimization conditions are given as:

$$\sigma_{\min} = 50\,\text{KN}; \quad \sigma_{\max} = 300\,\text{KN} \tag{5.34}$$

According to the description of the optimization design of the long-span cable-driven parallel mechanism based on the force characteristics in Sect. 5.2, the sensitivity of the parameters to be optimized on the force characteristics $W_{\sigma 1}$, $W_{\sigma 2}$, $W_{\sigma 3}$ of the mechanism can be obtained as shown in Figs. 5.3, 5.4 and 5.5. The effects of the five parameters to be optimized on the force characteristic $W_{\sigma 1}$ are shown in Fig. 5.3.

From Fig. 5.3, it can be found that:

(1) With the increase of θ_a/θ_b, the value of the force characteristic $W_{\sigma 1}$ of the six-cable driven parallel mechanism will become larger, i.e., its force characteristics become worse.
(2) When $\theta_a/\theta_b < 0.05$, as h/r_b increases, the value of the force characteristic $W_{\sigma 1}$ of the six-cable driven parallel mechanism will become smaller, i.e., its force characteristics become better. But when $\theta_a/\theta_b > 0.05$, $h/r_b = 0.91 \sim 0.97$, $W_{\sigma 1}$ is little changed, relatively, when $h/r_b = 0.97$, the change with θ_a/θ_b is more obvious.
(3) With the increase of r_a/r_b, the value of the force characteristic $W_{\sigma 1}$ of the six-cable driven parallel mechanism shows a decreasing trend, i.e., its force characteristics become better, but the change is small.

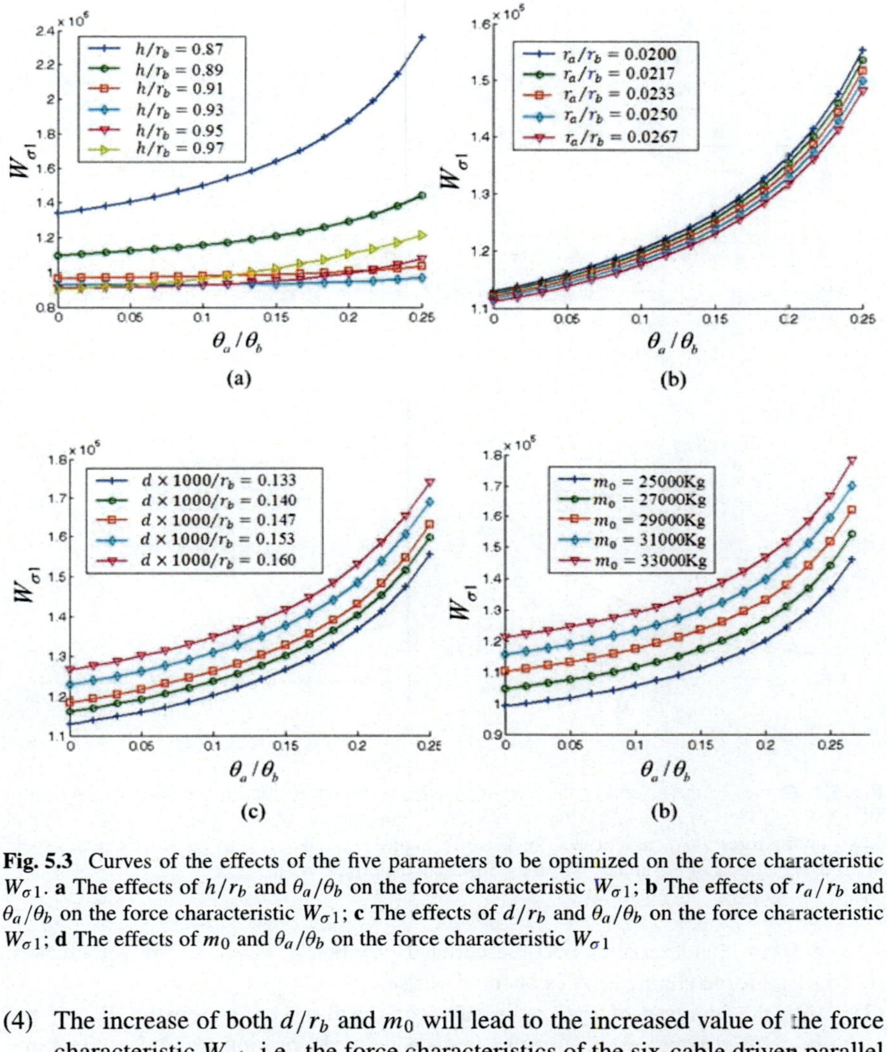

Fig. 5.3 Curves of the effects of the five parameters to be optimized on the force characteristic $W_{\sigma 1}$. **a** The effects of h/r_b and θ_a/θ_b on the force characteristic $W_{\sigma 1}$; **b** The effects of r_a/r_b and θ_a/θ_b on the force characteristic $W_{\sigma 1}$; **c** The effects of d/r_b and θ_a/θ_b on the force characteristic $W_{\sigma 1}$; **d** The effects of m_0 and θ_a/θ_b on the force characteristic $W_{\sigma 1}$

(4) The increase of both d/r_b and m_0 will lead to the increased value of the force characteristic $W_{\sigma 1}$, i.e., the force characteristics of the six-cable driven parallel mechanism become worse.

The effects of the five parameters to be optimized on the force characteristic $W_{\sigma 2}$ are shown in Fig. 5.4.

From Fig. 5.4, it can be found that:

(1) With the increase of θ_a/θ_b, the value of the force characteristic $W_{\sigma 2}$ of the six-cable driven parallel mechanism will become larger, i.e., its force characteristics become worse.
(2) When $h/r_b = 0.87 \sim 0.91$, as h/r_b increases, the value of the force characteristic $W_{\sigma 2}$ of the six-cable driven parallel mechanism will become smaller, i.e.,

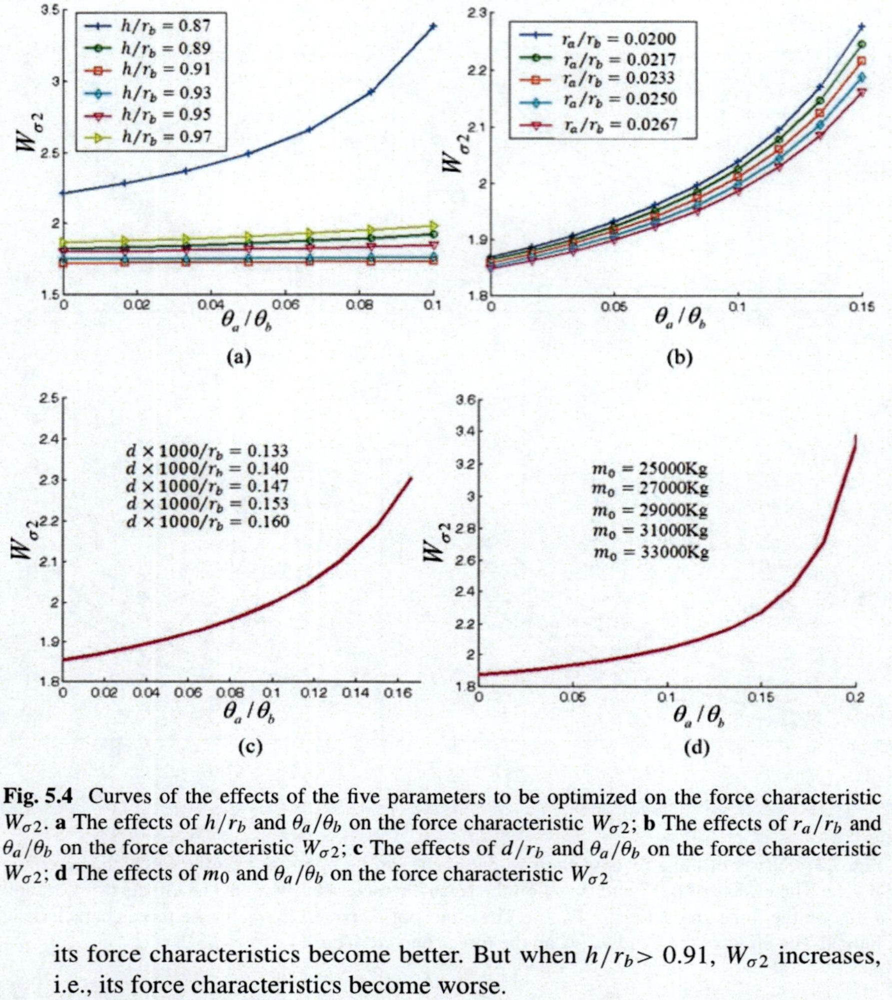

Fig. 5.4 Curves of the effects of the five parameters to be optimized on the force characteristic $W_{\sigma 2}$. **a** The effects of h/r_b and θ_a/θ_b on the force characteristic $W_{\sigma 2}$; **b** The effects of r_a/r_b and θ_a/θ_b on the force characteristic $W_{\sigma 2}$; **c** The effects of d/r_b and θ_a/θ_b on the force characteristic $W_{\sigma 2}$; **d** The effects of m_0 and θ_a/θ_b on the force characteristic $W_{\sigma 2}$

its force characteristics become better. But when $h/r_b > 0.91$, $W_{\sigma 2}$ increases, i.e., its force characteristics become worse.

(3) With the increase of r_a/r_b, the value of the force characteristic $W_{\sigma 2}$ of the six-cable driven parallel mechanism shows a decreasing trend, i.e., its force characteristics become better, but the effects are not great.

(4) The value of $W_{\sigma 2}$ is basically unaffected by the changes of d/r_b and m_0.

Assuming that the allowable maximum error at the end attitude of the long-span cable-driven parallel mechanism is $\left[100\,\text{mm}, 0.5^{\circ}\right]^{\text{T}}$, the effects of the five parameters to be optimized on the force characteristic $W_{\sigma 3}$ are shown in Fig. 5.5.

From Fig. 5.5, it can be found that:

(1) With the increase of θ_a/θ_b, the value of the force characteristic $W_{\sigma 3}$ of the six-cable driven parallel mechanism will become smaller, i.e., its force characteristics become worse.

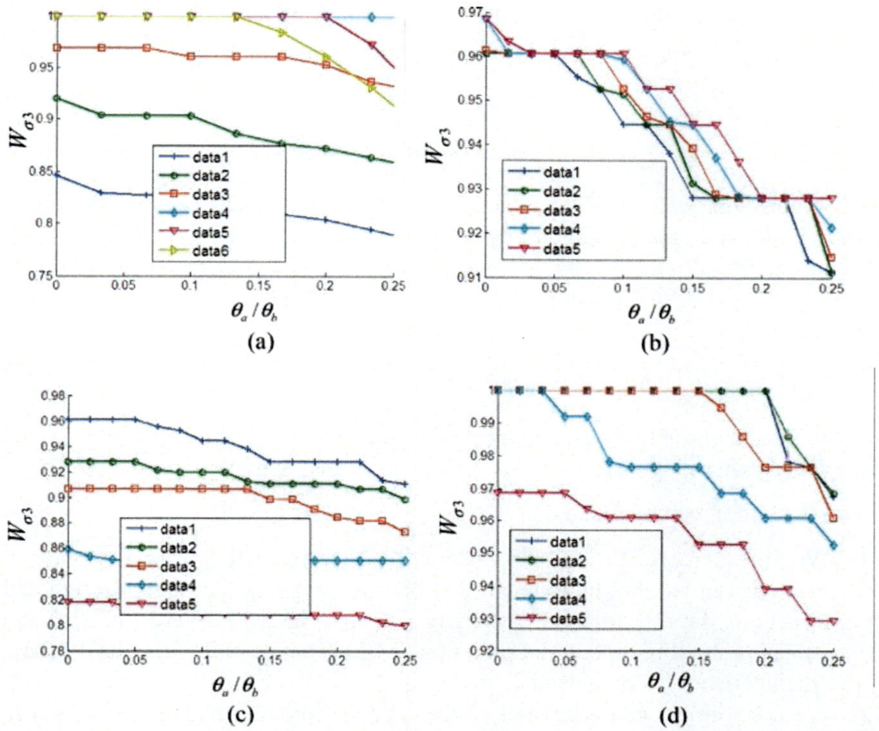

Fig. 5.5 Curves of the effects of the five parameters to be optimized on the force characteristic $W_{\sigma 3}$. **a** The effects of h/r_b and θ_a/θ_b on the force characteristic $W_{\sigma 3}$; **b** The effects of r_a/r_b and θ_a/θ_b on the force characteristic $W_{\sigma 3}$; **c** The effects of d/r_b and θ_a/θ_b on the force characteristic $W_{\sigma 3}$; **d** The effects of m_0 and θ_a/θ_b on the force characteristic $W_{\sigma 3}$

(2) When $h/r_b = 0.93$, the six-cable driven parallel mechanism has the best value of the force characteristic $W_{\sigma 3}$, i.e., it has the best force characteristics.
(3) With the increase of r_a/r_b, the value of the force characteristic $W_{\sigma 3}$ of the six-cable driven parallel mechanism shows an increasing trend, i.e., its force characteristics become better.
(4) The value of $W_{\sigma 3}$ tends to decrease as the values of d/r_b and m_0 increase, i.e., the force characteristics of the six-cable driven parallel mechanism will become worse.

5.4.2 *Dimensional Optimization Based on Stiffness Characteristics*

According to the optimization design function presented in Sect. 5.4.1, the optimization design function of the six-cable driven parallel mechanism based on stiffness characteristics can also be obtained according to the stiffness characteristics indices of the long-span cable-driven parallel mechanism in Sect. 5.3:

$$W_K = \Gamma\left(\frac{h}{r_a}; \frac{r_b}{r_a}; \frac{\theta_a}{\theta_b}; \frac{d}{r_a}\right) \tag{5.35}$$

The above parameters of the six-cable driven parallel mechanism are optimized to present its optimization conditions:

$$k_{x\,\min} = k_{y\,\min} = k_{z\,\min} = 1 \times 10^5\ \text{N/m} \tag{5.36}$$

$$k_{\alpha\,\min} = k_{\beta\,\min} = k_{\gamma\,\min} = 1 \times 10^6\ \text{N/m} \tag{5.37}$$

The effects of the four parameters to be optimized on the stiffness characteristic W_{K1} are shown in Fig. 5.6.

From Fig. 5.6, it can be found that:

(1) With the increase of θ_a/θ_b, the value of the force characteristic W_{K1} of the six-cable driven parallel mechanism will become larger in the three translational directions, i.e., its force characteristics become better, however, its stiffness values in the three rotational directions will become smaller, i.e., its stiffness characteristics become worse.
(2) For the stiffness characteristics of the six-cable driven parallel mechanism in the 3 translational directions, the values of the stiffness characteristics tend to become smaller as the value of h/r_b becomes larger, i.e., the stiffness becomes worse.
(3) With the increase of r_a/r_b and d/r_b, its stiffness values will become larger, i.e., its stiffness will become better. However, according to the stiffness values, the effects of the values of h/r_b and r_a/r_b on the stiffness characteristics are not obvious.
(4) The stiffness characteristics of the six-cable driven parallel mechanism in the three rotational directions tend to become larger as h/r_b, r_a/r_b and d/r_b become larger, i.e., its stiffness becomes better, where h/r_b has the least effects on the stiffness characteristics.

The effects of the four parameters to be optimized on the stiffness characteristic W_{K2} are shown in Fig. 5.7. It can be found that:

(1) With the increase of θ_a/θ_b, the value of the force characteristic W_{K2} of the six-cable driven parallel mechanism will become smaller in the three translational directions, i.e., its force characteristics become better, but its effects are not great. However, the stiffness values in the three rotational directions become significantly larger, i.e., its stiffness characteristics become worse.
(2) For the stiffness characteristics of the six-cable driven parallel mechanism in the three translational directions, the value of the stiffness characteristic W_{K2}

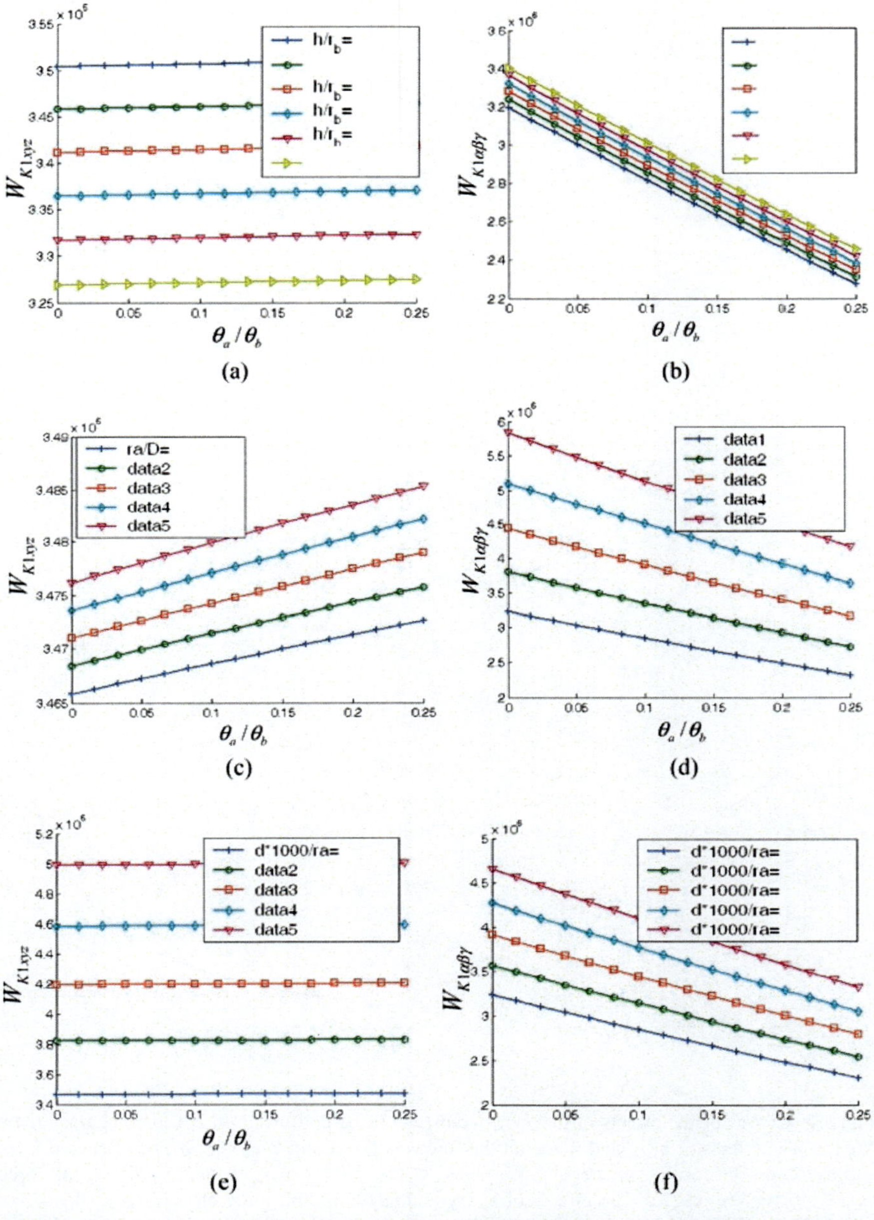

Fig. 5.6 Curves of the effects of the four parameters to be optimized on the force characteristic W_{K1}. **a** The effects of h/r_b and θ_a/θ_b on the force characteristic $W_{K1}xyz$; **b** The effects of h/r_b and θ_a/θ_b on the force characteristic $W_{K1}\alpha\beta\gamma$; **c** The effects of r_a/r_b and θ_a/θ_b on the force characteristic $W_{K1}xyz$; **d** The effects of r_a/r_b and θ_a/θ_b on the force characteristic $W_{K1}\alpha\beta\gamma$; **e** The effects of d/r_b and θ_a/θ_b on the force characteristic $W_{K1}xyz$; **f** The effects of d/r_b and θ_a/θ_b on the force characteristic $W_{K1}\alpha\beta\gamma$

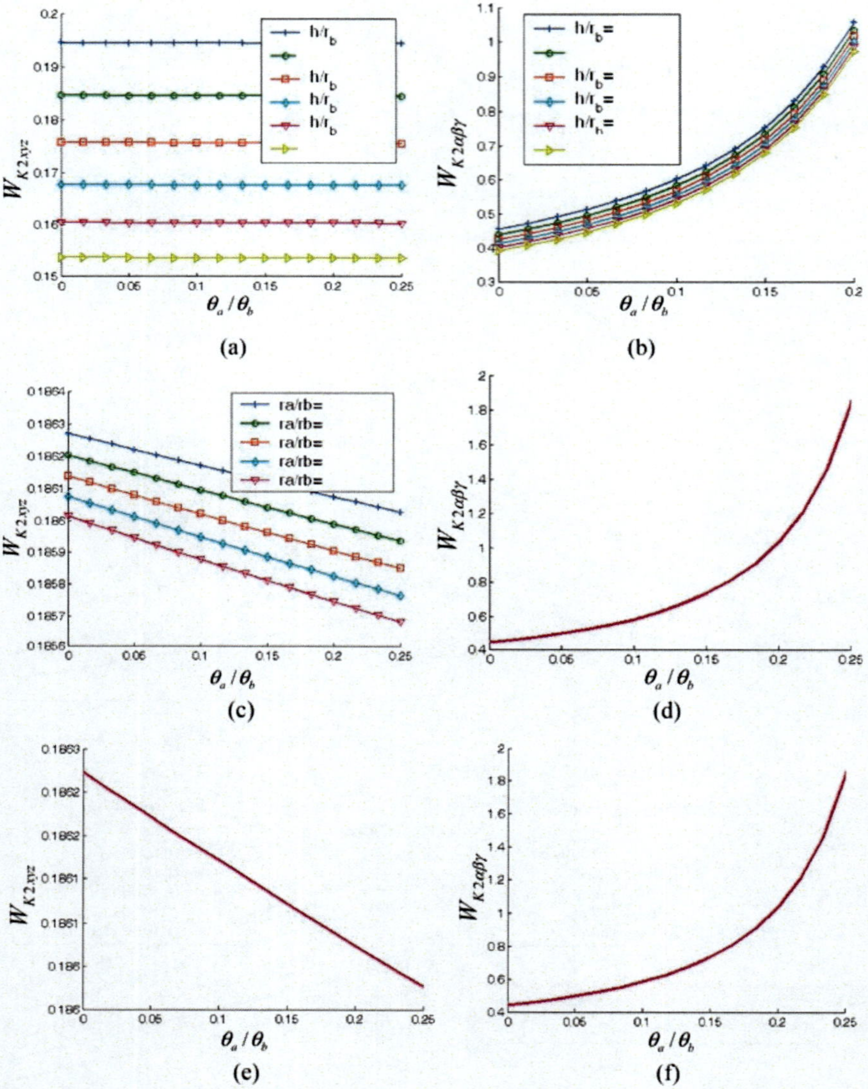

Fig. 5.7 Curves of the effects of the four parameters to be optimized on the force characteristic W_{K2}. **a** The effects of h/r_b and θ_a/θ_b on the force characteristic $W_{K2}xyz$; **b** The effects of h/r_b and θ_a/θ_b on the force characteristic $W_{K2}\alpha\beta\gamma$; **c** The effects of r_a/r_b and θ_a/θ_b on the force characteristic $W_{K2}xyz$; **d** The effects of r_a/r_b and θ_a/θ_b on the force characteristic $W_{K2}\alpha\beta\gamma$; **e** The effects of d/r_b and θ_a/θ_b on the force characteristic $W_{K2}xyz$; **f** The effects of d/r_b and θ_a/θ_b on the force characteristic $W_{K2}\alpha\beta\gamma$

tends to become smaller as h/r_b and r_a/r_b become larger, i.e., the stiffness characteristics become better; d/r_b has no significant effects on the stiffness characteristics.

(3) For the stiffness characteristics of the six-cable driven parallel mechanism in the three translational directions, the value of the stiffness characteristic W_{K2} tends to become smaller as h/r_b becomes larger, i.e., the stiffness characteristics become better; r_a/r_b and d/r_b have no significant effects on the stiffness characteristics.

5.4.3 Dimensional Optimization Based on the Tracking Angle at the Maximum Boundary

From Chap. 3, it can be found that the cable-driven parallel mechanism for FAST cannot meet the observation requirement of the tracking angle 40° when Z = 220 m, thus an A-B rotator is needed for angle compensation. To obtain a larger tracking angle for the cable-driven parallel mechanism, it can be deduced according to the optimization design function given in 5.4.1:

$$W_{\beta} = \Gamma\left(\frac{h}{r_b}; \frac{r_a}{r_b}; \frac{d}{r_b}; m_0\right) \tag{5.38}$$

The effects of the four parameters to be optimized on the maximum pitch angle at the boundary W_{β} are shown in Fig. 5.8. It can be found that:

(1) With the increase of h/r_b, W_{β} of the six-cable driven parallel mechanism at the boundary of the workspace becomes significantly larger.

(2) As r_a/r_b and d/r_b become larger, the tracking angle W_{β} of the six-cable driven parallel mechanism at the boundary of the workspace does not change significantly, i.e., these two dimension parameters have little effect on the tracking angle of the mechanism.

(3) The value of W_{β} shows an obvious decreasing trend as the value of m_0 increases, i.e., the tracking angle of the six-cable driven parallel mechanism at the boundary of the workspace becomes smaller.

5.4.4 Synthetic Optimization of Parameters

The objective of the synthetic optimization of the long-span cable-driven parallel mechanism is to choose better design parameters to make it have better force characteristics and stiffness characteristics. Based on the above analysis results, the effects of changing parameters on the mechanism performance and attitude can be obtained as shown in Table 5.1.

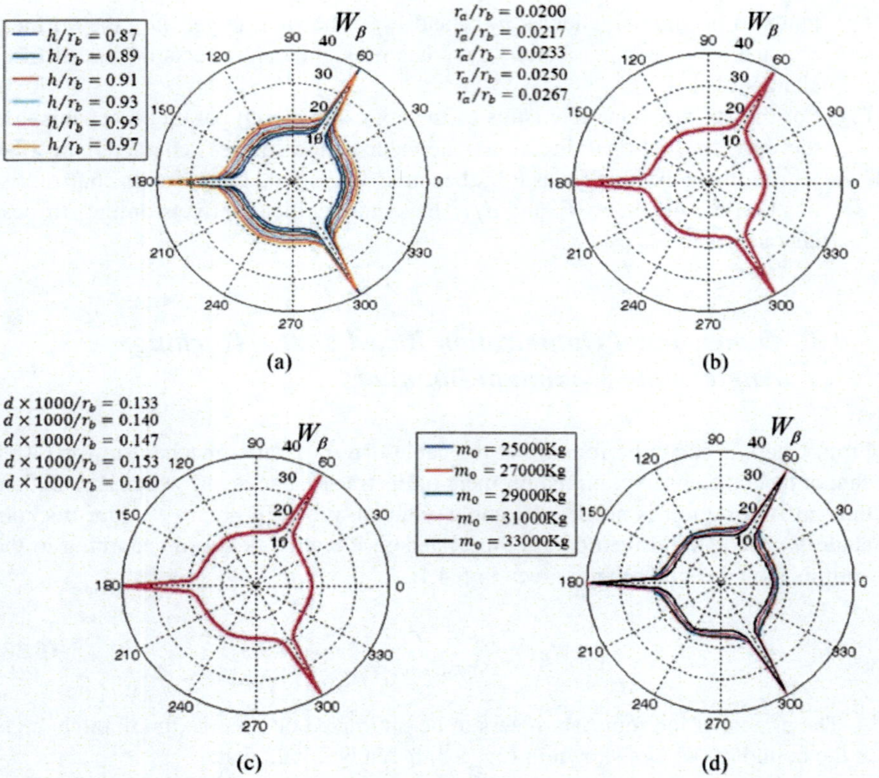

Fig. 5.8 Curves of the effects of the four parameters to be optimized on W_β (see color figures). **a** The effects of h/r_b on the maximum pitch angle W_β; **b** The effects of r_a/r_b on the maximum pitch angle W_β; **c** The effects of d/r_b on the maximum pitch angle W_β; **d** The effects of m_0 on the maximum pitch angle W_β

Through the above analysis, the law of effects of each parameter of the six-cable driven parallel mechanism for FAST on its force characteristics and stiffness characteristics can be obtained as follows:

(1) The analysis of θ_a/θ_b reveals that with the increase of this parameter value, the force characteristics and rotational stiffness characteristics of the mechanism show a trend of deterioration with an obvious change, except for the stiffness characteristics $W_{K1}xyz$ and $W_{K2}xyz$ of the mechanism in the translational direction, which show no change or little change. Based on this, the distribution of the joints of the moving platform of the mechanism is $\theta_a/\theta_b = 0$.
(2) The analysis of h/r_b reveals that the force characteristics, stiffness characteristics and the maximum tracking angle at the boundary of the mechanism basically show a trend of becoming better with the increase of parameter values. Among them, the stiffness characteristic $W_{K1}xyz$ becomes worse with little change, the

Table 5.1 The effects of changing parameters on the mechanism performance and attitude

Parameters	θ_a/θ_b	h/r_b	r_a/r_b	$d \times 1000/r_b$	m_0
Range of values	$0 \sim 0.25$	$0.87 \sim 0.97$	$0.02 \sim 0.0267$	$0.133 \sim 0.16$	25 ~ 33 KN
The force characteristic $W_{\sigma 1}$	↓↓↓	↑↑→	↑↑	↓	↓
The force characteristic $W_{\sigma 2}$	↓↓	↑↑→↓ The peak is 0.91	↑	→	→
The force characteristic $W_{\sigma 3}$	↓↓ / →↓↓	↑↑↑→ The peak is ≥ 0.93	↑	↓↓	→↓
The stiffness characteristic	→ / ↑	↓	↑	↑	/
The stiffness characteristic $W_{K1\alpha\beta\gamma}$	↓↓	↑	↑	↑	/
The stiffness characteristic W_{K2xyz}	→ / ↑	↑	↑	→	/
The stiffness characteristic $W_{K2\alpha\beta\gamma}$	↓↓	↑	→	→	/
The tracking angle W_β	/	↑	→	→	↓

Note ↑ indicates that the mechanism performance tends to become better with the change of parameters; ↓ indicates that the mechanism performance tends to become worse with the change of parameters; → indicates that the mechanism performance remain basically the same with the change of parameters. The more arrows, the more obvious the effects

force characteristic $W_{\sigma 2}$ becomes better and then worse with little change, while $W_{\sigma 3}$ becomes better and then stays the same with the peaks at $h/r_b = 0.91$ and $h/r_b \geq 0.93$ respectively. Based on this, to obtain better force characteristics and the tracking angle, the cable tower of the mechanism is set as $h/r_b = 0.93$, i.e., $h = 279$ m.

(3) The analysis of r_a/r_b reveals that the effects of the parameters on the force characteristics and stiffness characteristics of the mechanism basically tends to become better, thus $r_a/r_b = 0.0267$ is determined, i.e., $r_a = 8$ m.

(4) The analysis of d/r_b reveals that the force characteristics of the mechanism show a decreasing trend with the increase of parameter values, which especially has the greatest effects on the force characteristic $W_{\sigma 3}$, while the stiffness characteristics show a trend of becoming better, and the change of d/r_b has no obvious effects on W_β. Therefore, to make the mechanism have better force characteristics,

Table 5.2 The optimization design dimensions of the six-cable driven parallel mechanism for FAST

Symbols	Parameters	Values
r_a	The radius of the moving platform	8 (m)
r_b	The radius of the distribution circle of the cable tower	300 (m)
h	The height of the cable tower	279 (m)
d	The diameter of the cable	42 (mm)
ρ	The cable density	6.35 (Kg/m)
E	Elastic modulus of the cable	1.6×10^{11} (Pa)
m_0	The mass of the moving platform	25,000 (Kg)
θ_b	The distribution angle of the joints on the stationary platform	60°
θ_a	The distribution angle of the joints on the moving platform	0°

choosing $W_{\sigma 3} > 0.9$ and ensuring the advantages of the stiffness characteristics, thus $d/r_b = 0.14$ is determined, i.e., $d = 42$ mm.

(5) The analysis of m_0 reveals that the increase of the value of m_0 will make the mechanism have worse force characteristics and make the tracking angle at the boundary of the workspace become smaller, thus the value of m_0 should be reduced as much as possible. Here $\mathrm{m}_0 = 25000$ Kg is determined.

According to the law of effects of each design parameter on the force characteristics and stiffness characteristics of the mechanism, a set of dimensional optimization design results of the prototype of the six-cable driven parallel mechanism for FAST can be given, as shown in Table 5.2.

5.5 Conclusion

(1) The evaluation indices of the effects of the design parameters on the performance of the mechanism and the optimization design method are described.

(2) An optimization design method on the dimensions of the long-span cable-driven parallel mechanism based on the force characteristics and stiffness characteristics is established.

(3) Taking the six-cable driven parallel mechanism of the feed first-level support system for FAST as an example, the series optimization design functions are established based on the force characteristics, stiffness characteristics and the maximum tracking angle at the boundary; through synthetic analysis, the synthetic optimization of the basic dimension parameters, layout parameters and load-bearing parameters of the moving platform of the six-cable driven parallel mechanism are realized.

References

1. Tadokoro S, Nishioka, et al. On fundamental design of wire configuration of wire driven parallel manipulators with redundancy. In: Proceedings of the Japan- USA symposium on flexible automation. Boston, MA; 1996. pp. 151–58.
2. Zheng Y. Research on key theoretical issues of wire—driven parallel kinematic manipulatorsand the application to wind tunnel support systems [Ph.D. thesis]. Quanzhou: Huaqiao University; 2004.
3. Yaqing Z, Wenbai Z, Xiongwei L. Mechanism design of a wire-driven parallel suspension system for the feed cabin of fast. J Huaqiao Univ (Natl Sci). 2007;28(4):345–9.
4. Lafourcade P, Llibe M, Reboulet C. Design of a parallel wire-driven manipulator for wind tunnels. In: Proceeding of the workshop on Fundamental issues and future directions for parallel mechanisms and manipulators. Quebec City, Quebec; 2002. pp. 187–94.
5. Maeda K, Tadokoro S, Takamori T, et al. On design of a redundant wire-driven parallel robot WAPP manipulator. In: Proceeding of 1999 IEEE international conference on robotics and automation. Detroit, USA; 1999. pp. 895–900.
6. Liu S, Wu H. Design of new wire-driven parallel kinematic manipulator for wind tunnels. J Nanjing Univ Sci Technol. 2004;28(6):601–5.
7. Nan RD. Structure for supporting the feedback cabin of the FAST. National Astronomical Observatories Chinese Academy of Science; 2005.

Chapter 6
Rigid-Body Dynamic Modelling and Verification of the Fine-Tuning Stewart Platform

The feed support system of the FAST (Five-hundred-meter Aperture Spherical radio Telescope) adopts a three-level serial system, consisting of a cable-driven parallel robot (CDPR), an A-B rotator and a rigid Stewart parallel robot. The cable-driven parallel robot and the A-B rotator consist a motion platform (flexible support) with full DOFs, which provides the large workspace for the terminal receivers. The fine-tuning platform (or fine-tuning Stewart platform) is connected under the A-B rotator to improve the trajectory accuracy of the receivers. The base of the fine-tuning Stewart platform is not stationary, and moves with the cable-driven parallel robot and the A-B rotator. Therefore, the key in dynamic modelling of the fine-tuning platform is to consider the influence of the motion state of the base on the dynamics of the Stewart parallel robot, establishing a complete dynamic model containing the motion parameters of the base.

At present, most of the dynamic modelling on parallel robots focus on the robots with the stationary base (assuming that the base is stationary and rigidly supported), i.e., the inverse dynamics in the inertial coordinate system, and the dynamic model does not contain the motion parameters of the base. There are few research considering the motion of the base. This chapter takes the fine-tuning platform of the FAST feed support system as the research object, and completes dynamic modelling of the Stewart parallel robot with a motion base, considering the inertia force, centripetal force and gravity [1]. On the basis of the inverse dynamic modelling, the simulation analysis is implemented with the MATLAB software, and the accuracy of the dynamic model is verified with physical experiments [2]. The dynamic modelling of the fine-tuning Stewart platform lays a foundation for the theoretical study of the parallel robot and the construction of FAST project.

In this chapter, Sect. 6.1 overviews the existing dynamic modelling methods; Sect. 6.2 defines the coordinate systems, variables and constants required in modelling process based on the geometry of the fine-tuning Stewart platform and carries out the kinematic analysis; Sect. 6.3 adopts the Newton–Euler method to implement the dynamic modelling of the fine-tuning Stewart platform; Sect. 6.4

© Tsinghua University Press 2024

X. Tang et al., *Research and Application of Cable-Driven and Rigid Parallel Robots*,
https://doi.org/10.1007/978-981-99-7452-8_6

conducts the verification experiments through dynamic calibration and driving force measurement with the scale model of the fine-tuning Stewart platform.

The main contents of this chapter include:

(1) Dynamic modelling methods for parallel robots;
(2) Kinematic modelling of the fine-tuning Stewart platform;
(3) Dynamic modelling of the fine-tuning Stewart platform;
(4) Experimental verification of the dynamic model.

6.1 Dynamic Modelling Methods for Parallel Robots

The rigid body dynamics of parallel robots mainly include two types of problems: forward and inverse solutions. The former is mostly used for the motion simulation to adjust parameters of the control system. The latter is the basis of the driving force analysis, which is the theoretical foundation for the drive motor selection, transmission chain selection and dynamics control of the parallel mechanism. The inverse dynamics of the parallel robots is to deduce the driving force (or torque) of the active joint (drive joint) with the given motion parameters of the moving platforms (position, attitude, velocity and acceleration). For the parallel robots with special requirements, such as heavy loads, high speeds, high-frequency motions or high structural sensitivity, the inverse dynamics analysis is mandatory and unavoidable. Since the parallel robot adopts spatially closed kinematic chains and there are a large number of passive joints, its dynamic modelling is more complicated than that of the serial robot. The main methods for inverse dynamic modelling of parallel robots include: Newton–Euler method, Virtual work method, Lagrange method and Kane method. The derivation processes of these methods are different, and so are the forms of the deduced dynamic models. The method for inverse dynamic modelling needs to be selected according to the actual situation. The following sections will introduce these methods respectively.

1. Newton–Euler method

The Newton–Euler method establishes the dynamic model of parallel robots by combining Newton's mechanics and Euler equations. Since only adopts the most basic and simplest mechanical principles and concepts, the derivation process is explicit and the physical meaning is clear. Moreover, the internal forces of the joints can be calculated by establish the force and moment equations for each sub-module of the parallel robot. However, as a large number of mechanical equations need to be solved, the whole solving process is relatively complicated and the resulting dynamic equations are scattered in form. In the early days, Merlet [3] and Fichter [4] established mapping relationships between the workspace forces and the driving forces of joint space in the case of neglecting the mass of the branch chains. Do, Yang [5], Reboulet and Berthomieu [6] completed the dynamic modelling of the Stewart parallel mechanism with the Newton–Euler method. Subsequently, Ji [7], considering the inertia of branch chains, established the inverse dynamic model of the Stewart parallel mechanism and studied the inertia effect of branch chains on the driving force of the Stewart parallel robot, concluding that the inertia of branch chains has a large effect on the driving force, and cannot be neglected. Indian scholars Dasgupta et al. [8–10] conducted in-depth studies on the dynamic model of parallel robots with the Newton–Euler method, and meticulously deduced the complete dynamic model of Stewart parallel robots considering friction, gravity, Coriolis force and other factors. Zhang Lixin [11, 12] of Tsinghua University simplified the dynamic model of Stewart parallel robot deduced with the Newton–Euler method, significantly improved the computational efficiency of the inverse dynamics with limited error, and applied the simplified model to the dynamic feed-forward control.

2. Virtual work method

The virtual work method applies the principle of virtual work to deduce the dynamic model of parallel mechanisms. In the solving process, the joint internal forces do not work and therefore do not appear in the equations, avoiding the calculation of joint internal forces. The modelling process is simplified compared to the Newton–Euler method, and the deduced dynamic model of parallel robots is simpler in form. The focus of the virtual work method is the selection of key points, and appropriate key points can simplify the solving process of the inertia force and the dynamic equations. Zhang, Song [13], Wang and Gosselin [14] deduced the partial velocity and partial angular velocity matrix of the Stewart parallel mechanism, and established the dynamic inverse model with the virtual work method. Tsai [15] conducted an in-depth study on the Jacobi matrix of the parallel mechanism, proposing the link Jacobi matrix and standardizing the solution process of the virtual work method. Professor Staicu [17, 18] used the virtual work method to deduce the dynamic model of parallel robots with various configurations. Scholars of Tsinghua University such as Wu Jun [19] and Shao Hua [20] utilized the virtual work method to complete the dynamic modelling of the planar parallel machine tool and redundant planar parallel robots respectively.

3. Lagrange method

The Lagrange method is based on the work and energy (kinetic and potential energy) of the system to establish the dynamic model of the parallel mechanism through partial differential calculations. The modelling process of this method is clear and uniform, and the deduced dynamic equations are computationally efficient, but the burden of analytic solution of partial differentials is heavy. Abdellatif [21] used the Lagrange method to establish a computationally efficient inverse dynamics of the Stewart parallel platform by reasonably selecting the local coordinate system. Li [22] adopted the Lagrange method to deduce the dynamic equations of the 3-PRS parallel mechanism. Li Yuwen [23] applied the Lagrange method to obtain the dynamic equations of the hybrid three-degree-of-freedom parallel mechanism and carried out the optimization of counterweight.

4. Kane method

The mechanic principle of the Kane method is based on the Kane equation, which means that the sum of the generalized forces (including the generalized inertial forces) acting on a rigid body with respect to the generalized velocity is zero. This method does not need to calculate the constraining forces, and the calculation process only involves dot product and cross product of vectors, which is highly efficient. However, due to its obscure mechanical concepts, it has not been applied extensively. Yun [24] and Yang [25] employed the Kane method to establish the dynamic model of parallel robots and discussed the algorithm of dynamics control based on it. Liu [26] proposed a substructure-based Kane method, which simplified the solving process of the dynamic equations of the parallel mechanism.

Currently, the derivation of the inverse dynamics of parallel mechanism is carried out in the inertial coordinate system with the stationary base platform. Because of the motion of the base platform (with attitude, velocity and acceleration), the rigid-body dynamic model of the fine-tuning Stewart parallel robot needs to consider the motion of the base platform. As a result, the dynamic modelling process becomes particularly complex and there are few research and theoretical results in this area [27]. At the same time, the inverse dynamics considering the motion of the base platform is the basis for the theoretical analysis, simulation and control algorithm research of the FAST prototype, which must be deduced and studied.

6.2 Kinematic Analysis of the Fine-Tuning Platform

The FAST feed support system is a multi-level redundant hybrid mechanism. The motion description of parallel mechanism generally requires the establishment of coordinate systems on the base platform and the end effector (moving platform respectively. The base of the fine-tuning Stewart platform moves with the cable-driven parallel robot and the A-B rotator, and the coordinate system of the base platform becomes non-inertial. A global inertial coordinate system needs to be established separately. In addition, in order to describe the attitude of the cable-driven parallel robot and the A-B rotator, corresponding coordinate systems must be established on the moving platform of the cable-driven parallel robot and the A-B rotator. Finally, the coordinate systems of the FAST feed support system are established as shown in Fig. 6.1.

The global coordinate system $\{\boldsymbol{G}\}$: *O-XYZ* is established at the bottom of the active reflector, with the X-axis pointing to the cable tower C_1 and the Y-axis pointing to the midpoint of the line connecting cable towers C_2 and C_3. The coordinate system of cable platform $\{\boldsymbol{C}\}$: *O'-X'Y'Z'* is fixed to the geometric center of the plane determined by the hinge point of the steel cable of the moving platform of cable-driven parallel robot (referred to as the cable platform), and the direction of the coordinate system is shown in Fig. 6.1. The coordinate system of the moving platform of the fine-tuning Stewart platform $\{\boldsymbol{P}\}$: *o-xyz* is fixed to the geometric center of the plane determined by the rotation center of the spherical joints, with the *x*-axis pointing to the midpoint of the line connecting P_1 and P_6 and the *z*-axis is perpendicular to the plane defined

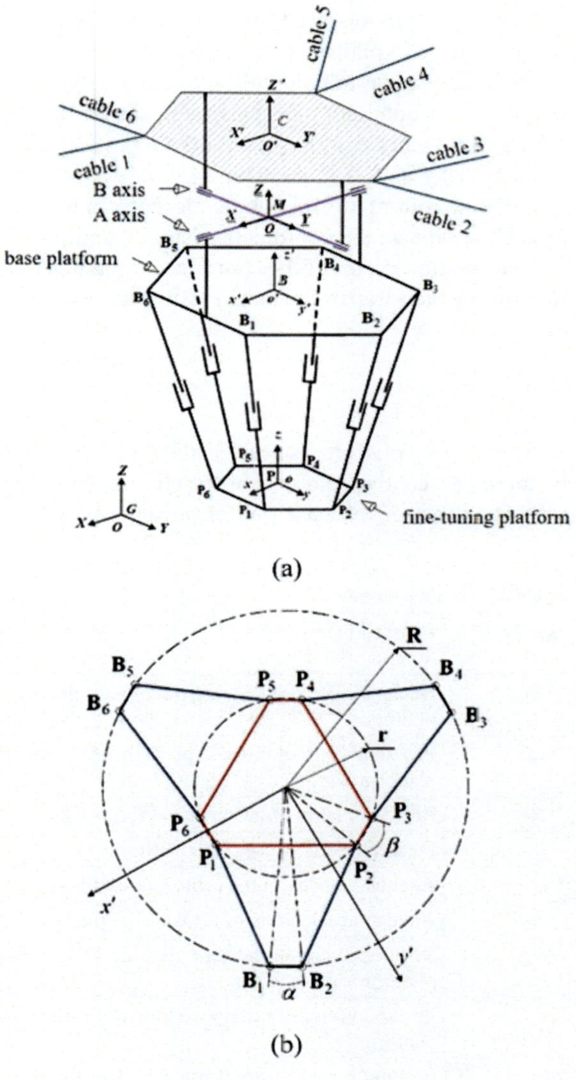

Fig. 6.1 Structure diagram of the FAST feed support system. **a** structure of the feed support system; **b** structure of the fine-tuning Stewart platform

by the center of the spherical joints. The coordinate system of the base of the fine-tuning Stewart platform{$\boldsymbol{B}$}: *o'-x'y'z'* is fixed to the geometric center of the plane determined by the rotation center of the Hooke joints, with the *x*'-axis pointing to the midpoint of the line connecting B_1 and B_6 and the *z*'-axis is perpendicular to the plane defined by the center of the Hooke joints. Among above coordinate systems, the global coordinate system is inertial, and others are non-inertial coordinate systems. In addition, a non-inertial translational coordinate system—the intermediate coordinate system {$\boldsymbol{M}$} is established, which is fixed at the intersection of the two rotational axes of the A-B rotator and is always parallel to the global coordinate system.

The main symbols used in the dynamic modelling are shown in Table 6.1, and others will be explained and illustrated where they first appear. The vectors and matrixes are shown in bold italics, and the numbers of the branch chain of the fine-tuning Stewart parallel mechanism are shown by the right subscript i ($i = 1, 2, \ldots, 6$); the cable number of the cable-driven parallel mechanism is denoted by the right subscript j ($j = 1, 2, \ldots, 6$). The right subscript of matrixes and vectors is used to indicate the information about the mechanism it describes, and the upper left corner is used to indicate the coordinate system it defines.

The position vector of the coordinate system origin of the moving platform of the fine-tuning Stewart parallel mechanism with respect to the base platform is described as:

$$^{\mathrm{B}}\boldsymbol{t}_{\mathrm{P}} = [x, y, z]^{\mathrm{T}} \tag{6.1}$$

The moving platform is described by the RPY angle, i.e., its position is described by three consecutive rotations of the coordinate system of the moving platform with respect to that of the base platform. Suppose at the beginning, $\{\boldsymbol{P}_3\}$: $o - x_3y_3z_3$

Table 6.1 Main symbols

Symbols	Physical meanings	Coordinate systems
$^{\mathrm{G}}\boldsymbol{R}_{\mathrm{B}}$	Rotation matrix of the base platform of the fine-tuning Stewart platform	$\{\boldsymbol{G}\}$
$^{\mathrm{G}}\boldsymbol{t}_{\mathrm{B}}$	Position vector of the base platform of the fine-tuning Stewart platform	$\{\boldsymbol{G}\}$
$^{\mathrm{G}}\boldsymbol{R}_{\mathrm{C}}$	Rotation matrix of the cable platform	$\{\boldsymbol{G}\}$
$^{\mathrm{G}}\boldsymbol{t}_{\mathrm{C}}$	Position vector of the cable platform	$\{\boldsymbol{G}\}$
$^{\mathrm{G}}\boldsymbol{\omega}_{\mathrm{C}}$	Angular velocity of the cable platform	$\{\boldsymbol{G}\}$
$^{\mathrm{G}}\boldsymbol{\varepsilon}_{\mathrm{C}}$	Angular acceleration of the cable platform	$\{\boldsymbol{G}\}$
$^{\mathrm{G}}\boldsymbol{R}_{\mathrm{B}}$	Rotation matrix of the base platform of the fine-tuning Stewart platform	$\{\boldsymbol{C}\}$
b_i	Position vector of the Hooke joint of the fine-tuning Stewart platform	$\{\boldsymbol{B}\}$
p_i	Position vector of the spherical joint of the fine-tuning Stewart platform	$\{\boldsymbol{P}\}$
$^{\mathrm{B}}\boldsymbol{R}_{\mathrm{P}}$	Rotation matrix of the moving platform of the fine-tuning Stewart platform	$\{\boldsymbol{B}\}$
$^{\mathrm{B}}\boldsymbol{t}_{\mathrm{P}}$	Position vector of the moving platform of the fine-tuning Stewart platform	$\{\boldsymbol{B}\}$
$^{\mathrm{B}}\boldsymbol{\omega}_{\mathrm{P}}$	Angular velocity of the moving platform of the fine-tuning Stewart platform	$\{\boldsymbol{B}\}$
$^{\mathrm{B}}\boldsymbol{\varepsilon}_{\mathrm{P}}$	Angular acceleration of the moving platform of the fine-tuning Stewart platform	$\{\boldsymbol{B}\}$

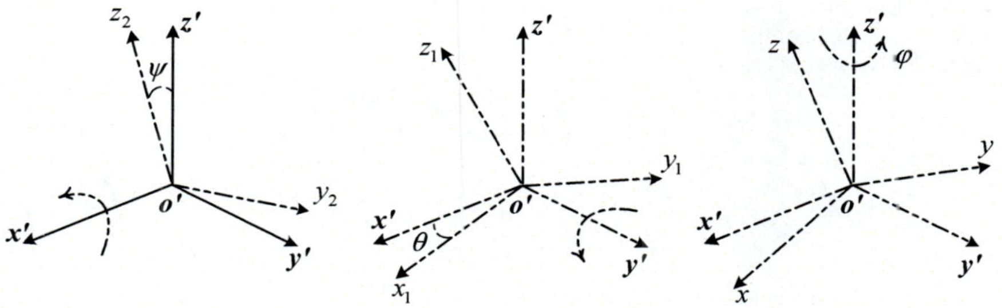

Fig. 6.2 The rotation matrix defined with RPY angles

coincides with $\{\boldsymbol{B}\}$. Firstly, $\{\boldsymbol{P}_3\}$ rotates ψ angle around the x'-axis of $\{\boldsymbol{B}\}$ to form $\{\boldsymbol{P}_2\}$: $o - x_2y_2z_2$, then $\{\boldsymbol{P}_2\}$ rotates θ angle around the y'-axis of $\{\boldsymbol{B}\}$ to form $\{\boldsymbol{P}_1\}$: $o - x_1y_1z_1$; finally $\{\boldsymbol{P}_1\}$ rotates ϕ angle around the z'-axis of $\{\boldsymbol{B}\}$ to form $\{\boldsymbol{P}\}$: $o - xyz$, which is illustrated in Fig. 6.2. The equation of the rotation matrix is:

$$\begin{aligned} {}^{\mathrm{B}}\boldsymbol{R}_{\mathrm{P}} &= \boldsymbol{R}(\psi, \theta, \phi) = \boldsymbol{R}(z', \phi)\boldsymbol{R}(y', \theta)\boldsymbol{R}(x', \psi) \\ &= \begin{bmatrix} c\phi c\theta & c\phi s\theta s\psi - s\phi c\psi & c\phi s\theta c\psi + s\phi s\psi \\ s\phi c\theta & s\phi s\theta s\psi + c\phi c\psi & s\phi s\theta c\psi - c\phi s\psi \\ -s\theta & c\theta s\psi & c\theta c\psi \end{bmatrix} \end{aligned} \tag{6.2}$$

where ψ is the roll angle, θ is the pitch angle, and ϕ is the deflection angle; c is the cosine operation and s is the sine operation. Since the three consecutive rotation angles of the Stewart parallel mechanism cannot reach 90 degrees, three rotation angles can be deduced reversely with the rotation matrix.

Since the base platform of the fine-tuning Stewart platform follows the motion of the cable-driven parallel mechanism and the A–B rotator, the coordinate system of the base platform $\{\boldsymbol{B}\}$ is non-inertial. A series of non-inertial translational coordinate systems parallel to the global coordinate system $\{\boldsymbol{G}\}$ are established, where the dynamic equations are combined. Therefore, each motion parameter needs to be deduced in the non-inertial translational coordinate systems. The linear accelerations (convective accelerations) of the non-inertial translational coordinate systems are deduced at the end. Since the branch chains of the Stewart parallel mechanism are identical, the dynamic deduction process of each branch chain is similar, and the right subscript i indicating the numbers of the branch chains can be omitted in the following dynamic deduction process.

The kinematic model of the branch chain of the fine-tuning Stewart parallel mechanism is shown in Fig. 6.3. The upper and lower parts of the branch chain are defined as local coordinate systems $\{\boldsymbol{L}\}$ and $\{\boldsymbol{U}\}$ respectively. $\{\boldsymbol{L}\}$ is fixed at Bi (the rotation center of the Hooke joint of the branch chain) with the x-axis pointing to Pi along the direction of the branch chain, and the cross product of the x-axis vector and the coordinate vector of Bi is defined as the y-axis of $\{\boldsymbol{L}\}$. $\{\boldsymbol{U}\}$ is fixed at the rotation center of the spherical joint of the branch chain, and the direction of each coordinate

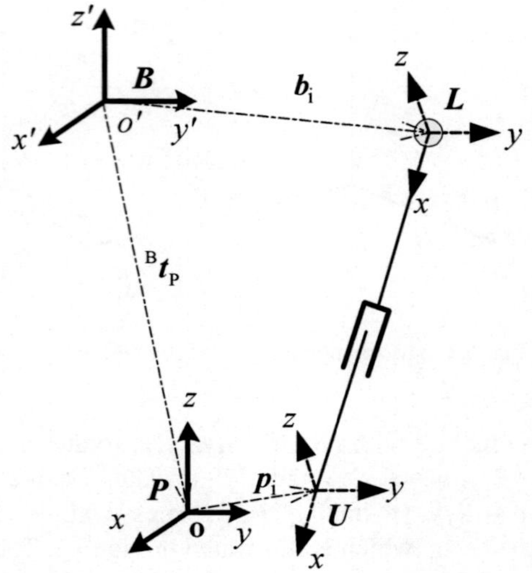

Fig. 6.3 Kinematic model of the branch chain of the fine-tuning Stewart platform

axis is parallel to $\{\boldsymbol{L}\}$. Two temporary non-inertial translational coordinate systems $\{\boldsymbol{L}'\}$ and $\{\boldsymbol{P}'\}$ (not marked in the figure) are established. They are fixed at the origin of $\{\boldsymbol{L}\}$ and $\{\boldsymbol{P}\}$ respectively, paralleling the global inertial coordinate system $\{\boldsymbol{G}\}$.

Based on the rotation matrix and the relative position vector, the position vector of the rotation center of the spherical joint Pi can be converted to the base coordinate system $\{\boldsymbol{B}\}$ and the global coordinate system $\{\boldsymbol{G}\}$ respectively as:

$$^{\mathrm{B}}\boldsymbol{P} = {}^{\mathrm{B}}\boldsymbol{R}_{\mathrm{P}}\boldsymbol{P} + {}^{\mathrm{B}}\boldsymbol{t}_{\mathrm{P}} \tag{6.3}$$

and

$$^{\mathrm{G}}\boldsymbol{P} = {}^{\mathrm{G}}\boldsymbol{t}_{\mathrm{B}} + {}^{\mathrm{G}}\boldsymbol{R}_{\mathrm{B}}{}^{\mathrm{B}}\boldsymbol{P} = {}^{\mathrm{G}}\boldsymbol{R}_{\mathrm{B}}\left({}^{\mathrm{B}}\boldsymbol{R}_{\mathrm{P}}\boldsymbol{P} + {}^{\mathrm{B}}\boldsymbol{t}_{\mathrm{P}}\right) \tag{6.4}$$

where ${}^{\mathrm{G}}\boldsymbol{t}_{\mathrm{B}}$ is the position vector of the geometric center of the base platform of the fine-tuning Stewart platform in the global coordinate system, which can be described by the attitude, position, and rotation angle of the A–B rotator. The rotation center of Hooke joint B_i can be expressed in $\{\boldsymbol{G}\}$ as:

$$^{\mathrm{G}}\boldsymbol{b} = {}^{\mathrm{G}}\boldsymbol{t}_{\mathrm{B}} + {}^{\mathrm{G}}\boldsymbol{R}_{\mathrm{B}}\boldsymbol{b} \tag{6.5}$$

The vector of the branch chain $\boldsymbol{S}$(pointing from the coordinate origin of the local coordinate system $\{\boldsymbol{L}\}$ to that of $\{\boldsymbol{U}\}$) can be expressed as the difference between position vectors of Bi and Pi. The vector of the branch chain obtained in the base coordinate system $\{\boldsymbol{B}\}$ and the global coordinate system $\{\boldsymbol{G}\}$ could be expressed as:

$$^{\mathrm{B}}\boldsymbol{S} = {}^{\mathrm{B}}\boldsymbol{P} - \boldsymbol{b} \tag{6.6}$$

and

$$^{\mathrm{G}}\boldsymbol{S}{=}^{\mathrm{G}}\boldsymbol{P}-{}^{\mathrm{G}}\boldsymbol{b}={}^{\mathrm{G}}\boldsymbol{R}_{\mathrm{B}}\left({}^{\mathrm{B}}\boldsymbol{P}-\boldsymbol{b}\right) \tag{6.7}$$

\varvec

Further, the length of the branch chain can be calculated, i.e., the inverse position equation of the Stewart parallel mechanism is:

$$L{=}\left\|{}^{\mathrm{G}}\boldsymbol{S}\right\|=\left\|{}^{\mathrm{B}}\boldsymbol{S}\right\| \tag{6.8}$$

The unit vector of the branch chain can be expressed as:

$$^{\mathrm{B}}\boldsymbol{s}={}^{\mathrm{B}}\boldsymbol{S}/L \tag{6.9}$$

and

$$^{\mathrm{G}}\boldsymbol{s}={}^{\mathrm{G}}\boldsymbol{S}/L={}^{\mathrm{G}}\boldsymbol{R}_{\mathrm{B}}\left({}^{\mathrm{B}}\boldsymbol{S}/L\right)={}^{\mathrm{G}}\boldsymbol{R}_{\mathrm{B}}{}^{\mathrm{B}}\boldsymbol{s} \tag{6.10}$$

When Newton–Euler equations are combined, the Euler equations of the branch chain are established around the rotation center of Hooke joint Bi. Therefore, it is necessary to deduce the description of the position vectors of the centroid of the upper and lower parts of the branch chain in {***L***'}. The rotation matrix of the branch chain from {***L***} to {***L***'} can be deduced from the position vector bi, as shown:

$$\begin{aligned}
\boldsymbol{x}&={}^{\mathrm{G}}\boldsymbol{R}_{\mathrm{B}}{}^{\mathrm{B}}\boldsymbol{s},\\
\boldsymbol{y}&={}^{\mathrm{G}}\boldsymbol{R}_{\mathrm{B}}\left({}^{\mathrm{B}}\boldsymbol{s}\times\boldsymbol{b}\right)/\left\|{}^{\mathrm{B}}\boldsymbol{s}\times\boldsymbol{b}\right\|,\\
\boldsymbol{z}&=\boldsymbol{x}\times\boldsymbol{y}.
\end{aligned}$$

Further, the rotation matrix can be obtained as:

$$^{\mathrm{G}}\boldsymbol{T}=\left[\boldsymbol{x}\ \boldsymbol{y}\ \boldsymbol{z}\right]={}^{\mathrm{G}}\boldsymbol{R}_{\mathrm{B}}{}^{\mathrm{B}}\boldsymbol{T} \tag{6.11}$$

The centroid vectors of the upper and lower parts of the branch chain are defined in their local coordinate systems {***L***} and {***U***} respectively, that is, $\boldsymbol{r}_{\mathrm{Lo}}$ and $\boldsymbol{r}_{\mathrm{Uo}}$, which can be expressed in {***L***'} as:

$$\boldsymbol{r}_{\mathrm{L}}={}^{\mathrm{G}}\boldsymbol{T}\boldsymbol{r}_{\mathrm{Lo}} \tag{6.12}$$

$$\boldsymbol{r}_{\mathrm{U}}={}^{\mathrm{G}}\boldsymbol{S}+{}^{\mathrm{G}}\boldsymbol{T}\boldsymbol{r}_{\mathrm{Uo}} \tag{6.13}$$

From the position analysis, it can be found that for the fine-tuning Stewart platform, the rotation matrix ${}^{\mathrm{G}}\boldsymbol{R}_{\mathrm{B}}$ of the base platform is included in the equations of the position vector of its components. That is, the position equations of the fine-tuning Stewart parallel mechanism can reflect the influence of the attitude of the base platform.

Deriving both sides of Eq. (6.7) simultaneously with respect to time, the velocity mapping equation of the branch chain in $\{G\}$ can be established as:

$$ {}^{G}\dot{\boldsymbol{S}}={}^{G}\omega_{B}\times{}^{G}\boldsymbol{R}_{B}\left({}^{B}\boldsymbol{R}_{P}\cdot\boldsymbol{P}+{}^{B}\boldsymbol{t}_{P}-\boldsymbol{b}\right)+{}^{G}\boldsymbol{R}_{B}\left({}^{B}\omega_{P}\times{}^{B}\boldsymbol{R}_{P}\boldsymbol{P}+{}^{B}\dot{\boldsymbol{t}}_{P}\right) \tag{6.14} $$

It can be further simplified as:

$$ {}^{G}\dot{\boldsymbol{S}}={}^{G}\omega_{B}\times{}^{G}\boldsymbol{S}+{}^{G}\boldsymbol{R}_{B}{}^{B}\dot{\boldsymbol{S}} \tag{6.15} $$

where,

$$ {}^{B}\dot{\boldsymbol{S}}={}^{B}\omega_{P}\times\left({}^{B}\boldsymbol{R}_{P}\boldsymbol{P}\right)+{}^{B}\dot{\boldsymbol{t}}_{P} $$

This equation can be obtained directly by deriving both sides of Eq. (6.6) simultaneously with respect to time.

Based on the analysis of the velocity vector ${}^{G}\dot{\boldsymbol{S}}$, it can be expressed in terms of the telescoping velocity $\dot{L}$ of the branch chain along its direction and the angular velocity ${}^{G}L$ along the vertical branch chain direction, i.e.:

$$ {}^{G}\dot{\boldsymbol{S}}=\dot{L}{}^{G}\boldsymbol{s}+L{}^{G}\boldsymbol{W}\times{}^{G}\boldsymbol{s} \tag{6.16} $$

Take the dot product with G$\boldsymbol{s}$ on each side of Eq. (6.16), the telescoping velocity $\dot{L}$ of the branch chain can be deduced as:

$$ \dot{L}={}^{G}\boldsymbol{s}\cdot{}^{G}\dot{\boldsymbol{S}}={}^{G}\boldsymbol{s}\cdot\left({}^{G}\omega_{B}\times{}^{G}\boldsymbol{S}\right)+\left({}^{G}\boldsymbol{R}_{B}{}^{B}\boldsymbol{s}\right)\cdot\left({}^{G}\boldsymbol{R}_{B}{}^{B}\dot{\boldsymbol{S}}\right) \tag{6.17} $$

Since ${}^{G}\boldsymbol{s}\left({}^{G}\omega_{B}\times^{G}\boldsymbol{S}\right)=0$, it can be simplified as:

$$ \dot{L}={}^{B}\boldsymbol{s}\cdot{}^{B}\dot{\boldsymbol{S}} $$

Take the cross product with ${}^{G}\boldsymbol{s}$ on each side of Eq. (6.16), the equation of the angular velocity of the branch chain can be deduced as:

$$ {}^{G}\boldsymbol{W}={}^{G}\boldsymbol{s}\times{}^{G}\dot{\boldsymbol{S}}/L \tag{6.18} $$

Substituting Eq. (6.15), can be simplified as:

$$ {}^{G}\boldsymbol{W}={}^{G}\omega_{B}-\left({}^{G}\boldsymbol{s}\cdot{}^{G}\omega_{B}\right){}^{G}\boldsymbol{s}+{}^{G}\boldsymbol{R}_{B}{}^{B}\boldsymbol{W} \tag{6.19} $$

where,

$$ {}^{B}\boldsymbol{W} = {}^{B}\boldsymbol{s} \times {}^{B}\dot{\boldsymbol{S}}/L $$

As can be found from the velocity analysis, the velocity equations of the fine-tuning Stewart parallel mechanism are more complex compared to the parallel mechanism with the stationary base. The velocity vector of each component is described in the global coordinate system, including parameters as attitude ${}^{G}\boldsymbol{R}_{B}$ and angular velocity ${}^{G}\boldsymbol{\omega}_{B}$ of the base platform, and reflects the influence of the base motion.

Similar to the velocity mapping equation, the acceleration mapping equation of the branch chain can be also expressed in two approaches. Deriving each side of Eq. (6.15) with respect to time gives:

$$ {}^{G}\ddot{\boldsymbol{S}} = {}^{G}\boldsymbol{\varepsilon}_{B} \times {}^{G}\boldsymbol{S} + {}^{G}\boldsymbol{\omega}_{B} \times {}^{G}\dot{\boldsymbol{S}} + {}^{G}\boldsymbol{\omega}_{B} \times \left({}^{G}\boldsymbol{R}_{B}{}^{B}\dot{\boldsymbol{S}}\right) + {}^{G}\boldsymbol{R}_{B}{}^{B}\ddot{\boldsymbol{S}} \tag{6.20} $$

where,

$$ {}^{B}\ddot{\boldsymbol{S}} = {}^{B}\boldsymbol{\varepsilon}_{P} \times {}^{B}\boldsymbol{R}_{P}\boldsymbol{P} + {}^{B}\boldsymbol{\omega}_{P} \times \left[{}^{B}\boldsymbol{\omega}_{P} \times \left({}^{B}\boldsymbol{R}_{P}\boldsymbol{P}\right)\right] + {}^{B}\ddot{\boldsymbol{t}}_{P} $$

An alternative expression for the acceleration of the branch chain can be obtained by deriving each side of Eq. (6.16) with respect to time:

$$ {}^{G}\ddot{\boldsymbol{S}} = \ddot{L}{}^{G}\boldsymbol{s} + {}^{G}\boldsymbol{W} \times \left({}^{G}\boldsymbol{W} \times {}^{G}\boldsymbol{S}\right) + 2{}^{G}\boldsymbol{W} \times \dot{L}{}^{G}\boldsymbol{s} + {}^{G}\boldsymbol{A} \times {}^{G}\boldsymbol{S} \tag{6.21} $$

And it can be simplified as:

$$ {}^{G}\ddot{\boldsymbol{S}} = \left(\ddot{L} - L{}^{G}\boldsymbol{W} \cdot {}^{G}\boldsymbol{W}\right){}^{G}\boldsymbol{s} + 2{}^{G}\boldsymbol{W} \times \dot{L}{}^{G}\boldsymbol{s} + {}^{G}\boldsymbol{A} \times {}^{G}\boldsymbol{S} \tag{6.22} $$

Both sides of Eq. (6.22) are dotted and crossed by the unit vector ${}^{G}\boldsymbol{s}$ of the branch chain, and the equations of the telescoping acceleration $\ddot{L}$ and the angular acceleration ${}^{G}\boldsymbol{A}$ of the branch chain are deduced as:

$$ \ddot{L} = L{}^{G}\boldsymbol{W} \cdot {}^{G}\boldsymbol{W} + {}^{G}\boldsymbol{s} \cdot {}^{G}\ddot{\boldsymbol{S}} \tag{6.23} $$

and

$$ {}^{G}\boldsymbol{A} = \left({}^{G}\boldsymbol{s} \times {}^{G}\ddot{\boldsymbol{S}} - 2\dot{L}{}^{G}\boldsymbol{W}\right)/L \tag{6.24} $$

The linear acceleration in the non-inertial translational coordinate system $\{\boldsymbol{L}'\}$ for the upper and lower centroids of the branch chain can be expressed as:

$$ \boldsymbol{a}_{L} = {}^{G}\boldsymbol{A} \times \boldsymbol{r}_{L} + {}^{G}\boldsymbol{W} \times \left({}^{G}\boldsymbol{W} \times \boldsymbol{r}_{L}\right) \tag{6.25} $$

$$ \boldsymbol{a}_{U} = {}^{G}\boldsymbol{A} \times \boldsymbol{r}_{U} + {}^{G}\boldsymbol{W} \times \left({}^{G}\boldsymbol{W} \times \boldsymbol{r}_{U}\right) + \ddot{L}{}^{G}\boldsymbol{s} + 2\dot{L}\left({}^{G}\boldsymbol{W} \times {}^{G}\boldsymbol{s}\right) \tag{6.26} $$

In the global coordinate system $\{\boldsymbol{G}\}$, the acceleration of the non-inertial translational coordinate system $\{\boldsymbol{L}'\}$ (translational convective acceleration with respect to the global coordinate system) can be expressed as:

$$\boldsymbol{a}_{\mathrm{Lr}} = {}^{\mathrm{G}}\ddot{\boldsymbol{t}}_{\mathrm{B}} + {}^{\mathrm{G}}\boldsymbol{\varepsilon}_{\mathrm{B}} \times \left({}^{\mathrm{G}}\boldsymbol{R}_{\mathrm{B}}\boldsymbol{b}\right) + {}^{\mathrm{G}}\boldsymbol{\omega}_{\mathrm{B}} \times \left({}^{\mathrm{G}}\boldsymbol{\omega}_{\mathrm{B}} \times {}^{\mathrm{G}}\boldsymbol{R}_{\mathrm{B}}\boldsymbol{b}\right) \tag{6.27}$$

The analysis shows that the equations of the acceleration parameters of the fine-tuning Stewart parallel mechanism contain the translational acceleration ${}^{\mathrm{G}}\ddot{\boldsymbol{t}}_{\mathrm{B}}$, the angular velocity of rotation ${}^{\mathrm{G}}\boldsymbol{\omega}_{\mathrm{B}}$, the angular acceleration ${}^{\mathrm{G}}\boldsymbol{\varepsilon}_{\mathrm{B}}$ and the attitude ${}^{\mathrm{G}}\boldsymbol{R}_{\mathrm{B}}$ of the base platform except for the relative position and the translational velocity of the base platform.

The velocity and acceleration parameters of the moving platform of the fine-tuning Stewart parallel mechanism are deduced below. A non-inertial translational coordinate system $\{\boldsymbol{P}'\}$ is introduced at the origin of the non-inertial coordinate system $\{\boldsymbol{P}\}$ of the moving platform. And $\{\boldsymbol{P}'\}$ is parallel to the global coordinate system.

Suppose $\boldsymbol{e}_{\mathrm{po}}$ is a position vector describing the centroid of the moving platform in $\{\boldsymbol{P}\}$, it can be expressed in $\{\boldsymbol{P}'\}$ as:

$${}^{\mathrm{G}}\boldsymbol{e}_{\mathrm{P}} = {}^{\mathrm{G}}\boldsymbol{R}_{\mathrm{B}}\left({}^{\mathrm{B}}\boldsymbol{R}_{\mathrm{P}}\boldsymbol{e}_{\mathrm{Po}}\right) \tag{6.28}$$

According to the superposition principle of rigid body motion, the angular velocity and angular acceleration of the moving platform of the fine-tuning Stewart parallel mechanism in the global coordinate system $\{\boldsymbol{G}\}$ can be expressed as:

$${}^{\mathrm{G}}\boldsymbol{\omega}_{\mathrm{P}} = {}^{\mathrm{G}}\boldsymbol{\omega}_{\mathrm{B}} + {}^{\mathrm{G}}\boldsymbol{R}_{\mathrm{B}}{}^{\mathrm{B}}\boldsymbol{\omega}_{\mathrm{P}} \tag{6.29}$$

$${}^{\mathrm{G}}\boldsymbol{\varepsilon}_{\mathrm{P}} = {}^{\mathrm{G}}\boldsymbol{\varepsilon}_{\mathrm{B}} + {}^{\mathrm{G}}\boldsymbol{R}_{\mathrm{B}}{}^{\mathrm{B}}\boldsymbol{\varepsilon}_{\mathrm{P}} + {}^{\mathrm{G}}\boldsymbol{\omega}_{\mathrm{B}} \times ({}^{\mathrm{G}}\boldsymbol{R}_{\mathrm{B}}{}^{\mathrm{B}}\boldsymbol{\omega}_{\mathrm{P}}) \tag{6.30}$$

Further, the convective acceleration of the non-inertial translational coordinate system $\{\boldsymbol{P}'\}$ of the moving platform of the fine-tuning Stewart parallel mechanism (translational acceleration of $\{\boldsymbol{P}'\}$ described in the global coordinate system) can be deduced as:

$$\begin{aligned}\boldsymbol{a}_{\mathrm{Pr}} = {}& {}^{\mathrm{G}}\ddot{\boldsymbol{t}}_{\mathrm{B}} + {}^{\mathrm{G}}\boldsymbol{R}_{\mathrm{B}}{}^{\mathrm{B}}\ddot{\boldsymbol{t}}_{\mathrm{P}} + {}^{\mathrm{G}}\boldsymbol{\varepsilon}_{\mathrm{B}} \times \left({}^{\mathrm{G}}\boldsymbol{R}_{\mathrm{B}}{}^{\mathrm{B}}\boldsymbol{t}_{\mathrm{P}}\right) + {}^{\mathrm{G}}\boldsymbol{\omega}_{\mathrm{B}} \\ & \times \left[{}^{\mathrm{G}}\boldsymbol{\omega}_{\mathrm{B}} \times \left({}^{\mathrm{G}}\boldsymbol{R}_{\mathrm{B}}{}^{\mathrm{B}}\boldsymbol{t}_{\mathrm{P}}\right)\right] \\ & + 2{}^{\mathrm{G}}\boldsymbol{\omega}_{\mathrm{B}} \times \left({}^{\mathrm{G}}\boldsymbol{R}_{\mathrm{B}}{}^{\mathrm{B}}\dot{\boldsymbol{t}}_{\mathrm{P}}\right)\end{aligned} \tag{6.31}$$

Since the Newton–Euler equations are combined in a non-inertial translational coordinate system (parallel to the global coordinate system), it is necessary to convert the inertia matrix of the components of the fine-tuning Stewart parallel mechanism defined in the local coordinate system with respect to the centroid to the global coordinate system. The rotation inertia $\boldsymbol{I}_{\mathrm{Lo}}$ of the upper segment of the branch chain is defined in $\{\boldsymbol{L}\}$. Converting it to $\{\boldsymbol{G}\}$ obtains:

$$\boldsymbol{I}_{\mathrm{L}} = {}^{\mathrm{G}}\boldsymbol{T}\boldsymbol{I}_{\mathrm{Lo}}{}^{\mathrm{G}}\boldsymbol{T}^{\mathrm{T}} \tag{6.32}$$

Similarly, the rotation inertia $\boldsymbol{I}_{\mathrm{Uo}}$ of the lower segment of the branch chain is defined in $\{\boldsymbol{U}\}$. Converting it to $\{\boldsymbol{G}\}$ obtains:

$$\boldsymbol{I}_{\mathrm{U}} = {}^{\mathrm{G}}\boldsymbol{T}\boldsymbol{I}_{\mathrm{Uo}}{}^{\mathrm{G}}\boldsymbol{T}^{\mathrm{T}} \tag{6.33}$$

The inertia matrix of the moving platform of the fine-tuning Stewart parallel mechanism in the local coordinate system $\{\boldsymbol{P}\}$ is $\boldsymbol{I}_{\mathrm{Po}}$. Converting it to $\{\boldsymbol{G}\}$ obtains:

$$\boldsymbol{I}_{\mathrm{P}} = \left({}^{\mathrm{G}}\boldsymbol{R}_{\mathrm{B}}{}^{\mathrm{B}}\boldsymbol{R}_{\mathrm{P}}\right)\boldsymbol{I}_{\mathrm{Po}}\left({}^{\mathrm{G}}\boldsymbol{R}_{\mathrm{B}}{}^{\mathrm{B}}\boldsymbol{R}_{\mathrm{P}}\right)^{\mathrm{T}} \tag{6.34}$$

Now, the kinematic and inertia parameters required for the dynamic equations of the fine-tuning Stewart parallel mechanism have been analyzed and deduced. The following sections will introduce the combination of the dynamic equations and the deduction of the driving force of the branch chain.

6.3 Dynamic Modelling of the Fine-Tuning Stewart Platform

Firstly, the Euler equation of the branch chain is established. In the non-inertial translational coordinate system $\{\boldsymbol{L}'\}$, the Euler equation of the branch chain is established with respect to the rotation center of Hooke joint B_i:

$$\begin{aligned}
&{}^{\mathrm{G}}\boldsymbol{S} \times \boldsymbol{F}_{\mathrm{S}} + M_{\mathrm{U}}{}^{\mathrm{G}}\boldsymbol{s} - C_{\mathrm{U}}{}^{\mathrm{B}}\boldsymbol{W} - C_{\mathrm{S}}\left({}^{\mathrm{B}}\boldsymbol{W} - {}^{\mathrm{B}}\boldsymbol{\omega}_{\mathrm{P}}\right)\\
&\quad = m_{\mathrm{L}}\boldsymbol{r}_{\mathrm{L}} \times \boldsymbol{a}_{\mathrm{L}} + m_{\mathrm{U}}\boldsymbol{r}_{\mathrm{U}} \times \boldsymbol{a}_{\mathrm{U}} + (\boldsymbol{I}_{\mathrm{L}} + \boldsymbol{I}_{\mathrm{U}})^{\mathrm{G}}\boldsymbol{A}\\
&\quad + {}^{\mathrm{G}}\boldsymbol{W} \times \left[(\boldsymbol{I}_{\mathrm{L}} + \boldsymbol{I}_{\mathrm{U}})^{\mathrm{G}}\boldsymbol{W}\right] + (m_{\mathrm{L}}\boldsymbol{r}_{\mathrm{L}} + m_{\mathrm{U}}\boldsymbol{r}_{\mathrm{U}})\\
&\quad \times \boldsymbol{a}_{\mathrm{Lr}} - (m_{\mathrm{L}}\boldsymbol{r}_{\mathrm{L}} + m_{\mathrm{U}}\boldsymbol{r}_{\mathrm{U}}) \times \boldsymbol{g}
\end{aligned} \tag{6.35}$$

where M_{U} is the torque modulus that applied by the Hooke joint to the branch chain to inhibit its rotation around itself; $\boldsymbol{F}_{\mathrm{S}}$ is the force applied by the spherical joint to the branch chain; C_{U} is the coefficient of friction at the Hooke joint; and M_{S} is the coefficient of friction at the spherical joint. Equation (6.35) can be simplified as:

$$ {}^{\mathrm{G}}\boldsymbol{S} \times \boldsymbol{F}_{\mathrm{S}} + M_{\mathrm{U}}{}^{\mathrm{G}}\boldsymbol{s} = \boldsymbol{D} \tag{6.36}$$

where,

$$\boldsymbol{D} = m_{\mathrm{L}}\boldsymbol{r}_{\mathrm{L}} \times \boldsymbol{a}_{\mathrm{L}} + m_{\mathrm{U}}\boldsymbol{r}_{\mathrm{U}} \times \boldsymbol{a}_{\mathrm{U}} + (\boldsymbol{I}_{\mathrm{L}} + \boldsymbol{I}_{\mathrm{U}})^{\mathrm{G}}\boldsymbol{A}$$

$$+{}^{G}\boldsymbol{W} \times \left[(\boldsymbol{I}_{L} + \boldsymbol{I}_{U}){}^{G}\boldsymbol{W} \right]$$
$$+ (m_{L}\boldsymbol{r}_{L} + m_{U}\boldsymbol{r}_{U}) \times \boldsymbol{a}_{Lr} - (m_{L}\boldsymbol{r}_{L} + m_{U}\boldsymbol{r}_{U})$$
$$\times \boldsymbol{g} + C_{U}{}^{B}\boldsymbol{W} + C_{S}\left({}^{B}\boldsymbol{W} - {}^{B}\boldsymbol{\omega}_{P}\right)$$

To eliminate M_U, both sides of Eq. (6.36) are crossed with the unit direction vector ${}^{G}\boldsymbol{s}$ of the branch chain respectively:

$$^{G}\boldsymbol{F}_{S} = \boldsymbol{C} + F_{Leg}{}^{G}\boldsymbol{s} \tag{6.37}$$

where,

$$\boldsymbol{C} = \frac{\boldsymbol{D} \times {}^{G}\boldsymbol{s}}{L}$$

$$^{G}\boldsymbol{s} \cdot {}^{G}\boldsymbol{F}_{S} = F_{Leg}$$

Now, the vector ${}^{G}\boldsymbol{F}_{S}$ containing three unknowns has been transformed into a one-dimensional scalar F_{Leg}.

To further establish dynamic equations of the moving platform of the fine-tuning Stewart parallel mechanism. In {***P'***}, the force balance equation can be established:

$$m_{P}\boldsymbol{g} + \boldsymbol{F}_{ext} - \sum_{i=1}^{6} {}^{G}\boldsymbol{F}_{Si} - m_{P}\boldsymbol{a}_{Pr} = m_{P}\boldsymbol{a}_{P} \tag{6.38}$$

The Euler equation of the moving platform can be established with respect to the origin of {***P'***}:

$$m_{P}{}^{G}\boldsymbol{e} \times \boldsymbol{g} + \boldsymbol{M}_{ext} - \sum_{i=1}^{6} \left[\left({}^{G}\boldsymbol{R}_{B}{}^{B}\boldsymbol{R}_{P}\boldsymbol{P}_{i}\right) \times {}^{G}\boldsymbol{F}_{Si} \right] - m_{P}{}^{G}\boldsymbol{e} \times \boldsymbol{a}_{Pr}$$
$$= m_{P}{}^{G}\boldsymbol{e} \times \boldsymbol{a}_{P} + {}^{G}\boldsymbol{\varepsilon}_{P}\boldsymbol{I}_{P} + {}^{G}\boldsymbol{\omega}_{p} \times \left(\boldsymbol{I}_{P}{}^{G}\boldsymbol{\omega}_{P}\right) \tag{6.39}$$

where $\boldsymbol{F}_{ext}$ and $\boldsymbol{M}_{ext}$ are the external force and moment acting on the moving platform (described in the global coordinate system). Substitute Eq. (6.37) into the Newton–Euler Eqs. (6.38) and (6.39) of the moving platform, and after simplification we can get:

$$\sum_{i=1}^{6} \left(F_{Leg_i}{}^{G}\boldsymbol{s}_i\right) = m_{P}\boldsymbol{g} + \boldsymbol{F}_{ext} - \sum_{i=1}^{6} \boldsymbol{C}_i - m_{P}\boldsymbol{a}_{Pr} - m_{P}\boldsymbol{a}_{P} \tag{6.40}$$

$$\sum_{i=1}^{6} \left[F_{Leg\,i}\left({}^{G}\boldsymbol{R}_{B}{}^{B}\boldsymbol{R}_{P}\boldsymbol{P}_{i}\right) \times {}^{G}\boldsymbol{s}_i \right] = m_{P}{}^{G}\boldsymbol{e} \times \left(\boldsymbol{g} - \boldsymbol{a}_{P} - \boldsymbol{a}_{pr}\right) + \boldsymbol{M}_{ext}$$

$$-{}^{\mathrm{G}}\boldsymbol{\varepsilon}_{\mathrm{P}}\boldsymbol{I}_{\mathrm{P}} - {}^{\mathrm{G}}\boldsymbol{\omega}_{\mathrm{P}} \times \left(\boldsymbol{I}_{\mathrm{P}}{}^{\mathrm{G}}\boldsymbol{\omega}_{\mathrm{P}}\right) - \sum_{i=1}^{6}\left[\left({}^{\mathrm{G}}\boldsymbol{R}_{\mathrm{B}}{}^{\mathrm{B}}\boldsymbol{R}_{\mathrm{P}}\boldsymbol{P}_i\right) \times \boldsymbol{C}_i\right] \tag{6.41}$$

The analysis shows that 6 equations containing 6 unknowns $F_{\mathrm{Leg}\,i}(i = 1, 2, \ldots 6)$ can be obtained by combining Eq. (6.40) and Eq. (6.41). A unique solution can be deduced with above equations. Equations (6.40) and (6.41) can be expressed in matrix form as:

$$\boldsymbol{H}\boldsymbol{F}_{\mathrm{Leg}} = \boldsymbol{N} \tag{6.42}$$

where,

$$\boldsymbol{H} = \begin{bmatrix} {}^{\mathrm{G}}\boldsymbol{s}_1 & {}^{\mathrm{G}}\boldsymbol{s}_2 & {}^{\mathrm{G}}\boldsymbol{s}_3 & {}^{\mathrm{G}}\boldsymbol{s}_4 & {}^{\mathrm{G}}\boldsymbol{s}_5 & {}^{\mathrm{G}}\boldsymbol{s}_6 \\ \boldsymbol{j}_1 & \boldsymbol{j}_2 & \boldsymbol{j}_3 & \boldsymbol{j}_4 & \boldsymbol{j}_5 & \boldsymbol{j}_6 \end{bmatrix}$$

where,

$$\boldsymbol{j}_i = \left({}^{\mathrm{G}}\boldsymbol{R}_{\mathrm{B}}{}^{\mathrm{B}}\boldsymbol{R}_{\mathrm{P}}\boldsymbol{P}_i\right) \times {}^{\mathrm{G}}\boldsymbol{s}_i \quad i = 1, 2, \ldots 6,$$
$$\boldsymbol{F}_{\mathrm{Leg}} = \left[F_{\mathrm{Leg1}}, F_{\mathrm{Leg2}}, F_{\mathrm{Leg3}}, F_{\mathrm{Leg4}}, F_{\mathrm{Leg5}}, F_{\mathrm{Leg6}}\right]^{\mathrm{T}},$$
$$\boldsymbol{N} = \begin{bmatrix} \boldsymbol{n}_1 \\ \boldsymbol{n}_2 \end{bmatrix}.$$

where,

$$\boldsymbol{n}_1 = m_{\mathrm{P}}\boldsymbol{g} + \boldsymbol{F}_{\mathrm{ext}} - \sum_{i=1}^{6}\boldsymbol{C}_i - m_{\mathrm{P}}\boldsymbol{a}_{\mathrm{Pr}} - m_{\mathrm{P}}\boldsymbol{a}_{\mathrm{P}}$$

$$\boldsymbol{n}_2 = m_{\mathrm{P}}{}^{\mathrm{G}}\boldsymbol{e} \times (\boldsymbol{g} - \boldsymbol{a}_{\mathrm{P}} - \boldsymbol{a}_{\mathrm{Pr}}) + \boldsymbol{M}_{\mathrm{ext}} - {}^{\mathrm{G}}\boldsymbol{\varepsilon}_{\mathrm{P}}\mathbf{I}_{\mathrm{P}} - \sum_{i=1}^{6}\left[\left({}^{\mathrm{G}}\boldsymbol{R}_{\mathrm{B}}{}^{\mathrm{B}}\boldsymbol{R}_{\mathrm{P}}\boldsymbol{P}_i\right) \times \boldsymbol{C}_i\right] - {}^{\mathrm{G}}\boldsymbol{\omega}_{\mathrm{P}} \times \left(\boldsymbol{I}_{\mathrm{P}}{}^{\mathrm{G}}\boldsymbol{\omega}_{\mathrm{P}}\right)$$

The force balance equation of the upper part of the branch chain is established:

$${}^{\mathrm{G}}\boldsymbol{F}_{\mathrm{S}} + m_{\mathrm{U}}\boldsymbol{g} + \boldsymbol{F}_{\mathrm{PJ}} - C_{\mathrm{p}}\dot{L}{}^{\mathrm{G}}\boldsymbol{s} = m_{\mathrm{U}}(\boldsymbol{a}_{\mathrm{U}} + \boldsymbol{a}_{\mathrm{Lr}}) \tag{6.43}$$

where Cp is the coefficient of friction of the prismatic joint; $\boldsymbol{F}_{\mathrm{PJ}}$ is the driving force of the prismatic joint. Both sides of Eq. (6.43) are dotted by ${}^{\mathrm{G}}\boldsymbol{s}$ simultaneously, then the driving force of the active joint (prismatic joint) of the branch chain can be deduced as:

$$F = U - F_{\mathrm{Leg}} \tag{6.44}$$

where,

$$F = {}^{\mathrm{G}}\boldsymbol{s} \cdot \boldsymbol{F}_{\mathrm{PJ}}$$

$$U = m_{\mathrm{U}}{}^{\mathrm{G}}\boldsymbol{s} \cdot \left[(\boldsymbol{a}_{\mathrm{U}} + \boldsymbol{a}_{\mathrm{Lr}}) - \boldsymbol{g}\right] + C_{\mathrm{p}}\dot{L}$$

The equation of the driving force of 6 branch chains can be expressed in matrix form as:

$$\boldsymbol{F} = \boldsymbol{F}_{\mathrm{Leg}} + \boldsymbol{U} \tag{6.45}$$

where,

$$\boldsymbol{F} = [F_1, F_2, F_3, F_4, F_5, F_6]^{\mathrm{T}}$$

$$\boldsymbol{U} = [U_1, U_2, U_3, U_4, U_5, U_6]^{\mathrm{T}}$$

Now, the deduction of the dynamic model of the fine-tuning Stewart parallel mechanism is completed. Compared with the traditional dynamic model of the Stewart parallel mechanism with a stationary base platform, it can reflect the influence of the motion parameters of the base platform on the driving force of the parallel mechanism. The above deduction process indicates that the position and linear velocity of the base platform in the global inertial coordinate system do not appear, that is, they do not influence the driving force.

6.4 Experimental Verification of the Dynamic Model

The idea of the experimental verification of the dynamic model is: the experimental parallel robot is controlled to move according to the given trajectory, and the actual driving forces of the branch chains are measured by the force sensors during the motion process. The data of the force sensor and the trajectory of the parallel mechanism is recorded simultaneously. The inertia parameters of the experimental parallel robot are obtained by calibration, and the theoretical driving force of the branch chain is deduced with the dynamic model in combination with the recorded trajectory. The accuracy of the dynamic model can be verified by comparing the theoretical and practical values of the driving forces of the branch chains. The keys are the simultaneous recording of the data of the trajectory and the force sensors, the calibration of such dynamic parameters as the inertia of the test bench, and the noise suppression of the collected data.

The 1:10 scale model of the fine-tuning Stewart platform is shown in Fig. 6.4. It includes three parts: a rigid Stewart parallel mechanism, an electrical control cabinet, and an Industrial Personal Computer (IPC). Each branch chain of the Stewart parallel mechanism is driven by a servo motor and ball screw, with a force sensor equipped in it. The signal acquisition card plugged in the IPC can read the data of the force

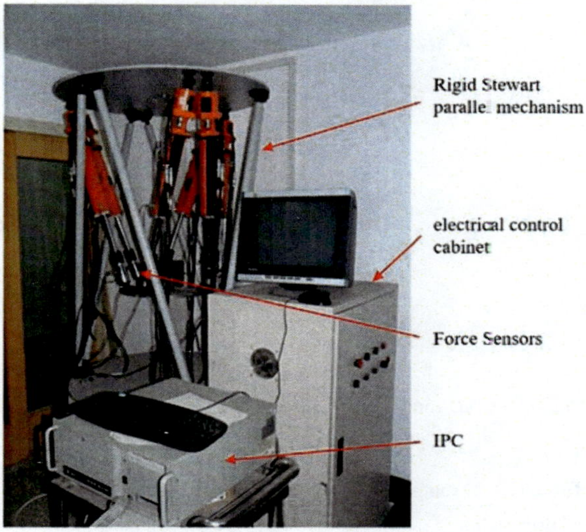

Fig. 6.4 The experimental robot (1:10 scale model of the fine-tuning Stewart platform)

sensors and collect the signal of the driving force of the branch chain of the Stewart parallel mechanism. The NC software enables high-speed simultaneous recording of the lengths of the branch chains and the data of the force sensors with a period of 2.2 ms.

The CAD and decomposition methods are combined to calibrate such dynamic parameters as mass, inertia matrix, and the coordinates of the centroid of the moving components of the scale model of the fine-tuning Stewart platform. The specific steps are: firstly, using UG software to establish the detailed CAD model of the experimental object, in which the explosive view of the single branch chain is shown in Fig. 6.5. Then, the 1:10 scale model of the fine-tuning platform is physically disassembled and measured to deduce the actual masses of all components. The mass properties of each component in the UG model are modified to the corresponding weighed-out masses. Finally, dynamic parameters, such as the position of the centroid and the inertia matrix of each moving component, are obtained with the UG software. The dynamic parameters of the 1:10 scale model of the fine-tuning Stewart platform obtained with the above calibration method are shown in Table 6.2. The parameters of the centroid and inertia matrix are described in the local coordinate system of each component.

To describe the experimental trajectory and the dynamics, the following coordinate systems are established in the 1:10 scale model of the fine-tuning platform. The global coordinate system *O-XYZ* is established at the geometric center of the base platform, and the direction of *X*-axis and *Y*-axis as well as the position of the joint point are shown in Fig. 6.6. According to the right-hand rule, the positive direction of the *Z*-axis is perpendicular to the paper outward. After the experimental parallel robot returning to zero, the initial length of the branch chains is 967.25 mm, and the position of the moving platform in the global coordinate system are (0, 0, −

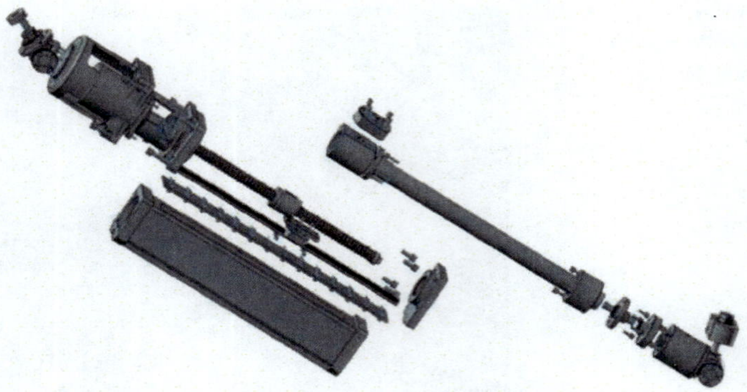

Fig. 6.5 CAD model of the branch chain of the 1:10 scale model of the fine-tuning platform

Table 6.2 Dynamic parameters of the 1:10 scale model of the fine-tuning Strewart platform

Components	Mass/kg	Centroid coordinates/m	Inertia matrix/kg.m^2
Upper part	7.470	$\begin{bmatrix} 0.2538 \\ 0.0002 \\ 0.0051 \end{bmatrix}$	$\begin{bmatrix} 0.0075 & -0.0001 & 0.0057 \\ -0.0001 & 0.2175 & 0 \\ 0.0057 & 0 & 0.2167 \end{bmatrix}$
Lower part	3.355	$\begin{bmatrix} -0.2752 \\ 0.0001 \\ -0.0007 \end{bmatrix}$	$\begin{bmatrix} 0.0011 & 0.0001 & 0.0001 \\ 0.0001 & 0.1446 & 0 \\ 0.0001 & 0 & 0.1444 \end{bmatrix}$
The moving platform	6.513	$\begin{bmatrix} 0 \\ 0 \\ 0.0089 \end{bmatrix}$	$\begin{bmatrix} 0.0865 & 0 & 0 \\ 0 & 0.0865 & 0 \\ 0 & 0 & 0.1494 \end{bmatrix}$

887)mm with a horizontal attitude. The radii of the joint points of the base platform and the moving platform are 450 mm and 250 mm respectively. The workspace of the moving platform of the scale model is a solid sphere with a center at (0, 0, −967) mm and a radius of 100 mm.

By writing the corresponding NC program, the moving platform of the 1:10 scale model of the fine-tuning platform can be controlled to make the reciprocating motion with constant acceleration and deceleration along the X-axis and Y-axis respectively on the maximum horizontal cross-section of the workspace (i.e. the horizontal plane of Z = −967 mm) with a feed speed of 2000 mm/min and an acceleration of 1 m/s^2. The distance of reciprocating motion is −50~+50 mm. The velocity curve of the trajectory of the moving platform is shown in Fig. 6.7, where t_A is the time of acceleration or deceleration, and t_D is the time of a motion segment including acceleration, deceleration and constant velocity.

Fig. 6.6 Structure diagram of the 1:10 scale model of the fine-tuning platform

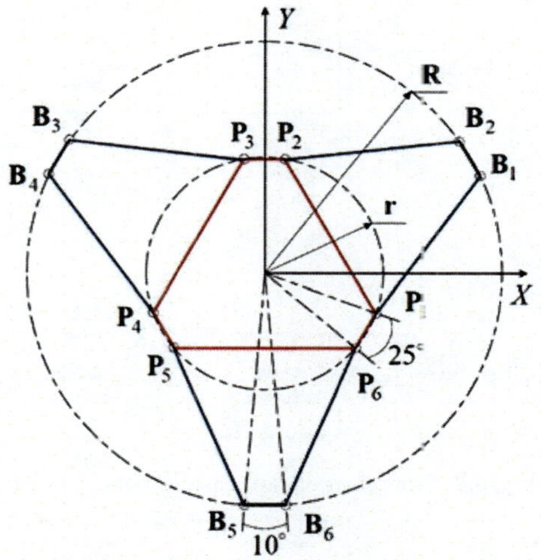

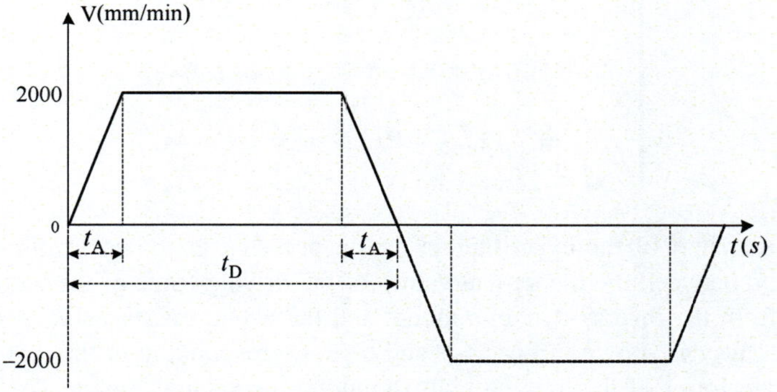

Fig. 6.7 Velocity curve of the motion trajectory

The simultaneous recording cycle of the force sensor data and the branch chain lengths is 2.2 ms. The directly recorded data is the branch chain lengths of the 1:10 scale model of the fine-tuning Stewart platform, and the terminal attitude of the moving platform at each recording moment is deduced with the positive position solution. The velocity and acceleration of the moving platform are calculated with the differential operation. Finally, the calculated motion parameters are substituted into the inverse dynamic model of the fine-tuning Stewart platform to deduce the theoretical driving force of each branch chain.

The signals obtained through the high-frequency acquisition are mostly interferential. In this experiment, resampling (5 times period resampling) and data smoothing (five-point linear moving average method) are used to process the measured branch chain lengths and the driving forces to suppress the noise signals and obtain the real data. The formula of the five-point linear moving average method can be written as:

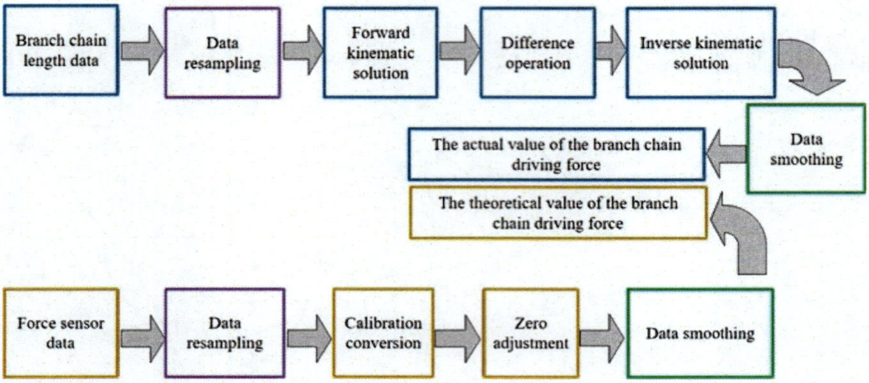

Fig. 6.8 Flow chart of experimental data processing

$$\begin{cases} y_1 = \frac{1}{5}(3x_1 + 2x_2 + x_3 - x_4) \\ y_2 = \frac{1}{10}(4x_1 + 3x_2 + 2x_3 + x_4) \\ \ldots \\ y_i = \frac{1}{5}(x_{i-2} + x_{i-1} + x_i + x_{i+1} + x_{i+2}) \\ \ldots \\ y_{n-1} = \frac{1}{10}(x_{n-3} + 2x_{n-2} + 3x_{n-1} + 4x_n) \\ y_n = \frac{1}{5}(-x_{n-3} + x_{n-2} + 2x_{n-1} + 3x_n) \end{cases}$$

The flow chart of the experimental data processing is shown in Fig. 6.8. The comparison between the theoretical values of the driving force of the branch chain deduced from the inverse dynamic model and the actual measured values through the force sensor is shown in Figs. 6.9 and 6.10, corresponding to the reciprocating motion trajectories of the moving platform along X-axis and Y-axis. The blue line illustrate the practical values while the red line indicate the theoretical values.

The driving force curve of the branch chain for the reciprocating motion along the X-axis of the 1:10 scale model of the fine-tuning platform can be divided into three segments:

The first one is the fast-feed motion of the moving platform along the *Z*-axis from the machine origin to the starting point of the given trajectory (the center of the workspace). Because of the symmetry of the Stewart parallel mechanism, driving forces of the 6 branch chains are the same in this segment.

The second one is a linear reciprocating motion of ±50 mm along the *X*-axis at a feed rate of 2000 mm/min from the starting point. The process starts at 4.3 s, undergoes a reciprocating motion, and lasts until 10.1 s;

The third one reveal the static driving forces of the branch chains that is stationary at the center of the workspace after the motion is finished.

The driving force curve of the reciprocating motion along the *Y*-axis can also be divided into 3 segments. The first and third ones are exactly the same as the driving force curve for the reciprocating motion along the *X*-axis, while in the second

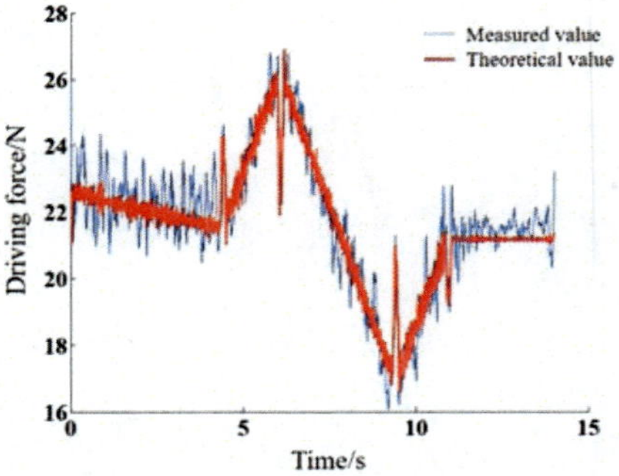

(a) Driving force comparison of branch chain 1

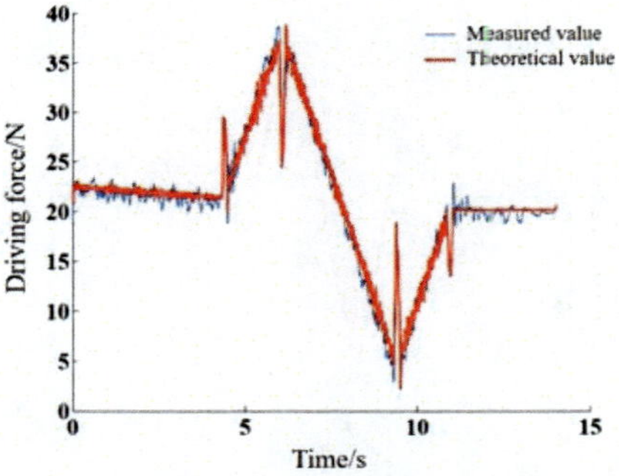

(b) Driving force comparison of branch chain 2

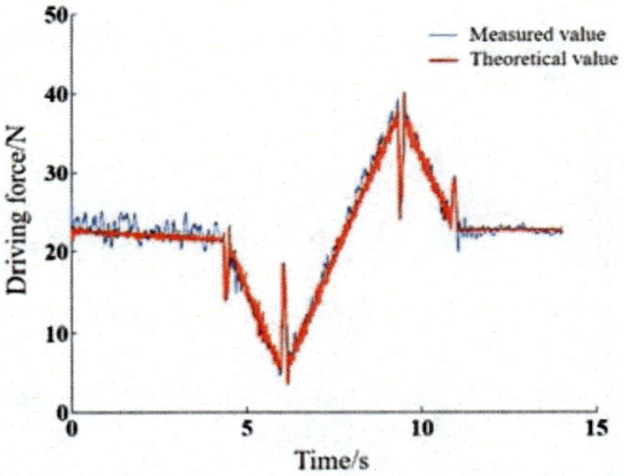

(c) Driving force comparison of branch chain 3

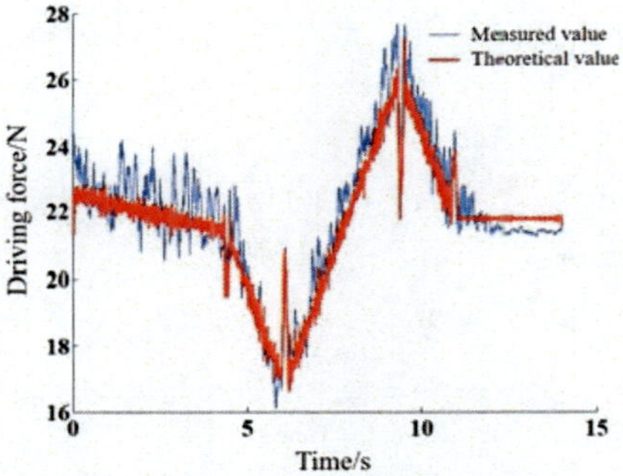

(d) Driving force comparison of branch chain 4

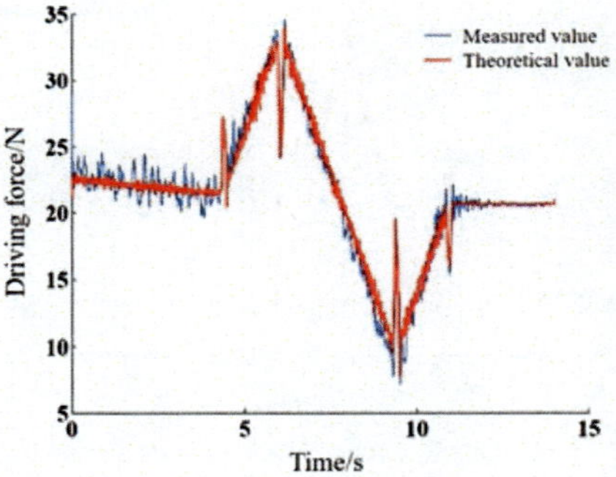

(e) Driving force comparison of branch chain 5

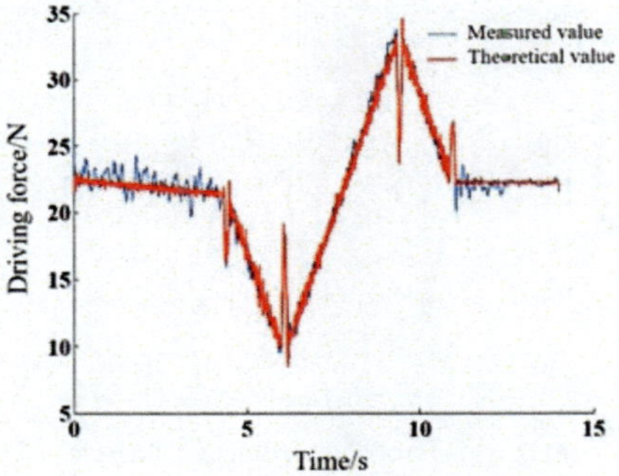

(f) Driving force comparison of branch chain 6

Fig. 6.9 Experiment results of the reciprocating motion along the X-axis

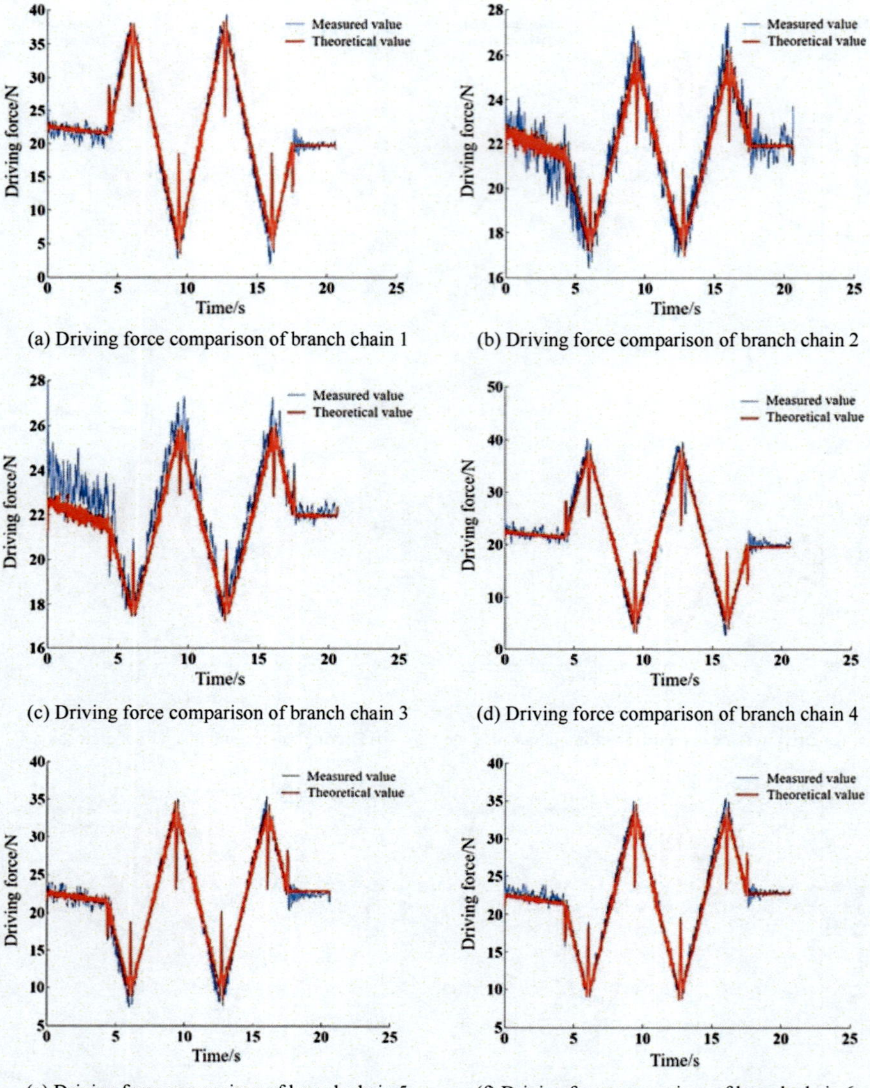

(a) Driving force comparison of branch chain 1
(b) Driving force comparison of branch chain 2
(c) Driving force comparison of branch chain 3
(d) Driving force comparison of branch chain 4
(e) Driving force comparison of branch chain 5
(f) Driving force comparison of branch chain 6

Fig. 6.10 Experiment results of the reciprocating motion along the *Y*-axis

segment, the moving platform implements two reciprocating motions along the *Y*-axis from 4.3 s to 17.5 s, as shown in Fig. 6.10.

The root-mean-square (RMS) error and the maximum error between the theoretical value and the practical value of the driving force of the 1:10 scale model of the fine-tuning Stewart platform in the experiments are shown in Table 6.3. Due

Table 6.3 Deviation between practical and theoretical values

	Branch chain 1	Branch chain 2	Branch chain 3	Branch chain 4	Branch chain 5	Branch chain 6
RMS deviation of the motions along *X*-axis(N)	0.696	1.148	1.242	0.793	1.007	0.8016
Max deviation of the motions along *X*-axis(N)	3.268	5.650	6.150	2.968	6.220	4.216
RMS deviation of the motions along *Y*-axis(N)	1.377	0.746	0.926	1.516	1.097	1.043
Max deviation of the motions along *Y*-axis(N)	5.364	2.753	6.591	6.310	4.671	4.960

to the friction and data sampling noise, there are some differences between actual measured values and theoretical calculation values, with a maximum deviation of 6.6 N and 1.5 N in the RMS. In general, the actual measured values of the driving force of the branch chain coincide with the theoretical values deduced by the inverse dynamic model, and they fit each other well. In particular, the theoretical values and the practical values can precisely coincide with each other during the acceleration and deceleration phases, which verifies the accuracy of the established dynamic model with the Newton–Euler method.

6.5 Conclusion

(1) Based on the non-inertial translational coordinate system, dynamic modelling of the fine-tuning Stewart platform in the FAST is established with the Newton–Euler method, which reflects the influence of the motion of the base platform on the dynamics of the rigid parallel mechanism.
(2) Position and linear velocity of the base platform do not affect the driving force of the fine-tuning Stewart parallel mechanism.
(3) Based on the dynamic calibration, verification of the dynamic model is implemented by synchronously measuring and recording the motion state and the driving force of the branch chain.

References

1. Shao ZF, Tang X, Wang LP, Chen X. Driving force analysis for the secondary adjustable system in FAST. Robotica.; 29(6):903–15.
2. Shao ZF, Tang X, Wang LP. Dynamics verification experiment of the Stewart parallel manipulator. Int J Adv Robotic Syst.; 12(10):144.
3. Merlet JP. Parallel manipulators part 1: theory, design, kinematics, dynamics and control, INRIA report; 1987.
4. Fichter EF. A Stewart platform-based manipulator: general theory and practical construction. Int J Robot Res. 1986;5(2):157–82.
5. Do W, Yang D. Inverse dynamic analysis and simulation of a platform type of robot. J Robot Syst. 1988;5:29–227.
6. Reboulet C, Berthomieu T. Dynamic models of a six degree of freedom parallel manipulators. Proceeding of the 5th international conference on advanced robotics. Pise, Italy, 1991. p. 1153–157.
7. Ji Z. Study of the effect of leg inertia in Stewart platforms. Proceedings IEEE international conference on robotics and automation. Atlanta, USA. 1993;1:121–26.
8. Dasgupta B, Mruthyunjaya TS. Closed-form dynamic equations of the general Stewart platform through the Newton-Euler approach. Mech Mach Theory. 1998;33(7):993–1012.
9. Dasgupta B, Choudhury P. A general strategy based on the Newton-Euler approach for the dynamic formulation of parallel manipulators. Mech Mach Theory. 1999;34(6):801–24.
10. Dasgupta B, Mruthyunjaya TS. A Newton-Euler formulation for the inverse dynamics of the Stewart platform manipulator. Mech Mach Theory. 1998;33(8):1135–52.
11. Zhang LX, Wang JS, Wang LP, et al. Simplified rigid body dynamics model for a parallel kinematic machine under uniform motion. J Tsinghua Univ. 2003;43(8):1041–4.
12. Zhang LX, Wang JS, Wang LP. Simplification of the rigid body dynamic model for a 6-UPS parallel kinematic machine under the acceleration motion and the deceleration motion. J Mech Eng. 2003;39(11):117–22.
13. Zhang CD, Song SM. An efficient method for inverse dynamics of manipulators based on the virtual work principle. J Robot Syst. 1993;10(5):605–27.
14. Wang J, Gosselin CM. A new approach for the dynamic analysis of parallel manipulators. Multibody SysDyn. 1998;2(3):317–34.
15. Tsai LW. The Jacobian analysis of a parallel manipulator using reciprocal screws. In: Proceeding of the 6th international symposium on recent advances in robot kinematics; 1998; Salzburg, Austia. p. 327–36.
16. Tsai LW. Solving the inverse dynamics of a Stewart-Gough manipulator by the principle of virtual work. ASME J Mech Des. 2000;122(1):3–9.
17. Staicu S. Dynamic analysis of the Star parallel manipulator. Robot Auton Syst. 2009;57(11):1057–64.
18. Staicu S. Dynamics of the 6–6 Stewart parallel manipulator. Robot Comput-Integr Manipulator. 2011;27(1):212–20.
19. Wu J, Wang JS, Li TM, Wang LP. Dynamic analysis of the 2-DOF planar parallel manipulator of a heavy duty hybrid machine tool. Int J Adv Manuf Technol. 2007;34(3–4):413–20.
20. Shao H, Wang LP, Guan LW, Wu J. Dynamic manipulability and optimization of a redundant three DOF planar parallel manipulator. In: ASME international conference on reconfigurable mechanisms and robots; 2009; London, UK. p. 302–08.
21. Abdellatif H, Heimann B. Computational efficient inverse dynamics of 6-DOF fully parallel manipulators by using the Lagrangian formalism. Mech Mach Theory. 2009;44(1):192–207.
22. Li YM, Xu QS. Kinematics and inverse dynamics analysis for a general 3-PRS spatial parallel mechanism. Robotica. 2005;23(2):219–29.
23. Li YW, Wang JS, Liu XJ, Wang LP. Dynamic performance comparison and counterweight optimization of 3-DOF parallel manipulators for a new hybrid machine tool. Mech Mach Theory. 2010;45(11):1668–80.

24. Yun Y, Li YM, Xu QS. Active vibration control of a 3-DOF parallel platform based on Kane's dynamics method. In: Proceeding of SCIE annual conference; 2008; Chofu, Japan. p. 2667–672.
25. Yang CF, Huang QT, He JF, Jiang HZ, Han JW. Model-based Control for 6-DOF parallel manipulator. In: International Asia conference on informatics in control, automation, and robotic; 2009; Bangkok, Thailand. p. 81–4.
26. Liu MJ, Tian YT, Li CX. Dynamics of parallel manipulator using sub-structure Kane method. J Shanghai Jiaotong Univ (Chin Ed). 2001;35(7):1032–5.
27. Lopes AM. Complete dynamic modeling of a moving base 6-dof parallel manipulator. Robotica. 2009;28(5):781–93.

Chapter 7
Dynamic Modeling of the Rigid-Flexible Series Coupling System

The cable-driven parallel robot in the FAST feed support system is characterized by long span and low stiffness, and is a flexible manipulator. The fine-tuning Stewart platform is a rigid robot, and is the flexibly supported parallel robot. They are connected in series, and form a typical flexibly supported robot system, which can also be called the rigid-flexible coupling system. Because of the low stiffness, the long-span CDPR is sensitive to the wind disturbance and external forces and is easy to deform and vibrate, affecting the terminal accuracy. To improve the terminal trajectory accuracy of the FAST feed support system and ensure its comprehensive observation performance, its accuracy characteristics and control method must be studied. The dynamic modeling of the flexibly supported parallel robot is the theoretical foundation for the study of the force characteristics and vibration suppression control of the FAST feed support system, which is also a prerequisite for engineering problems, such as the construction of the driving system and the structural check. Therefore, this chapter carries out the study of the dynamic modeling of the flexibly supported parallel robot, focusing on the inverse dynamics to deduce the driving forces of the flexibly supported parallel robot with the given geometrical dimensions, the inertia parameters, and the motion parameters (position, attitude, velocity, and acceleration) of the base and the moving platform. The kernel is to combine rigid and flexible dynamic models to efficiently establish the complete dynamic model of the rigid-flexible series coupling [1, 2].

This chapter introduces the modeling and simulation methods for the rigid-flexible coupling system in FAST. The flexibly supported robot system is briefly introduced in Sect. 7.1. The elastic dynamic model of the cable-driven parallel robot is established in Sect. 7.2. In Sect. 7.3, a wind disturbance model is established according to the topographic conditions of the working environment of FAST, which is applied to the numerical simulation of the feed support system to obtain the error curves of the cable platform under the wind disturbance. In Sect. 7.4, the simulation model of the

© Tsinghua University Press 2024

X. Tang et al., *Research and Application of Cable-Driven and Rigid Parallel Robots*,
https://doi.org/10.1007/978-981-99-7452-8_7

feed support system is established, and the coupling characteristics of the fine-tuning Stewart platform, the A-B rotator and the cable-driven parallel robot are analyzed.

The Main Contents of This Chapter Are

(1) Flexibly supported robot and its dynamics;
(2) Elastic dynamics of the CDPR;
(3) The FAST wind load model;
(4) Analysis of rigid-flexible coupling characteristics and models combination.

7.1 Flexibly Supported Robot and Its Dynamics

With the development of space technology, robotic systems with a large workspace, light weight, low power consumption and high accuracy are needed for satellite recovery and repair. To satisfy this requirement, the concept of flexibly supported robot emerges, also known as the compliant base manipulator, which is a rigid robot attached to the base that is passively compliant during the motion. The major difference between the flexibly supported robot and the conventional robot is the flexible base (Fig. 7.1). The flexibly supported robot is characterized by a large workspace, low cost and light weight [1–3], showing great potential for applications in aerospace and aviation industries. The flexibly supported robot also refers to a system that contains a flexible robot as the base and a rigid robot, which is generally referred to as a flexibly supported robot system. The reactive force of the rigid robot can cause deformation of the flexible base, which will further introduce terminal errors. There is the coupling of forces and motion between the rigid robot and the flexible support.

The research on the flexibly supported robot began in the 1980s. In 1981, NASA sent the serial Canadarm manipulator with six DOFs into space (see Fig. 7.2) for space

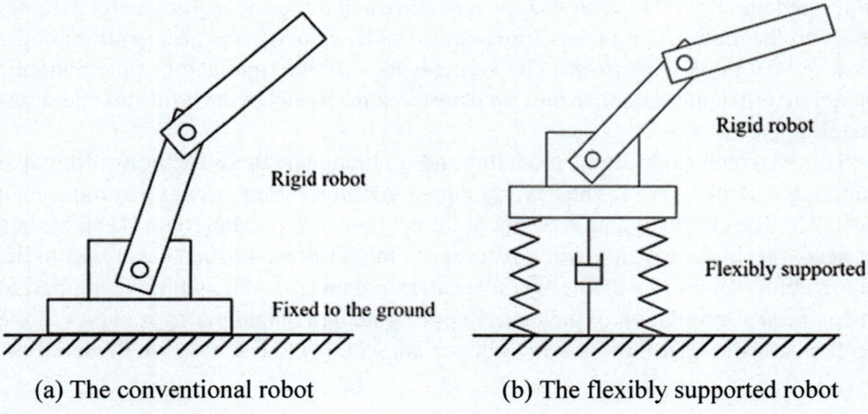

Fig. 7.1 The flexibly supported robot and the conventional robot

transport. The manipulator adopts a human-like arm structure, and the wrist robot at the end is a flexibly supported robot. According to the different structures of the rigid robot connected in series on the flexible support/base, the flexibly supported robot can be divided into the flexible-supported serial robot and the flexibly supported parallel robots. The former is simple in structure and appears earlier. Its related research on dynamic modeling is sufficient. Its modeling methods contain two categories: in the early days, the rigid robot and the flexible support were considered uniformly, and Lagrange equation and hypothetical modality were used to establish the dynamic model of the flexibly supported serial manipulator [4]. The obtained equations were complex, long, and difficult to simplify. Scholar Lew gave a general form of the dynamic coupling equations for the flexibly supported serial manipulator, which transformed the dynamic modeling into the parameter identification problem and was widely applied to the dynamic modeling of the flexibly supported serial manipulator with simple structures [5]. However, the dynamic modeling of the rigid-flexible coupling manipulator with complex structures is still at the forefront of theoretical research [6, 7].

The flexibly supported parallel manipulator is complex in structure, and the two modeling methods mentioned above are not applicable. We can establish the separated models first, i.e., the dynamic model of the parallel manipulator with the moving base and the elastic dynamic model of the flexible support, and combine separated models according to the coupling characteristics of the system. At present, although the research on the flexible dynamics of the cable-driven parallel robot is not mature, there has been a large amount of research on the dynamic modeling of single cable in the engineering industry. There are various studies on dynamic modeling of the

Fig. 7.2 Canadarm manipulator (flexibly supported robot)

rigid parallel robot [8–10], but studies on dynamic modeling for the parallel manipulator with a moving base are few. Lopes [11] used the general momentum theory to obtain the dynamic model of the Stewart parallel robot with a fixed base and further deduced the dynamic model considering the base motion. However, the study on coupling characteristics is insufficient and constrains the solution of the dynamic model of the flexibly supported parallel manipulator.

7.2 Elastic Dynamics of the CDPR

The first level of the FAST feed support system is a long-span cable-suspended parallel robot with lower stiffness, which is prone to deform elastically under the wind disturbance. In the dynamic modelling, the cables are appropriately simplified and equated to a spring damping system. The cable mass is concentrated at both ends of the cable, neglecting the friction at the cable anchor points and the cable exit points. Replacing six driving cables with the spring damping models, the physical model of the feed support system is illustrated in Fig. 7.3. The elastic dynamics of the large cable-driven parallel robot is established with the following assumptions: the driving cables are simplified to the ideal spring damping models; mass of each cable is concentrated at both ends; the cable exit points of the cable tower and the cable anchor points of the cable platform can be considered as ideal Hooke joints and spherical joints.

Next, the elastic dynamics of the cable-driven parallel robot will be conducted based on the spring damping model.

In the global coordinate system, the unit directional vector of the j-th driving cable (from the exit point C_j to the anchor point D_j) can be expressed as:

$$ {}^{G}\mathbf{s}_{Cj} = {}^{G}\mathbf{S}_{Cj} / L_{Cj} \tag{7.1} $$

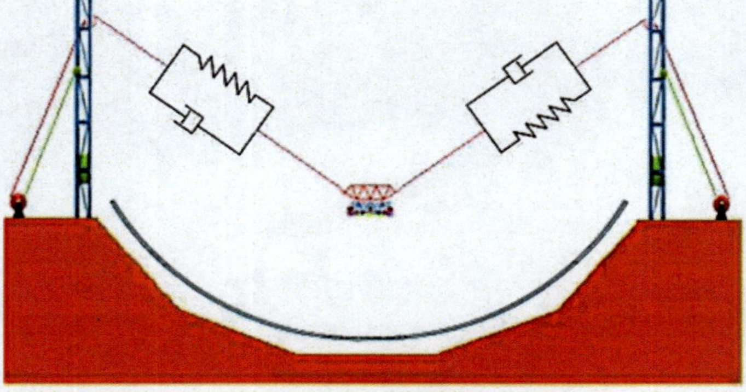

Fig. 7.3 Feed support system with the equivalent spring damping models

where:

$$ {}^{G}\boldsymbol{S}_{Cj} = {}^{G}\boldsymbol{t}_{C} + {}^{G}\boldsymbol{R}_{C}\boldsymbol{d}_{j} - \boldsymbol{c}_{j}, L_{Cj} = \| {}^{G}\boldsymbol{S}_{Cj} \| $$

Deriving on both sides of Eq. (7.1) with respect to time obtains the equation for the velocity of the cable anchor point as:

$$ {}^{G}\dot{\boldsymbol{S}}_{Cj} = {}^{G}\dot{\boldsymbol{t}}_{C} + {}^{G}\dot{\boldsymbol{\theta}}_{C} \times \left({}^{G}\boldsymbol{R}_{C}\boldsymbol{d}_{j} \right) \tag{7.2} $$

The telescopic velocity of the driving cable is the component of the velocity of the anchor point along the direction of cable length, which can be written as:

$$ \dot{L}_{Cj} = {}^{G}\boldsymbol{s}_{Cj} \cdot {}^{G}\dot{\boldsymbol{S}}_{Cj} \tag{7.3} $$

The tension of the *j*-th driving cable acting on the cable platform can be derived with the spring damping model, and is a function of the elastic modulus, damping and telescopic velocity, i.e.

$$ F_{j} = K_{j}(L_{Cj} - L_{Cj0}) + C_{j}\dot{L}_{Cj} \tag{7.4} $$

where K_j is the elastic coefficient of the *j*-th cable; A is the cross-sectional area; E is the elastic modulus; C_j is the damping coefficient; L_{C0j} is the initial length of the cable in the ideal state. According to the Newton–Euler equation, the elastic dynamic model of the cable-driven parallel robot can be obtained as:

$$ \begin{bmatrix} \sum_{j=1}^{6} F_{j}{}^{G}\boldsymbol{s}_{Cj} \\ \sum_{j=1}^{6} F_{j}\left[\left({}^{G}\boldsymbol{R}_{C}\boldsymbol{d}_{j} \right) \times {}^{G}\boldsymbol{s}_{Cj} \right] \end{bmatrix} + \begin{bmatrix} \boldsymbol{M}{}^{G}\ddot{\boldsymbol{t}}_{C} \\ {}^{G}\dot{\boldsymbol{\theta}}_{C} \times \left(\boldsymbol{I}{}^{G}\dot{\boldsymbol{\theta}}_{C} \right) + \boldsymbol{I}{}^{G}\ddot{\boldsymbol{\theta}}_{C} \end{bmatrix} = \begin{bmatrix} \boldsymbol{F}_{e} \\ \boldsymbol{N}_{e} \end{bmatrix} \tag{7.5} $$

where $_{M}$ and $_{I}$ are the mass of the feed cabin and the inertia matrix relative to the centroid; $\boldsymbol{F}_{\mathbf{e}}$ and $\boldsymbol{N}_{\mathbf{e}}$ are the external force and moment acting on the feed cabin; ${}^{\mathrm{G}}\ddot{\boldsymbol{t}}_{\mathrm{C}}$ and ${}^{\mathrm{G}}\ddot{\boldsymbol{\theta}}_{\mathrm{C}}$ are the linear and angular accelerations of the cable platform in the global coordinate system.

Since the analytical solution of Eq. (7.5) is difficult to deduce, the numerical solution is considered to carry out the simulation, which is usually based on either the numerical integration or the discrete analog method. Here, the numerical-integration based simulation method is adopted, and the Fourth-Order Runge–Kutta Method is used specifically. Assuming that $\boldsymbol{X} = \left[{}^{\mathrm{G}}t_{\mathrm{C}}; {}^{\mathrm{G}}\theta_{\mathrm{C}} \right]$, the elastic dynamic model of the cable-driven parallel robot can be written as:

$$\begin{cases} \dot{X} = V \\ \dot{V} = f(t, X, V) \\ X(\mathrm{t}_0) = X_0 \\ V(\mathrm{t}_0) = \dot{X}_0 \end{cases} \tag{7.6}$$

The Fourth-Order Runge–Kutta Method uses the first four derivatives of the Taylor expansion, and its recursive formula can be expressed as:

$$\begin{cases} X_{\mathrm{n}+1} = X_{\mathrm{n}} + hV_{\mathrm{n}} + \frac{h^2}{6}(\eta_1 + \eta_2 + \eta_3) \\ V_{\mathrm{n}+1} = V_{\mathrm{n}} + \frac{h}{6}(\eta_1 + 2\eta_2 + 2\eta_3 + \eta_4) \end{cases} \tag{7.7}$$

where h is the time step of the simulation, which determines the sampling time of the calculation, and other unknown parameters in the above equation are listed as:

$$\eta_1 = f(\mathrm{t_n}, \mathbf{X}_{\mathrm{n}}, V_{\mathrm{n}}),$$

$$\eta_2 = f(t_{\mathrm{n}} + \frac{h}{2}, X_{\mathrm{n}} + \frac{h}{2}V_{\mathrm{n}}, V_{\mathrm{n}} + \frac{h}{2}\eta_1),$$

$$\eta_3 = f(t_{\mathrm{n}} + \frac{h}{2}, X_{\mathrm{n}} + \frac{h}{2}V_{\mathrm{n}} + \frac{h^2}{4}\eta_1, V_{\mathrm{n}} + \frac{h}{2}\eta_2)$$

$$\eta_4 = f(t_{\mathrm{n}} + h, X_{\mathrm{n}} + hV_{\mathrm{n}} + \frac{h^2}{2}\eta_2, V_{\mathrm{n}} + h\eta_3).$$

7.3 The FAST Wind Load Model

The FAST feed support system, with its long span and lightweight, is vulnerable to external disturbances, and it is susceptible to large position displacements under the wind load. Before conducting error and control studies, the wind disturbance response of the FAST feed support system is analyzed. The wind load model is established based on the topographic conditions at the site of the FAST prototype.

The airflow on the surface forms the wind. When the wind encounters an obstacle, it generates wind pressure on the windward side of the obstacle, forcing it to shift its position and generating vibrations around the shifted position. The atmosphere near the ground and affected by the surface friction is called the friction layer. The height of the friction layer is not fixed because of the different surface topography, but is generally in the range of 300~1000 m above the ground. The feed cabin of FAST is in the friction layer.

The main wind-bearing side of the FAST feed support system is the shield wrapped around the feed cabin. The wind force is mainly determined by the wind velocity. According to the Davenport wind spectrum, wind velocity can be expressed as a function of time t and height z, which mainly composed of two parts: the mean velocity and the pulse velocity.

$$v(z, t) = v_m(z) + v_g(z, t) \tag{7.8}$$

where $v_m(z)$ stands for the average wind velocity, which is only related to height, and $v_g(z, t)$ stands for the pulse wind velocity, which is a function changing with height and time.

In plain terrain, the average wind velocity increases with the height, and the relation is called the average wind velocity gradient (or wind profile), which can be expressed as:

$$v_m(z) = v_{10} \cdot \left(\frac{z}{10}\right)^{\alpha} \tag{7.9}$$

where z is the height of the feed cabin center from the ground, and according to the actual structural parameters of the FAST prototype, z = 145 m; v_{10} is the mean velocity measured at 10 m above the ground; α is the terrain factor reflecting the roughness of the ground. Based on the physical environment of the FAST site, the value of 0.16 for the coefficient of the ground surface can be determined.

The design requirements of the FAST prototype include: (1) the FAST feed support system can work normally when the average wind velocity at a height of 10 m is not greater than 4 m/s (maximum working wind velocity); (2) when the avearage wind velocity at a height of 10 m is not greater than 8 m/s (maximum safety wind velocity), the feed cabin does not need to fall back to the safety maintenance cabin on the ground. According to Eq. (7.9), the variation curve of the average wind velocity with height can be derived, as shown in Fig. 7.4. Converting the wind velocity at 10 m height from the ground to the height of the feed cabin, i.e. the 145 m height, the maximum working velocity can be obtained as 6.14 m/s and the maximum safe velocity as 12.27 m/s.

The realistic pulse wind is a three-dimensional wind turbulence, including downwind, crosswind and vertical directions. The downwind turbulence plays a decisive role. The pulse wind is a stable random process that conforms to the Gaussian distribution with a mean of zero. The pulse wind cannot be repeated, and the pulse wind velocity is analyzed with the random vibration theory.

The power spectrum (probability density function) of the pulse wind includes the self-power spectrum and the mutual power spectrum. Since the scale of the feed cabin is relatively small, this section focuses on the former. The Davenport spectrum is widely applied, and can be expressed as:

$$S_v(f) = \frac{4\kappa v_m^2 \beta}{f(1+\beta^2)^{4/3}} \tag{7.10}$$

where $\beta = L_v f / v_m$, and L_v is the scale of wind turbulence; f is the frequency of pulse wind; κ is the roughness coefficient of the ground ($0.005 \leq \kappa \leq 0.015$).

Fig. 7.4 Variation curves of the average wind velocity with height

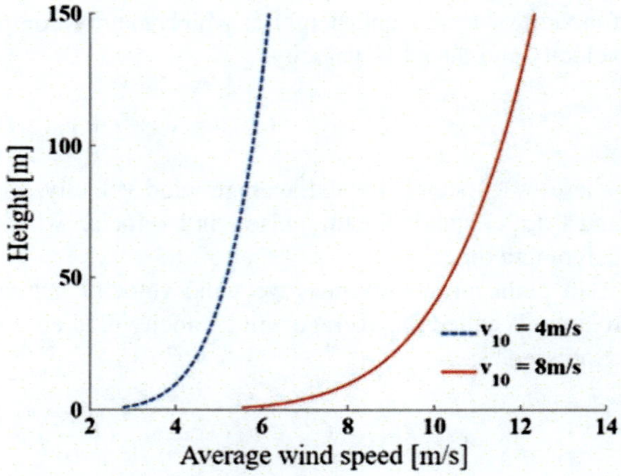

Based on the actual conditions of the FAST prototype, $L_v = 1200\,\text{m}$ and $\kappa = 0.01$ are adopted.

The pulse wind samples can be obtained by inputting a white-noise signal satisfying the unit standard deviation into the filter constructed based on the Davenport spectrum. When the transfer function of the filter reaches the third order, the obtained filter will be consistent well with the Davenport spectrum within the bandwidth range of [10^{-4}, 10] Hz by adopting reasonable parameters, as shown in Fig. 7.5. The transfer function of the established filter is

$$TF = \frac{1.869s^2 + 1.147s + 0.002927}{s^3 + 1.646s^2 + 0.1057s + 0.0008963} \tag{7.11}$$

Fig. 7.5 Fitting effects of the filter

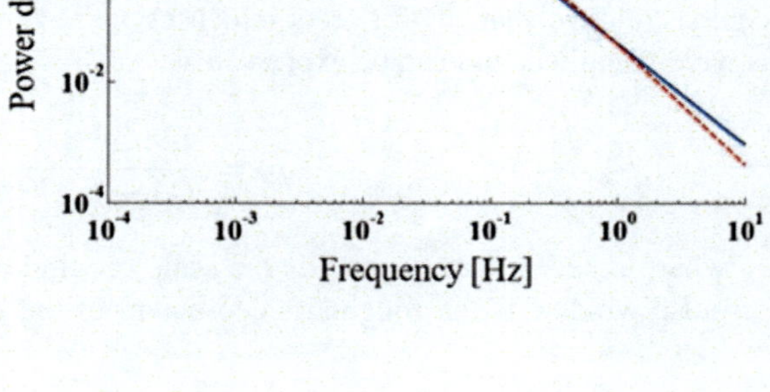

It can be seen from Fig. 7.5 that the main power components of the pulse wind are concentrated on the lower frequency bandwidth range, i.e., $10^{-4} \sim 0.1$ Hz. The disturbance errors in this frequency band are easy to be implemented with the vibration suppression control of the fine-tuning Stewart platform because of the long period and low frequency.

The actual wind velocity can be obtained by adding the pulse wind velocity and the average wind velocity. To calculate the magnitude of the wind force on the feed cabin, the wind pressure acting on the surface of the feed cabin shield needs to be calculated first. According to the Bernoulli Equation, the wind pressure can be expressed as:

$$p(z,t) = \frac{1}{2}\rho v(z,t)^2 = \frac{1}{2}\rho\left[v_m(z) + v_g(z,t)\right]^2 \tag{7.12}$$

where ρ is the air density, and $\rho = 1.225\ \mathrm{kg/m^3}$ at the location of the FAST prototype.

The wind force acting on the object is equal to the product of the wind pressure and the effective wind area. The effective wind area of the feed cabin is about $150\mathrm{m}^2$. According to the simulation process of the wind load described above, the simulation block diagram of the wind load of the feed cabin is shown in Fig. 7.6. Further, a set of wind samples at the maximum working wind velocity of 100 s can be generated through numerical simulation and shown in Fig. 7.7.

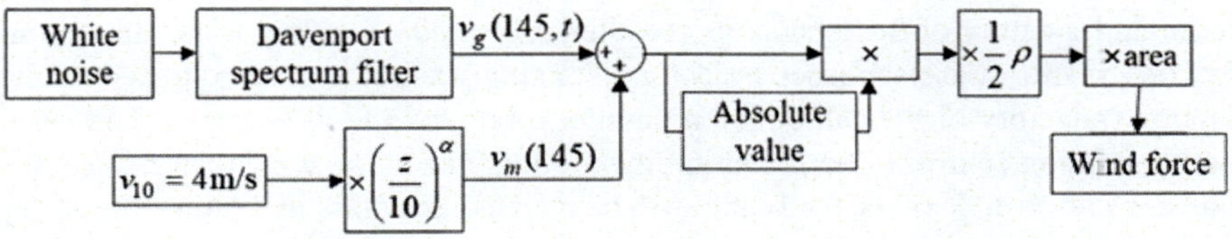

Fig. 7.6 Simulation block diagram of the wind load of the FAST feed cabin

Fig. 7.7 A set of wind samples of the feed cabin under the maximum working wind

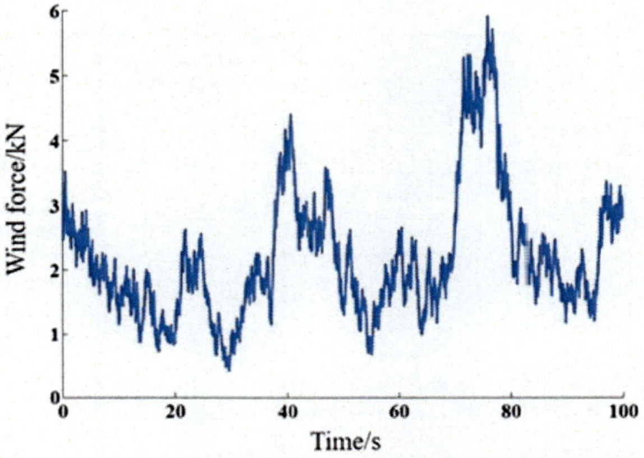

7.4 Analysis of Rigid-Flexible Coupling Characteristics and Models Combination

The rigid-body dynamic model of the Stewart parallel robot with the moving base, the elastic dynamic model of the cable-driven parallel robot and the wind load samples of the feed cabin have been derived. The integrated dynamic model of the FAST feed support system can be obtained by combining them together. The interaction between the first two models is mainly reflected in the coupling of motion and force. The fine-tuning Stewart platform and the A-B rotator are fixed under the cable platform and move with the cable-driven parallel robot. At the same time, the wind force and the reactive force of the fine-tuning Stewart platform and the A-B rotator will disturb the cable-driven parallel robot, generating acceleration, velocity and displacement in turn. The relation between the fine-tuning Stewart platform, the A-B rotator and the cable-driven parallel robot can be transformed into a pose follow-up system with force feedback, as illustrated in Fig. 7.8. The part in the dashed box reflects the follow-up relation and force transfer relation between the cable-driven parallel manipulator, the A-B rotator and the Stewart fine-tuning platform. The wind force, as an external disturbance, can be directly introduced into the complete simulation model of the feed support system. In addition, the model takes the vibration control strategy into account, which can be used for subsequent simulation experiments of the control system.

The overall simulation process can be described as follows. According to the terminal trajectory of the feed support system obtained according to the astronomical observation, the established trajectory planning strategy is adopted to deduce the motion trajectory of the cable-driven parallel robot and the A-B rotator, and carry out the motion control of both. During the motion, due to the wind disturbance, the position and attitude of the feed cabin will be affected, resulting in terminal trajectory errors. With the vibration suppression control strategy, the motion trajectory of the fine-tuning Stewart platform in each control cycle is figured out on the basis of the

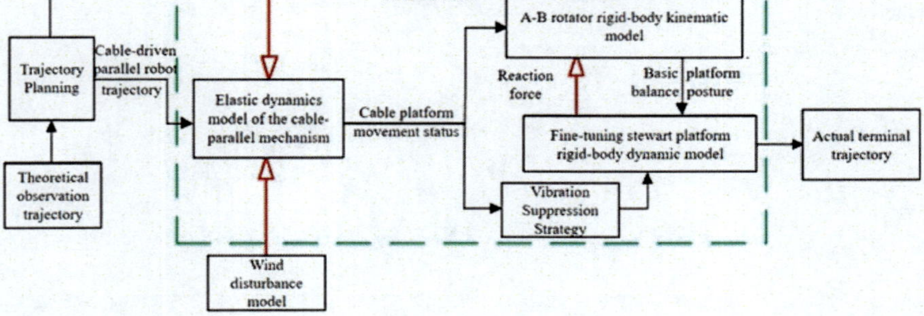

Fig. 7.8 Numerical simulation model of the feed support system

motion state of the cable platform collected by sensors and complete the vibration suppression control. Meanwhile, the fine-tuning Stewart platform will generate reactive force, which can be transmitted to the cable platform through the A-B rotator, forming a combined disturbance force with the wind load.

At the center of mass of the FAST feed cabin, the concentrated pulse force disturbance is applied along the positive direction of the X, Y and Z axes of the global coordinate system, and the concentrated pulse moment disturbance is applied along that of the A, B and C axes of the global coordinate system. In addition, the established simulation model of the feed support system is used to deduce the position and attitude step response curves of the cable platform and conduct the natural frequency and stiffness analysis of the feed support system of the FAST prototype. Results of the position and attitude response of the FAST feed support system are shown in Fig. 7.9.

The resonant frequency of the feed support system along the Z-axis direction is about 0.55 Hz, which is higher than the resonant frequency along the X-axis (about 0.24 Hz) and Y-axis (about 0.3 Hz) directions. The torsional vibration frequency around the C-axis is about 0.84 Hz, which is higher than the torsional vibration frequency around the A-axis (about 0.3 Hz) and the B-axis directions (about 0.24 Hz). From the simulation results, we can tell that the lowest natural frequency of the feed support system is about 0.24 Hz. The low resonant frequencies along X-axis and B-axis are basically the same, while the low resonant frequencies of Y-axis and A-axis are basically the same. The feed cabin is prone to compound vibration.

In above simulation experiments, the step force amplitude is 500N and the step moment amplitude is $500\,\mathrm{N} \cdot \mathrm{m}$. Based on the displacements along three coordinate axes, the translational stiffness of the feed support system can be determined with the order of 10^5N/m, and the stiffness in the Z-axis direction is slightly higher than that in the X-axis and Y-axis directions. The torsional stiffness of the feed support system around three coordinate axes is of order of $10^6\mathrm{N} \cdot \mathrm{m/rad}$. The torsional stiffness of the C-axis is about one order of magnitude higher than that of the A-axis and B-axis. In general, the stiffness of the feed support system is relatively low, and it is prone to generate large position and attitude errors under the wind disturbance.

The obtained wind force samples (Fig. 7.7) are applied to the geometric center of the feed cabin along the positive direction of the X-axis of the global coordinate system. Through numerical simulation, position and attitude errors of the cable platform can be derived as shown in Figs. 7.10a, b, and the attitude errors are described by the RPY Euler angle. The terminal position error of the feed support system are shown in Fig. 7.11.

The position error direction of the cable platform is consistent with the wind disturbance direction, along the X-axis direction of the global coordinate system. There is an initial distance of 7 m between the terminal of the feed support system and the cable platform. The angular error of the cable platform along the B-axis is amplified to a significant terminal position error of the feed support system. The instantaneous error direction of the cable platform and the terminal are opposite. It is mainly due to the superimposed effects of translational and torsional vibrations (translational vibration of the cable platform along the X-axis and torsional vibration

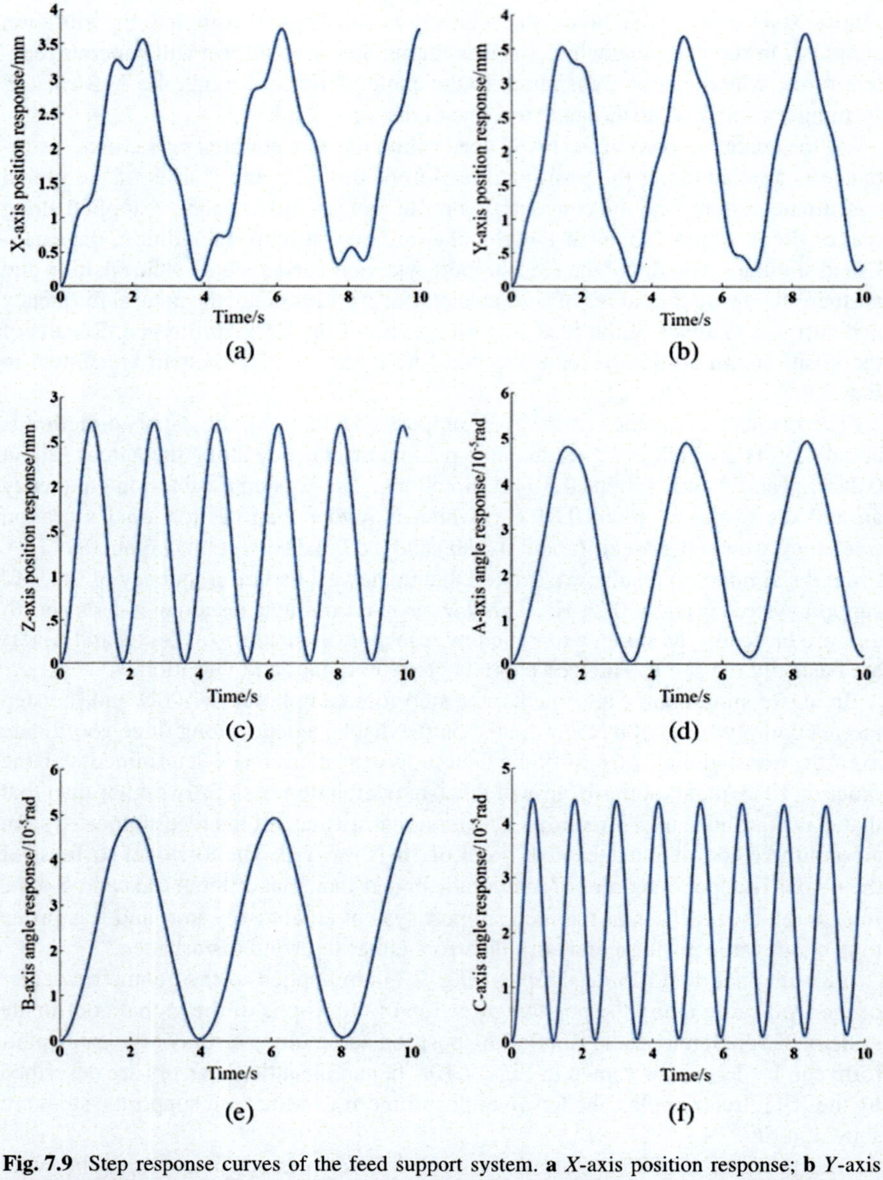

Fig. 7.9 Step response curves of the feed support system. **a** *X*-axis position response; **b** *Y*-axis position response; **c** *Z*-axis position response; **d** *A*-axis angle response; **e** *B*-axis angle response; **f** *C*-axis angle response

around the Y-axis). The maximum position error of the geometric center of the cable platform is 31.16 mm, while the maximum terminal position error of the feed support system is 82.14 mm, which cannot satisfy the design requirements of the feed support system (within 10 mm root-mean-square error). An effective vibration suppression control strategy should be adopted. As shown in Fig. 7.10b, the angle error of the cable platform is small, thus it can satisfy the attitude accuracy requirements of 0.2° root-mean-square. Therefore, vibration suppression control is mainly used to reduce the terminal position errors of the feed support system.

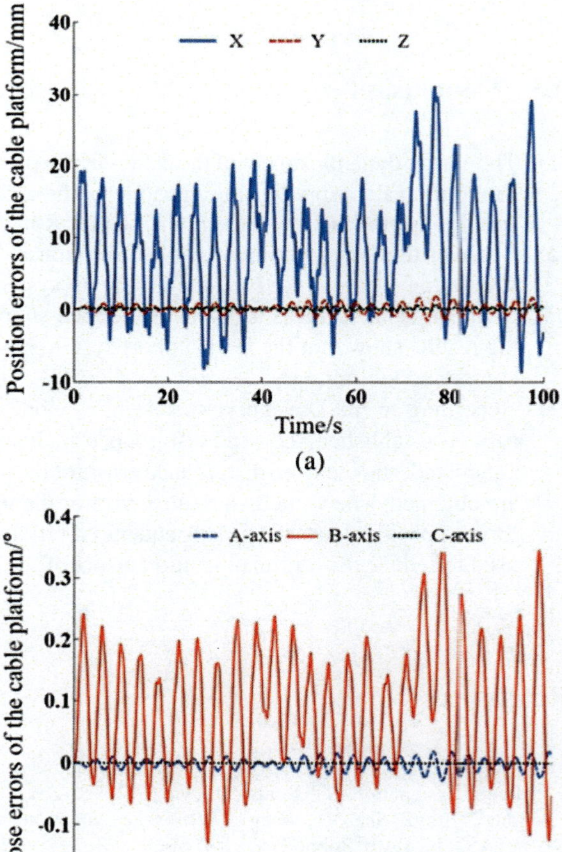

Fig. 7.10 Trajectory errors of the cable platform under the wind disturbance. **a** Position errors; **b** Attitude errors

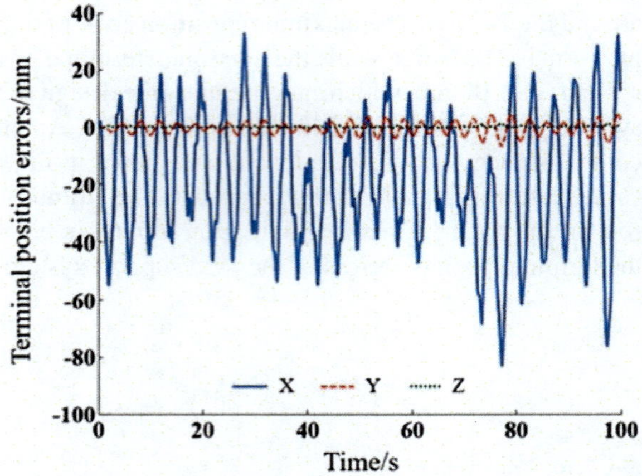

Fig. 7.11 Terminal position errors of the feed support system under the wind disturbance

7.5 Conclusion

(1) The elastic dynamic model of the cable-driven parallel robot is established based on the equivalent spring damping model of the cable. And the numerical iterative solution is proposed with the Fourth-Order Runge–Kutta Method.
(2) Based on the coupling relation between motion and force, the dynamic numerical simulation model of the FAST feed support system is established. The resonant frequency and stiffness analysis of the feed support system is completed, and the results show that the feed support system has low stiffness and is prone to compound vibration.
(3) According to the Davenport spectrum, a wind load model of the FAST feed cabin is established, the wind disturbance analysis of the feed support system is conducted, and the wind disturbance errors of the cable platform and the terminal are obtained. The wind disturbance error of the terminal is large, mainly in the form of position errors. The subsequent vibration suppression control is mainly used to reduce the terminal position errors of the feed support system.

References

1. Shao ZF, Tang X, Wang LP, Chen X. Dynamic modeling and wind vibration control of the feed support system in FAST. Nonlinear Dyn. 67(2):965–85.
2. Tang X, Liu Z, Shao ZF, Wang L. Self-excited vibration analysis for the feed support system in FAST. Int J Adv Robot Syst. 11(4):63.
3. Glassell RL, Burks BL, Glover WH. System review of the modified light duty utility arm after the completion of the nuclear waste removal from seven underground storage tanks at Osk Ridge National Laboratory. the Providence group, Knoxville, TN; 2000.

4. Miguel A T, Dubowsky S. Path planning for elastically constrained space manipulator systems. In: Proceedings of the IEEE international conference on robotics and automation. Atlanta, GA; 1993(1):812–17.
5. Yoshida K, Nenchev DN, Uchiyama M. Vibration suppression and zero reaction maneuvers of flexible space structure mounted manipulators. J Smart Mater Struct. 1999;8(6):847–56.
6. Obergfell K. End-Point Sensing and Control of Flexible Multi-Link Manipulators. Ph. D. Dissertation, Department of Mechanical Engineering, Georgia Institute of Technology, Atalanta, GA, 1998.
7. Lin J, Huang ZZ. A novel PID control parameters tuning approach for robot manipulators mounted on oscillatory base. Robotica. 2007;25(4):467–77.
8. Zhang XP, Yu YQ. A new spatial robot beam element for modeling manipulators with links and joint flexibility. Mech Mach Theory. 2000;35(3):403–21.
9. Liu SZ, Yu YQ, Zhu ZC, et al. Dynamic modeling and analysis of 3-RRS parallel manipulator with flexible links. J Cent South Univ Technol. 2010;17(2):323–31.
10. Staicu S. Dynamics of the 6–6 Stewart parallel manipulator. Robot Comput-Integr Manuf. 2011;27(1):212–20.
11. Lopes AM. Complete dynamic modeling of a moving base 6-dof parallel manipulator. Robotica. 2010;28(5):781–93.

Chapter 8
Vibration Control of Flexibly Supported Parallel Robots

The simulation in the Chap. 7 shows that it is necessary to implement the vibration suppression control (or vibration control) to reduce terminal position error of the FAST feed support system caused by the wind disturbance. This chapter mainly discusses the vibration control carried out by the fine-tuning Stewart platform to improve the terminal accuracy. The Stewart parallel robot is supported by the flexible cable-driven parallel robot, and is a typical flexibly supported parallel robot. In Sect. 8.1, the vibration control methods of flexibly supported robots are discussed in general. The trajectory planning method of the FAST feed support system is presented in Sect. 8.2, which lays a foundation for the subsequent research on the vibration control. Trajectory compensation vibration control and internal force vibration control are illustrated and applied to the FAST feed support system in Sects. 8.3 and 8.4. Specifically, Sect. 8.3 discusses the trajectory compensation vibration control strategy, and the Stewart parallel robot compensates the terminal error through real-time trajectory planning. Section 8.4 discusses the internal force vibration control strategy, which adopts the motion reaction force of the Stewart platform to offset the external disturbance.

The main contents of this chapter are:

(1) Vibration control of flexibly supported robots
(2) Trajectory planning of the FAST feed support system
(3) Trajectory compensation vibration control
(4) Internal force vibration control.

8.1 Vibration Control Method of Flexibly Supported Robot

Due to rigid-flexible coupling and force-position coupling in the flexibly supported robot system, it is prone to vibrate under the influence of external force or reaction force of the rigid robot, leading to the loss of terminal precision. Vibration

© Tsinghua University Press 2024

X. Tang et al., *Research and Application of Cable-Driven and Rigid Parallel Robots*,
https://doi.org/10.1007/978-981-99-7452-8_8

control is the key and difficulty problems in the research and application of flexibly supported robots, which can be generally divided into two categories: passive and active vibration controls.

Passive vibration control means that when a large vibration appears, the flexibly supported robot system stops working, and uses the damping and friction of the system to dissipate the vibration energy. After the vibration gradually disappears, the robot system continues to operate. The simplest way is to lock all joints and wait for the vibration energy gradually running out [1]. This method requires a long waiting time. The dissipation of vibration energy can be accelerated by adding appropriate dampers [2]. Van [3], Nenchev [4] and Torres [5] proposed a mode to accelerate the dissipation of the vibration energy. When the vibration of the flexibly supported robot goes severe, the energy dissipation mode is activated to reduce the system vibration. Passive vibration control is hard to realize the continuous motion of the flexibly supported robot, and it is suitable for the point control and the scope of application is limited.

Active vibration control is the main control method of flexibly supported robot, which adopts continuous control to curb the target vibration within the allowable range. Active vibration control methods can be divided according to the implementation object, as shown in Fig. 8.1. Due to the low stiffness and limited control bandwidth of the flexible support, its ability to suppress vibration is limited. Even if the control bandwidth of the flexible support is sufficient, the use of flexible support for vibration control requires large energy consumption which is relatively dangerous [6, 7]. At present, the vibration control is mainly implemented by the rigid robot for flexibly supported robot systems. The vibration control carried out by the rigid robot can be subdivided into three categories: trajectory compensation vibration control, trajectory planning vibration control, and internal force vibration control.

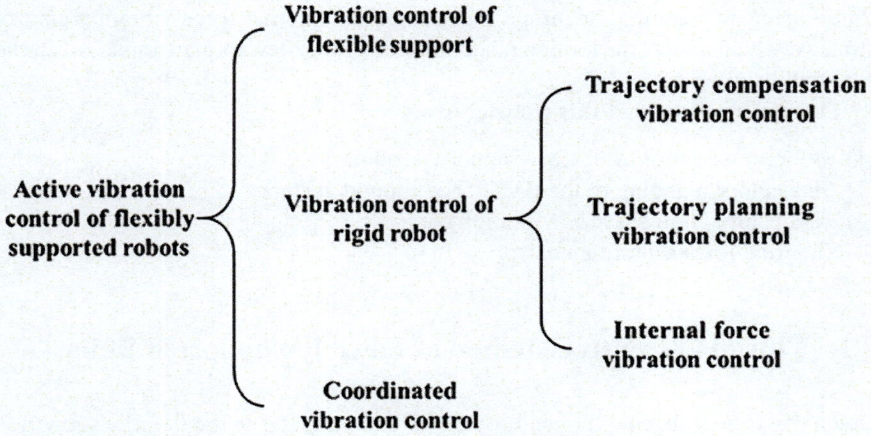

Fig. 8.1 Classification of active vibration control for flexibly supported robots

The trajectory compensation vibration control is to calculate the terminal error of the system in the global coordinate system (or directly measure the terminal error of the system) by measuring the pose error of the flexible support in real time, and convert the error to the joint coordinate system of the rigid robot for compensation. Staffetti [8] and Yoshikawa [9] used kinematically redundant micro robot to compensate the elastic deformation of the moving base with low stiffness, and applied the principle of minimum compensation to the trajectory planning of the flexibly supported robot system. Cho et al. [10] used cameras to measure the terminal position of flexible support/base, and the rigid micro robot was used to complete the trajectory compensation. Finally, the automatic assembly of automobile tires was realized.

Trajectory planning vibration control is to ensure the vibration of flexible support as small as possible through reasonably trajectory planning of the rigid robot. Nenchev [11] took advantage of the redundancy of the flexibly supported robot to select the trajectory with the least system vibration among the infinite set of joint space solutions. Torres [12] avoided large vibration of the system by reasonably planning of the trajectory between the given start and end points.

Internal force vibration control is to counteract the vibration of flexible support through the reaction force of the rigid robot, improving the stability and terminal accuracy of the whole system. Xie et al. [13] installed sensors on the flexible support, and control the rigid robot to creates a force against the acceleration of the support, according to the sensor feedback. Lew [14] conducted the vibration control experiment with industrial series robots considering internal forces. Taking advantage of the kinematic redundancy of the flexibly supported robot, Parsa [15] established the algorithm of internal force vibration control, and verified the performance on the planar flexibly supported robot with simulation.

However, researches on vibration control of flexibly supported parallel mechanism (robot) is limited. A typical example is the study on the Large Adaptive Reflector (LAR) radio telescope in Canada (Fig. 1.2). The project is the Canadian Square Kilometer Array Telescope (SKA) [16], began in 2004, and a number of universities are involved in the project, such as McGill, Laval and Columbia University. LAR's feed support system is divided into two layers: the first layer is a large-span flexible parallel robot driven with eight cables, and the second layer is a small parallel robot driven with eight cables inside the feed cabin. LAR's feed support system is a typical flexibly supported parallel robots. Taghirad and Nahon [17, 18] simplified the two 8-cable spatial parallel robots into 4-cable planar parallel robots, analyzed the kinematics and singularity of the flexible supported parallel mechanism, and studied the trajectory planning and dynamic control of the redundant mechanism. It is a pity that this project only carried out theoretical analysis and simulation, without physical experiments of prototype construction.

The other important application of flexibly supported parallel robots is the FAST project. The preliminary model of the FAST feed support system is proposed by academician Baoyan Duan of the Xidian University as illustrated in Fig. 2.2b, which consisted of a parallel robot driven with six cables and a rigid Stewart parallel robot connected in series. The CDPR independently controls the terminal position and posture [19–23]. Subsequently, Gexue Ren and Hui Zhang of Tsinghua University proposed the three-layer series structure of the feed support system that composed of four-cable parallel robot, A-C rotator and Stewart parallel robot, and built the 50-m scale model (Fig. 2.2a). The CDPR, composed of bearing cables and counterweight cable, implements the spatial 3 DOFs translation motions. The kinematic modeling and analysis of the Stewart parallel robot were carried out. And the differential velocity prediction algorithm was proposed with the research on the stability and structure optimization of the feed support system [24–27]. Based on the above research, authors of this book proposed the final configuration of the FAST feed support system, which is composed of six-cable parallel robot, A-B rotator, and the rigid Stewart parallel robot in series. The research on vibration control is introduced as follows.

8.2 Trajectory Planning of the Feed Support System

The FAST feed support system consists of the CDPR, A-B rotator and rigid Stewart platform in series. The CDPR adopts Stewart configuration with 6 DOFs (degrees of freedom), the A-B rotator has 2 DOFs, and the fine-tuned Stewart platform has 6 DOFs. Theoretically, the 5 DOFs spatial trajectory tracking required by the terminal receiver can be well realized only with the CDPR. The feed support system possesses redundant DOFs. The trajectory of the feed support system should be planned before the study of vibration control of the rigid Stewart platform. The terminal trajectory of the feed support system should be reasonably and uniquely decomposed into motions of three layers. Trajectory planning is the basis of FAST feed support system research [28] and the premise of vibration control of fine-tuned Stewart platform. In order to complete the trajectory planning of the redundant mechanism, the optimization objective and constraints are added to ensure that the unique trajectory planning results can be deduced. To facilitate the realization of the control function of the FAST feed support system, two constraints are put forward, respectively:

(1) The kinematic control model of the CDPR should be simple with good motion accuracy.
(2) The fine-tuned Stewart platform is used to carry out the vibration control. Under ideal conditions (without external disturbance), the Stewart platform can be stationary while the trajectory accuracy of the feed support system still meet the requirements.

As shown in Fig. 8.2, the terminal of the FAST feed support system mainly moves on the feed receiving surface, which is a spherical crown surface. At each point of the surface, the terminal receiver of the feed support system needs to maintain a certain attitude angle to ensure that the normal vector of the receiver passes through the sphere center of the reflector. As the receiver moves from the center to the boundary on the feed receiving surface, the absolute value of the attitude angle gradually changes from 0° to 40°. According to the analysis in Chaps. 2 and 3 of this book [29, 30],

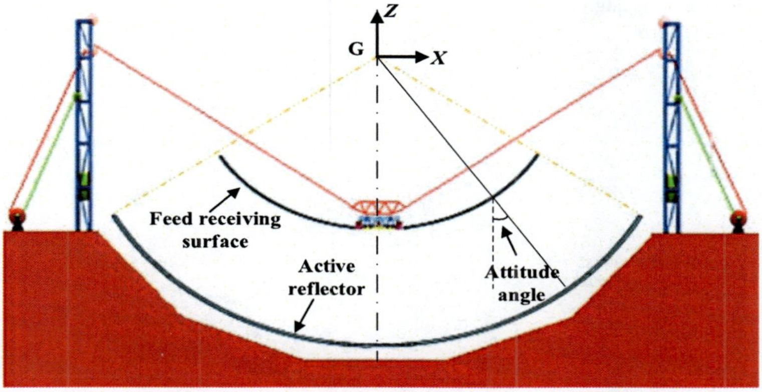

Fig. 8.2 Workspace of the FAST feed support system

the cable-driven parallel robot can meet the above constraint (1) by homogenizing the cable tension in the whole workspace and narrowing down the variation range of cable forces. The driving cables adopt the linear model, and the control accuracy of the CDPR is good. According to the results of cable force optimization, the attitude coupling requirements on the cable-driven parallel robot are established. The cable platform of the CDPR must maintain a natural dip angle at different positions on the feed receiving surface. There is a mathematical relationship between the dip angle of the cable platform and the required observation attitude angle of the terminal receiver, which can be simplified as:

$$\varphi_{\mathrm{Cable}} = (3/8)\varphi_{\mathrm{Receiver}} \tag{8.1}$$

where φ_{Cable} is the natural dip angle of the cable platform (feed cabin), and $\varphi_{\mathrm{Receiver}}$ is the required observation attitude angle of the receiver. An intuitive description of their relationship is shown in Fig. 8.3. When the CDPR satisfies the above restriction, each position of the workspace will correspond to a certain natural dip angle of the cable platform. The natural dip angel can only achieve 3/8 of the required observation attitude angle, and the remaining 5/8 is implemented by the A-B rotator.

To sum up, in order to simplify the kinematic control model of the cable-driven parallel robot, it is necessary to ensure the uniformity conditions of the cable force and limit the variation range. The cable force optimization determines that a certain natural dip angle must be guaranteed at each position of the cable platform (feed cabin) on the feed receiving surface. Since the attitude and position of the cable-driven parallel robot are coupled, it is equivalent to a 3-DOF mechanism with rotational concomitant motion. The A-B rotator has 2-DOF rotation and is responsible for compensating the difference between the observation attitude angle and the natural dip angle. Under ideal conditions, the CDPR and the A-B rotator realize the 5-DOF spatial trajectory tracking of the terminal receiver, and the fine-tuned Stewart

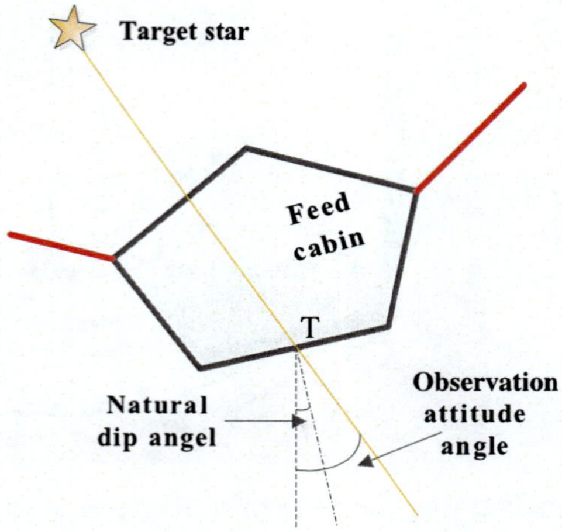

Fig. 8.3 Natural dip angle of the cable platform

platform does not need to move and meets the restriction condition (2). The trajectory planning of the FAST feed support system is derived below.

As shown in Fig. 8.2, a global inertial coordinate system is established in the sphere center of the active reflector $\{\boldsymbol{G}\} : O - XYZ$. The target attitude is described relative to the terminal receiver, which is the geometric center of the moving platform of the rigid Stewart platform. The target pose is given as ${}^{G}\boldsymbol{X}_{\mathrm{P}} = [x_t, y_t, z_t, \alpha_t, \beta_t, \gamma_t]^{\mathrm{T}}$, which can be converted into the pose of the CDPR ${}^{G}\boldsymbol{X}_{\mathrm{C}} = [x_{\mathrm{C}}, y_{\mathrm{C}}, z_{\mathrm{C}}, \alpha_{\mathrm{C}}, \beta_{\mathrm{C}}, \gamma_{\mathrm{C}}]^{\mathrm{T}}$ and angles of A-B rotator $\boldsymbol{\theta} = [\theta_{\mathrm{A}}, \theta_{\mathrm{B}}]^{\mathrm{T}}$.

According to Chap. 6, it can be concluded that in the global coordinate system $\{\boldsymbol{G}\}$, the expression of the position and attitude of the feed support system is:

$$ {}^{G}\boldsymbol{t}_{P} = {}^{G}\boldsymbol{t}_{C} + {}^{G}\boldsymbol{R}_{C}{}^{C}\boldsymbol{t}_{M} + {}^{G}\boldsymbol{R}_{B}\left(-{}^{B}\boldsymbol{t}_{M}\right) + {}^{G}\boldsymbol{R}_{B}{}^{B}\boldsymbol{t}_{P} \tag{8.2} $$

and

$$ {}^{G}\boldsymbol{R}_{P} = {}^{G}\boldsymbol{R}_{C}{}^{C}\boldsymbol{R}_{B}{}^{B}\boldsymbol{R}_{P} \tag{8.3} $$

where ${}^{\mathrm{C}}\boldsymbol{t}_{\mathrm{M}}$ represents the position vector of point M (the origin of the intermediate coordinate system) in the cable platform coordinate system; ${}^{\mathrm{B}}\boldsymbol{t}_{\mathrm{M}}$ represents the position vector of point M in the coordinate system $\{\boldsymbol{B}\}$; ${}^{\mathrm{B}}\boldsymbol{R}_{\mathrm{P}}$ is the identity matrix; ${}^{\mathrm{B}}\boldsymbol{t}_{\mathrm{P}}$ is the position vector of the moving platform of Stewart platform in the base coordinate system $\{\boldsymbol{B}\}$ under the initial state.

The natural dip angle of the cable platform (feed cabin) can be described as a function of the position of the cable platform, and the expression is written as:

$$ \varphi_{\mathrm{Cable}} = \frac{3}{8}\varphi_{\mathrm{Receiver}} = \frac{3}{8}\arccos\left(\frac{-z_t}{\sqrt{x_t^2 + y_t^2 + z_t^2}}\right) \tag{8.4} $$

In order to simplify the subsequent partial differential calculation, the rotation matrix of the cable platform is described in the form of quaternion [28] and expressed as the function of ${}^{\mathrm{G}}\boldsymbol{t}_{\mathrm{C}}$:

$$ {}^{\mathrm{G}}\boldsymbol{R}_{\mathrm{C}} = f\left({}^{\mathrm{G}}\boldsymbol{t}_{\mathrm{C}}\right) = \begin{bmatrix} \cos\left(\frac{\varphi_{\mathrm{Cable}}}{2}\right) \\ \frac{y}{\sqrt{x_{\mathrm{C}}^2+y_{\mathrm{C}}^2}}\sin\left(\frac{\varphi_{\mathrm{Cable}}}{2}\right) \\ -\frac{x}{\sqrt{x_{\mathrm{C}}^2+y_{\mathrm{C}}^2}}\sin\left(\frac{\varphi_{\mathrm{Cable}}}{2}\right) \\ 0 \end{bmatrix} \tag{8.5} $$

Equation (8.5) is substituted into Eq. (8.2), and Newton's iteration method is used to convert it into an equation with the independent variable ${}^{\mathrm{G}}\boldsymbol{t}_{\mathrm{C}}$ as:

$$ \begin{aligned} F\left({}^{\mathrm{G}}\boldsymbol{t}_{\mathrm{C}}\right) = {}^{\mathrm{G}}\boldsymbol{t}_{\mathrm{C}} + f\left({}^{\mathrm{G}}\boldsymbol{t}_{\mathrm{C}}\right){}^{\mathrm{C}}\boldsymbol{t}_{\mathrm{M}} + {}^{\mathrm{G}}\boldsymbol{R}_{\mathrm{B}}\left(-{}^{\mathrm{B}}\boldsymbol{t}_{\mathrm{M}}\right) \\ + {}^{\mathrm{G}}\boldsymbol{R}_{\mathrm{B}}{}^{\mathrm{B}}\boldsymbol{t}_{\mathrm{P}} - {}^{\mathrm{G}}\boldsymbol{t}_{\mathrm{P}} = 0 \end{aligned} \tag{8.6} $$

Further, the above equation can be rearranged into the following form of linear equations, namely:

$$ \boldsymbol{K}\,{}^{\mathrm{G}}\boldsymbol{t}_{\mathrm{C}} = 0 \tag{8.7} $$

where

$$ F\left({}^{\mathrm{G}}\boldsymbol{t}_{\mathrm{C}}\right) = \begin{bmatrix} f_1(x_{\mathrm{C}}, y_{\mathrm{C}}, z_{\mathrm{C}}) \\ f_2(x_{\mathrm{C}}, y_{\mathrm{C}}, z_{\mathrm{C}}) \\ f_3(x_{\mathrm{C}}, y_{\mathrm{C}}, z_{\mathrm{C}}) \end{bmatrix}, \quad \boldsymbol{K} = \begin{bmatrix} \partial f_1/\partial x_{\mathrm{C}} & \partial f_1/\partial y_{\mathrm{C}} & \partial f_1/\partial z_{\mathrm{C}} \\ \partial f_2/\partial x_{\mathrm{C}} & \partial f_2/\partial y_{\mathrm{C}} & \partial f_2/\partial z_{\mathrm{C}} \\ \partial f_3/\partial x_{\mathrm{C}} & \partial f_3/\partial y_{\mathrm{C}} & \partial f_3/\partial z_{\mathrm{C}} \end{bmatrix} $$

Once ${}^{\mathrm{G}}\boldsymbol{t}_{\mathrm{C}}$ is deduced, ${}^{\mathrm{G}}\boldsymbol{R}_{\mathrm{C}}$ can be determined with Eq. (8.5), and ${}^{\mathrm{C}}\boldsymbol{R}_{\mathrm{B}}$ can be deduced with Eq. (8.3). The cable lengths and the rotation angles of A-B rotator can be figured out with the inverse kinematic equation and Euler angle of the CDPR.

According to the optimization results of cable forces, the CDPR can be considered as the rigid parallel mechanism in control. Therefore, the inverse kinematic solution is basically the same as the solution process in Chap. 6. Cable lengths are deduced as:

$$ L_j = \left\| {}^{G}\boldsymbol{S}_{Cj} \right\| \tag{8.8} $$

$$ {}^{G}\boldsymbol{S}_{Cj} = {}^{G}\boldsymbol{t}_{C} + {}^{G}\boldsymbol{R}_{C}\boldsymbol{d}_j - \boldsymbol{c}_j \tag{8.9} $$

where j is the cable number, $\boldsymbol{d}_j$ is the position vector of the j-th cable connection point on the cable platform, and $\boldsymbol{c}_j$ is the position vector of the j-th cable exit point on the cable tower.

Taking 1:15 scale model (40-meter scale model) of the FAST feed support system as example, the mechanical structure of A-B rotator is shown in Fig. 8.4. Axis A is directly driven by the servo motor after passing through the reducer. The inverse equation is simple as:

$$n_A = \frac{\eta\theta_A}{360} \tag{8.10}$$

where η is the deceleration ratio of the reducer, θ_{A} is the rotation angle of axis A, and n_{A} is the number of cycles of motor rotation. The B axis is driven by telescopic branch chains, and its inverse equation can be easily obtained with the law of cosine:

$$h_B = \sqrt{a^2 + b^2 - 2ab\cos(\theta_B + \theta_{B0})} \tag{8.11}$$

Among them, $h_B = \|O_2O_3\|$, $a = \|O_1O_2\|$, $a = \|O_1O_3\|$, θ_{B0} is the angle between O_1O_2 and O_1O_3 in the initial state of the A-B rotator.

The trajectory planning process of the FAST feed support system is represented with a flow chart, as shown in Fig. 8.5. ε is the iterative accuracy, and is selected as 10^{-6} m. In order to verify the trajectory planning method of the feed support system and the computational efficiency of the iterative algorithm, numerical simulation experiments are carried out in the MATLAB software. Parameters of 1:15 scale model of the feed support system were adopted in the simulation. In the global coordinate system, the terminal trajectory of the feed support system is described as:

$$\begin{cases} x_t = r \times \sin(\omega t) \\ y_t = r \times \cos(\omega t) \\ z_t = -9.6 \end{cases}$$

where t is simulation time, $r = 2$ m, and ω is the angular velocity of circular motion, $\omega t \in [0, 360°)$. According to the given terminal position, the rotation matrix of the terminal receiver is represented by the quaternion:

$$^G\boldsymbol{R}_P = \begin{bmatrix} \cos\left(\frac{\psi}{2}\right) \\ \frac{y_t}{\sqrt{x_t^2 + y_t^2}} \sin\left(\frac{\psi}{2}\right) \\ -\frac{x_t}{\sqrt{x_t^2 + y_t^2}} \sin\left(\frac{\psi}{2}\right) \\ 0 \end{bmatrix} \quad \psi = \arccos\left(\frac{-z_{\mathrm{t}}}{\sqrt{x_{\mathrm{t}}^2 + y_{\mathrm{t}}^2 + z_{\mathrm{t}}^2}}\right)$$

According to the proposed trajectory planning method, the trajectory of the cable platform and angles of A-B rotator can be calculated directly as shown in Fig. 8.6a and b. When $\varepsilon = 10^{-6}$ m, the average number of iterations of the Newton iteration method

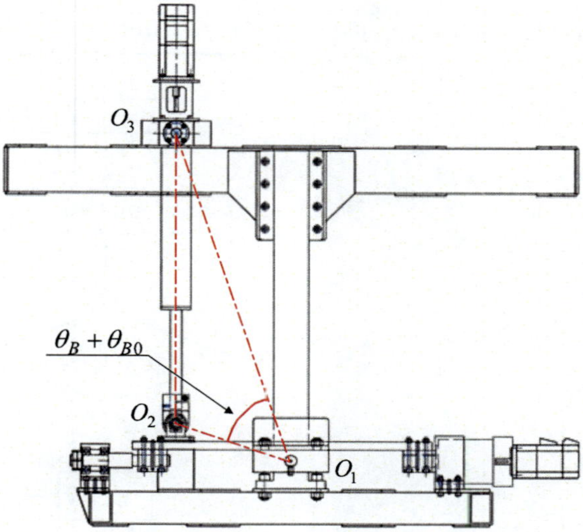

Fig. 8.4 Mechanical structure of the A-B rotator

is 4, and the maximum number of iterations is 10. The proposed method converges fast and can be used for real-time trajectory calculation of the feed support system.

8.3 Trajectory Compensation Vibration Control

The CDPR in the FAST feed support system has the characteristics of large span and low stiffness. It is sensitive to the disturbance of wind and external forces, which is prone to produce deformation and vibration increasing the terminal error. In order to improve the trajectory accuracy of FAST feed support system, it is necessary to implement effective vibration control through the fine-tuning Stewart platform, which is a key theoretical and practical problem in FAST project.

There are mainly three types of vibration controls implemented by rigid robots: (1) trajectory planning vibration control, (2) trajectory compensation vibration control, (3) internal force vibration control. This section focuses on trajectory compensation vibration control, and Sect. 8.4 focuses on internal force vibration control.

In this section, the fine-tuning Stewart platform is taken as the research object, and the fuzzy PD controller is proposed to realize the trajectory compensation vibration control. Finally, the vibration control experiment is implemented. The experiment shows that the terminal trajectory accuracy of the 1:15 (40 m) scale model of the feed support system meets the design requirements, and the effect of the trajectory compensation vibration control is remarkable.

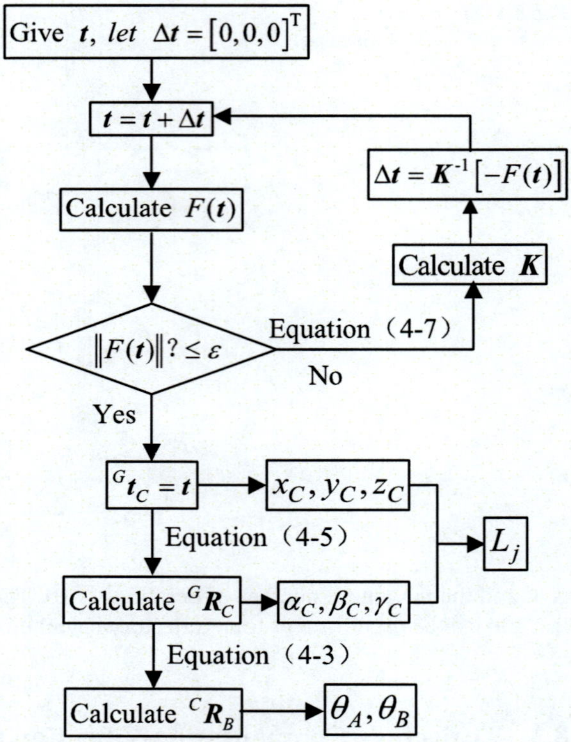

Fig. 8.5 Flow chart of trajectory planning for FAST feed support system

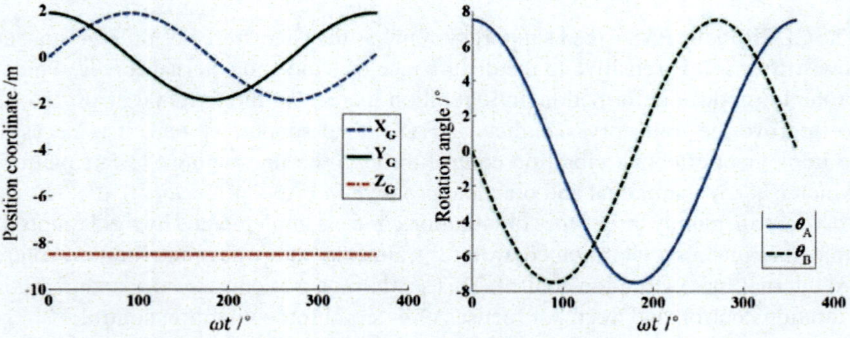

(a) Cable platform trajectory. (b) Angles of A-B rotator.

Fig. 8.6 Simulation results of trajectory planning of the feed support system

8.3.1 Trajectory Compensation Vibration Control Principle

Based on the previous trajectory planning, the fine-tuned Stewart platform is used to carry out the vibration control of the FAST feed support system, which is a typical flexibly supported parallel robot. The flexible support consists of cable-driven parallel robot and the A-B rotator, which will introduce elastic deformation and vibration under the influence of wind or other factors. The fine-tuned Stewart platform adopts the trajectory compensation vibration control to ensure the terminal accuracy of the feed support system, avoiding the influence of deformation and vibration of flexible support on terminal accuracy. Under the trajectory compensation vibration control, the control system structure of the Stewart platform is shown in Fig. 8.7.

The actual terminal position and attitude of the FAST feed support system are jointly determined by the cable-driven parallel robot, A-B rotator and fine-tuning

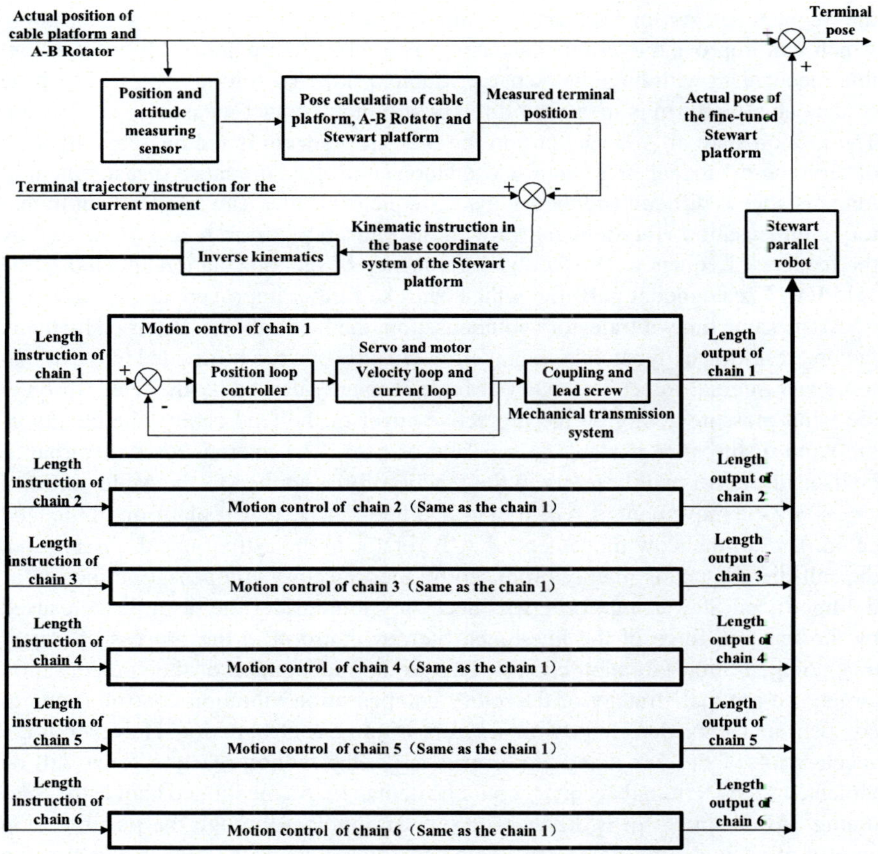

Fig. 8.7 Block diagram of the trajectory compensation vibration control

Stewart platform. The trajectory compensation vibration control calculates the actual terminal position and attitude of the feed support system through real-time detection of the position and attitude of the cable platform. According to the theoretical terminal trajectory given by the astronomical observation requirement, the terminal error is deduced, and the fine-tuning Stewart platform is controlled to compensate the error in real time.

The trajectory compensation vibration control of the fine-tuning Stewart platform is a kind of dynamic tracking control without a predetermined full trajectory. After the trajectory compensation vibration control is implemented on the fine-tuned Stewart platform, the actual terminal error of the feed support system is mainly affected by three factors: (1) Frequency of vibration control; (2) Disturbance of the feed cabin (cable platform); (3) Vibration control logic. We will discuss them separately.

The control frequency is often determined by the feedback frequency. In the trajectory compensation vibration control, the error compensated in the current cycle is calculated with the feedback of the previous cycle. Therefore, the higher the frequency of sensor feedback, the smaller delay of the trajectory compensation, which will improve the terminal accuracy. The cable-driven parallel robot is a flexible mechanism with huge workspace, which is full closed loop control. The pose of the cable platform is measured directly with non-contact measuring equipment. The motion state of A-B rotator and the Stewart platform is fed back by encoders of servomotors to realize the semi-closed loop control. Non-contact measurement of long distance is difficult and the feedback frequency is low. The frequency of trajectory compensation vibration control with the Stewart platform is mainly limited by the feedback frequency. The feedback frequency of the total station adopted in the 1:15 FAST scale model is 10 Hz, which can't be further improved temporarily.

At the same time of trajectory compensation, the feed cabin is disturbed by many factors, resulting in dynamic terminal error. Disturbance can be divided into external force and internal force. The external force is mainly caused by the wind, through the wind pressure acting on the protective cover of the feed cabin. The energy of the wind disturbance is stable in a period of time. The internal force disturbance (self-disturbance) mainly refers to the reaction force applied to the cable platform caused by the movement of A-B rotator and fine-tuned Stewart platform in the feed cabin. Due to the slow movement of A-B rotator in the astronomical observation, the self-disturbance is mainly produced by the trajectory compensation motion of the fine-tuned Stewart platform. It is necessary to consider the disturbance caused by the reaction force of the fine-tuned Stewart platform in the process of trajectory compensation vibration control. Due to the randomness of the compensation target, the optimal strategy of trajectory compensation vibration control seems to compensate for the detected terminal errors within a control period. However, if the acceleration of the fine-tuned platform is large, the motion reaction force will be violent, and the terminal accuracy could be damaged. As mentioned above, the fine-tuning Stewart platform is driven with the servomotors through the position loop control of each chain. The control logic and parameters of the branch chain of the Stewart platform must be determined reasonably in order to improve the terminal trajectory accuracy and system stability.

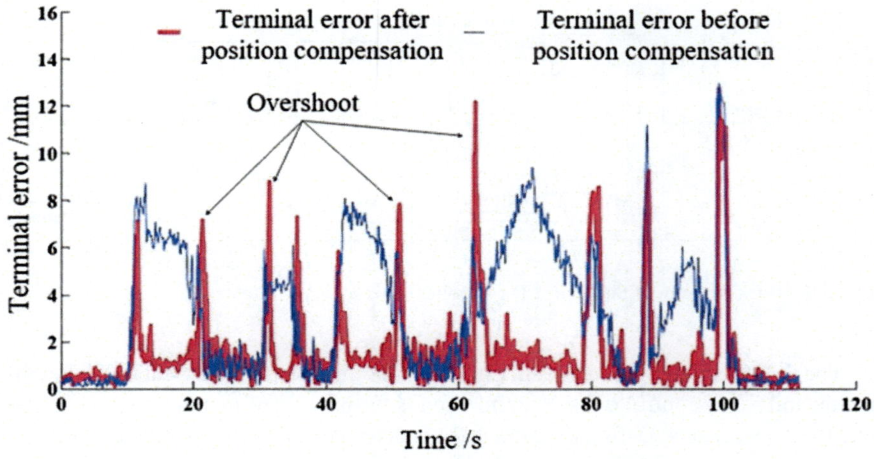

Fig. 8.8 Trajectory compensation vibration control results with PD controller

The PD control logic is considered first. During the process of adjusting the proportional and differential parameters, it is found that the terminal accuracy of the feed support system is significantly improved with large proportional gain and small differential gain, but the Stewart platform will produce position overshoots, which could excite the vibration of the cable-driven parallel robot. When small proportional gain and large differential gain are used, the compensation effect of the fine-tuning Stewart platform is diminished, and the required terminal trajectory accuracy cannot be achieved. Experiment was carried out with the 1:15 scale model of the feed support system. (According to the kinematic similarity, the required terminal accuracy of the 1:15 scale model of the feed support system is determined as root-mean-square (RMS) 2 mm and 0.2°). A group of experimental results obtained with PD controller are shown in Fig. 8.8. The terminal position error is RMS 1.912 mm, and meets the accuracy requirements, but the position overshoot is obvious.

Considering the complexity (fuzziness) of the reaction force effect on the accuracy of the feed support system and the uncertainty of the random wind disturbance, fuzzy control logic is introduced into the position loop of the Stewart platform to boost the trajectory compensation vibration control. Fuzzy control is a control method based on the natural language, which adopts fuzzy logical reasoning and does not need to establish a systematic mathematical model [31]. Fuzzy control is effective for the control of complex, nonlinear, hysteretic and coupled systems. The core of fuzzy control is fuzzy rules. This book is mainly obtained through the experiment of 1:15 scale model of the feed support system. After adopting fuzzy PD control logic, the control block diagram of single branch chain of the Stewart platform is shown in Fig. 8.9 [32].

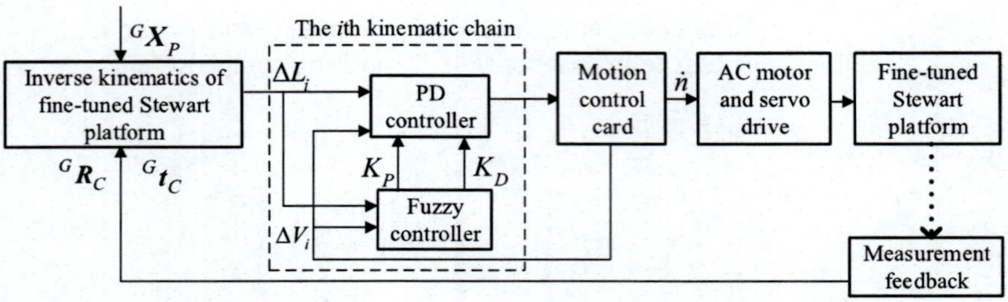

Fig. 8.9 Block diagram of the fuzzy PD controller

The design of fuzzy controller mainly includes three parts: fuzzification, defuzzification and fuzzy control rules. Fuzzification is the process of mapping input parameters to fuzzy subsets. Defuzzification is the process of converting fuzzy output generated by fuzzy control rules into deblurring quantities. The input quantities of fuzzy controller are the length variation of chain $|\Delta L_i|$ and the velocity variation $|\Delta V_i|$, the outputs are proportional gain K_P and differential gain K_D of the PD controller.

All fuzzy subsets adopt triangular functions, and variation domains of the chain length and velocity are divided into three membership functions respectively, namely "Small", "Medium" and "Large", as illustrated in Figs. 8.10 and 8.11.

Fig. 8.10 Fuzzification of input variables

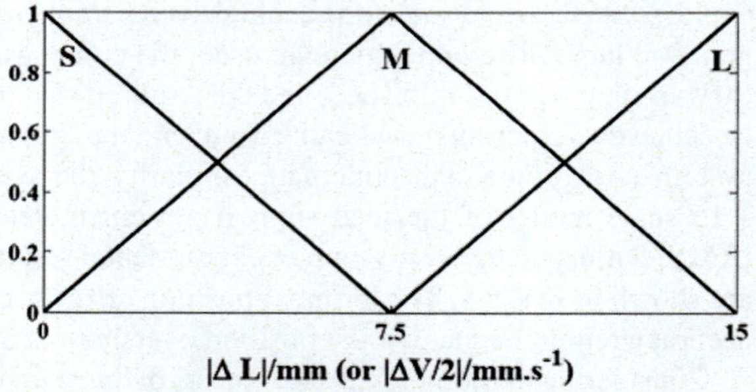

Fig. 8.11 Fuzzification of output variables

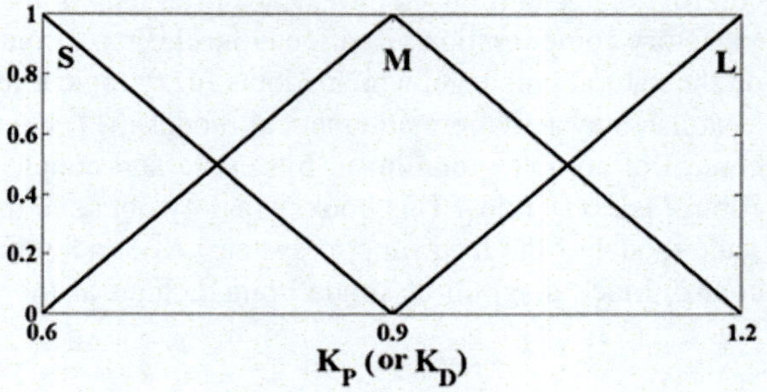

Table 8.1 Fuzzy control rules of proportional gain

$\|\Delta V\|$	$\|\Delta L\|$		
	L	M	S
L	S	S	M
M	M	L	L
S	L	L	L

Table 8.2 Fuzzy control rules of differential gain

$\|\Delta V\|$	$\|\Delta L\|$		
	L	M	S
L	S	M	S
M	M	M	S
S	L	L	S

Since length and velocity of the chain have three membership functions, a total of 9 rules are required, as shown in Tables 8.1 and 8.2. The fuzzy rules are obtained from the experimental results of 1:15 scale model of the feed support system. Specific steps are as follows: Firstly, a large number of PD control experiments were carried out with different proportions and differential parameters; Then the segmented PD control was studied according to the length and velocity variations of the kinematic chain; Finally, the segmented PD controller is refined to obtain fuzzy control logic, and adjusted until meets the requirements. When the velocity variation and length error of the branch chain is large, the compensation speed should be reduced appropriately to avoid overshoot. When the length error and velocity variation are medium, the fast compensation is carried out to improve the accuracy. When the error is small, the velocity fluctuation should be reduced.

The result of fuzzy derivation cannot be directly used for control, and it must be transformed by defuzzification. The main methods of clarity include Maximum Subordination Method, Mamdani Method and Gravity Method [32]. This book adopts the Mamdani Method. The method intercepts the lower half of each output membership function at the corresponding membership value (membership value is calculated by fuzzy rules), combines it into a new region after "or" operation, and gets the center of gravity of the new region, which is the deblurring value of fuzzy output.

By adopting fuzzy PD controller, the fine-tuned Stewart platform can not only ensure that the terminal trajectory accuracy meets the requirements, but also avoid position overshoot, effectively reduce the reaction force impact of the fine-tuned Stewart platform on the flexible cable-driven parallel robot, and improve the stability of the feed support system. The concrete experimental procedure and results are given in the Sect. 8.3.2.

8.3.2 Vibration Control Experiment on the FAST 1:15 Scale Model

The experiment of trajectory compensation vibration control is carried out on the 1:15 scale model of the FAST feed support system, as shown in Fig. 8.12. The experiment focuses on the terminal accuracy and system stability of the trajectory compensation vibration control. The cable-driven parallel robot is simplified in the mechanical structure, and the scale model of the feed cabin is suspended below the fixed support through three steel cables. Two non-contact measuring systems are used in the experiment. Three targets of Leica TC2003 total station are installed on the edge of the feed cabin, and evenly distributed with the angle of 120°. Based on the feedbacks of three total stations, position and attitude of the cable platform are calculated for the real-time vibration control. In addition, a laser tracker of API company is used to measure the actual position of the moving platform of the fine-tuned Stewart mechanism and to verify the terminal accuracy. The sampling frequency of the total station TC2003 is 5–10 Hz, the resolution is 0.01 mm, the static measurement accuracy is 1 mm + 1 ppm (part per million), and the tracking measurement accuracy is 1 mm + 2 ppm; The max sampling frequency of Tracker 3 is 300 Hz, the resolution is 1 μm, the static measurement accuracy is 5 ppm, and the tracking measurement accuracy is 10 ppm. One laser tracker measures the terminal position without the attitude. TC2003 total station and Tracker 3 laser tracker are shown in Fig. 8.13.

First of all, kinematic accuracy of the fine-tuned Stewart platform is verified. The laser tracker records the terminal motion when the fine-tuned Stewart platform executes the given trajectory, checking the machining and assembling precision and the control accuracy. The moving platform of the fine-tuned Stewart manipulator keeps horizontal in the workspace center as the initial state, moves in the positive direction of the *X*-axis until the workspace boundary, and returns to the workspace

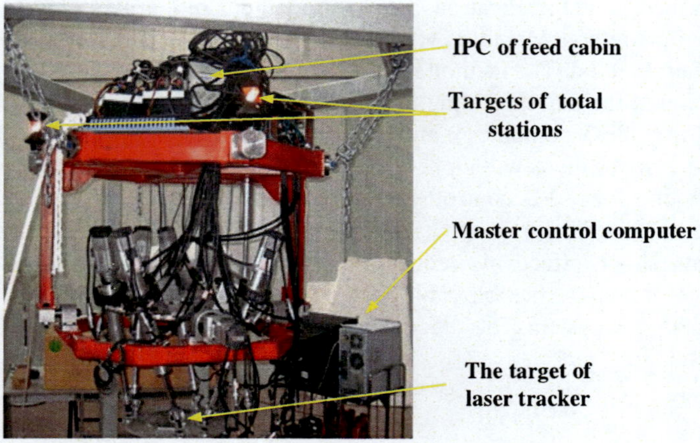

Fig. 8.12 1:15 scale model of the FAST feed support system

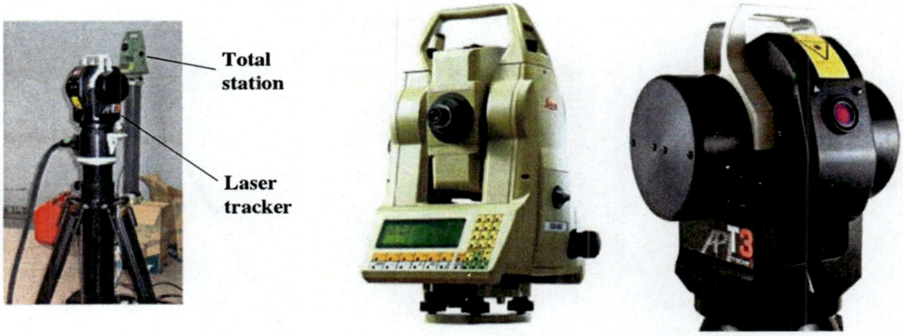

(a) Photo of working condition. (b) Leica TC2003. (c) API Tracker 3.

Fig. 8.13 Measuring equipment

center. The terminal motion is recorded with the laser tracker as shown in Fig. 8.14. The abscissa is the number of sampling points of the laser tracker. The sampling frequency of the laser tracker is 35 Hz. The ordinate is the terminal displacement. Compared with the command track recorded in the control system, the maximum terminal error of the fine-tuned Stewart platform is 0.06 mm, which well meets the design requirements.

The tracking ability of the fine-tuned Stewart platform for sinusoidal signals of the specific frequency is also tested. The simulation of the FAST prototype shows that the low-order resonance frequency of the feed support system is about 0.2 Hz. The sinusoidal signal with a frequency of 0.2 Hz is adopted as the command trajectory, the actual terminal track of the moving platform measured by the laser tracker is shown in Fig. 8.15. The terminal position error is RMS 0.104 mm.

Fig. 8.14 Terminal trajectory of the fine-tuned Stewart platform

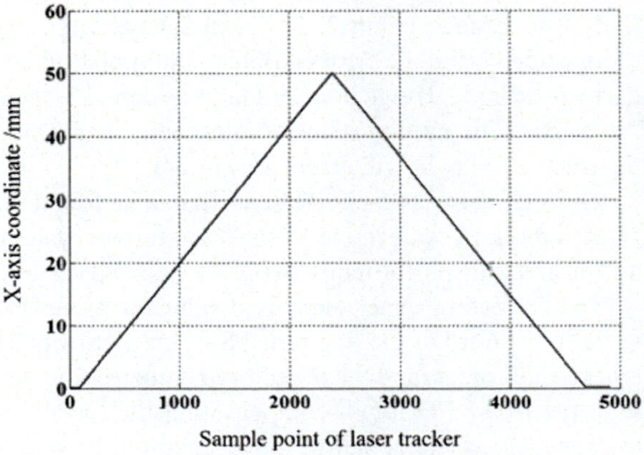

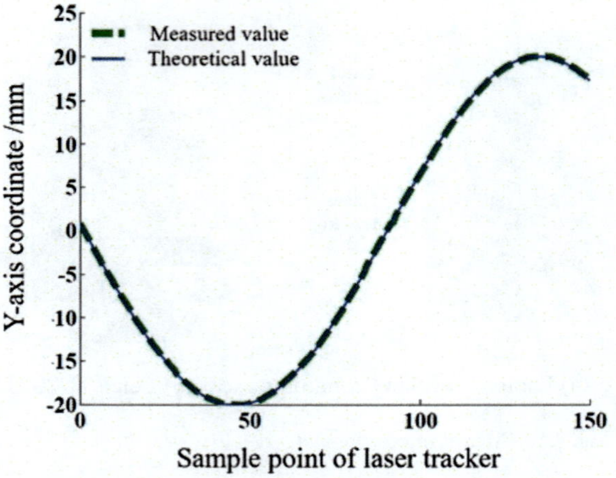

Fig. 8.15 Tracking experiment of 0.2 Hz sinusoidal signal

Due to the limitation on the feedback frequency of the total stations, the control frequency of the fine-tuning Stewart platform is 10 Hz. There are fewer interpolation points for the compensation trajectory. In addition, the network communication is adopted between the main control computer and the industrial computer in the feed cabin. Network delay will have adverse effects on tracking errors, which is generally in the order of milliseconds. The industrial computer of the feed cabin receives a set of instruction signals every 0.1 s, and the network delay introduces an error of several percent per unit moving distance (moving distance within 0.1 s). Due to the above two reasons, the tracking error of the fine-tuned Stewart platform increased slightly, but it still has good tracking accuracy and meets the requirements of trajectory compensation.

When the Stewart platform tracks 0.2 Hz sinusoidal signal, the length error of each telescopic chain is shown in Fig. 8.16. The maximum length error is about 0.05 mm. Branch chain 2, 3, 4 and 5 have large extensional variation with high motion speed, and the error caused by interpolation error and network delay increase correspondingly. Hinge points of the fine-tuned Stewart platform is symmetric about the Y-axis. The motion states of telescopic chains symmetrical about the Y-axis are the same and the length errors are similar.

Through above two experiments, it can be found that the fine-tuned Stewart platform of the 1:15 scale model of the feed support system has good kinematic accuracy, and the tracking performance of 0.2 Hz sinusoidal signal fully meets the requirements.

The following experiment is conducted to verify the trajectory compensation vibration control of the fine-tuned Stewart platform. The cable platform is disturbed by external forces, and the trajectory compensation vibration control is implemented with the fuzzy PD logic, compensating the terminal error in real time. The laser tracker records the actual terminal position of the feed support system, while the position and posture of the cable platform are measured by the total stations to realize the control. The terminal error of the feed support system without trajectory

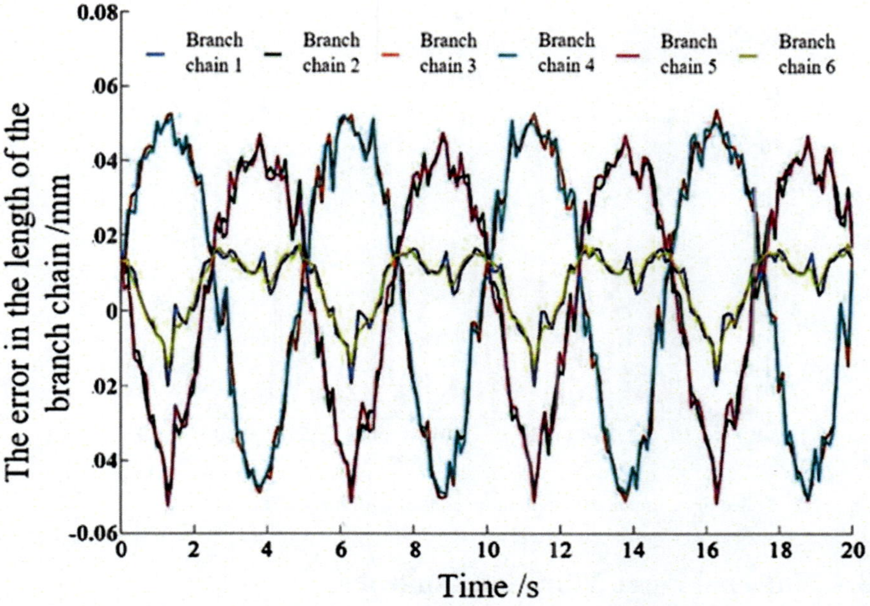

Fig. 8.16 Chain length errors of tracking experiment with 0.2 Hz sinusoidal signal

compensation is obtained through the positive kinematics according to measured actual pose of the cable platform. The experiment results are shown in Fig. 8.17 (the blue line represents the terminal error without the vibration control, and the red line represents the terminal error with the vibration control). With the trajectory compensation vibration control, the terminal error of the feed support system is effectively reduced to RMS 1.861 mm, which meets the requirement of RMS 2 mm. At the same time, there is no position overshoot in the experiment, which verifies that the fuzzy PD logic can effectively improve the stability of the feed support system.

In the process of vibration control, the terminal position offsets of the feed support system along each axis of the global coordinate system are shown in Fig. 8.18. At the step points of the square waves, the feed support system has a large terminal position offset. This is because the vibration control is implemented through the trajectory compensation, which belongs to the tracking control, and the control system can't compensate for newly generated errors in the current cycle. Besides, the control cycle of 0.1 s is long, and the instantaneous position tracking error is unavoidable.

The astronomical observation of FAST requires the dynamic accuracy of the terminal which is described with the root-mean-square error, and the instantaneous position deviation will not affect the overall observation effect. The experiment results show that when the trajectory compensation vibration control is applied to the fine-tuned Stewart platform, the terminal position error of each coordinate axis will be stabilized near zero rapidly. The RMS accuracy meets the requirements.

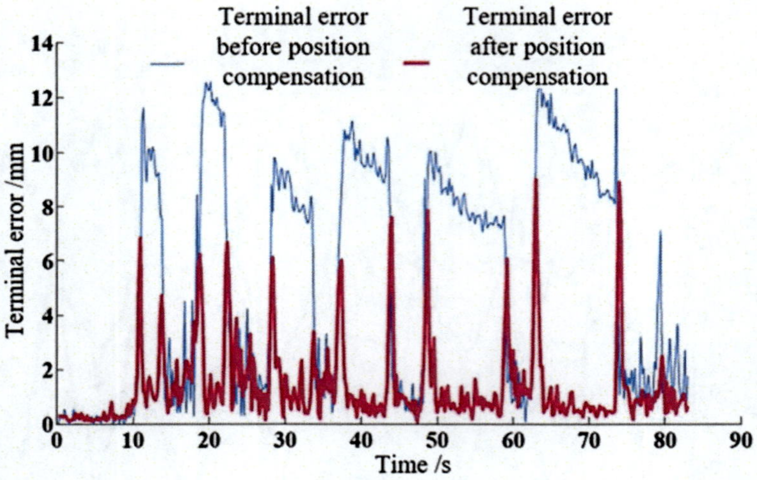

Fig. 8.17 Trajectory compensation vibration control with fuzzy PD controller

8.4 Internal Force Vibration Control

In Sect. 8.3, the fuzzy PD controller is established to realize the trajectory compensation vibration control with the fine-tuned Stewart platform, and the 1:15 scale model of the FAST feed support system can reach the required terminal trajectory accuracy. However, the reaction force caused by acceleration and deceleration of the fine-tuned Stewart platform can't be eliminated. In this section, the internal force vibration control of the flexible supported parallel mechanism is studied, and the vibration of the feed support system is counteracted by the reaction force of the fine-tuned Stewart platform to improve the terminal trajectory accuracy.

A notable feature of the FAST fine-tuned Stewart platform is the large mass of the moving platform. Its prototype reaches three tons, and is a typical heavy-duty parallel robot. Slight acceleration and deceleration of the moving platform of the Stewart platform can produce obvious reaction force to the feed support system. Objective conditions are available for using the motion reaction force to suppress the vibration of cable-driven parallel robot and improve the terminal accuracy and stability of the feed support system.

8.4.1 Internal Force Vibration Control Principle

The essence of the internal force vibration control with the flexibly supported parallel robot is to make full use of the reaction force of the rigid parallel robot to suppress the vibration of the flexible support and improve the terminal trajectory accuracy and stability of the whole system [30]. Compared with the trajectory compensation

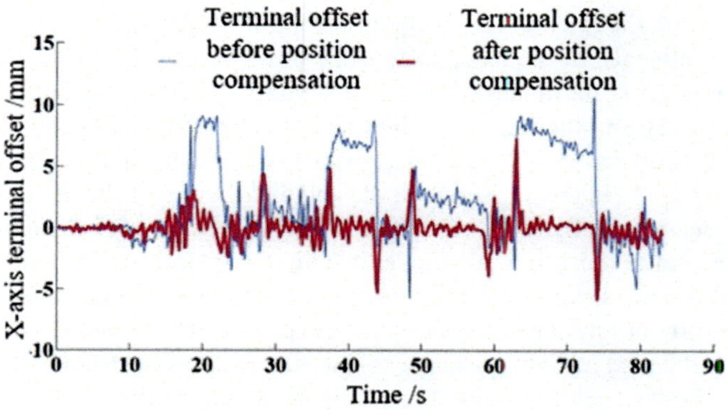

(a) Terminal position offset along the X-axis.

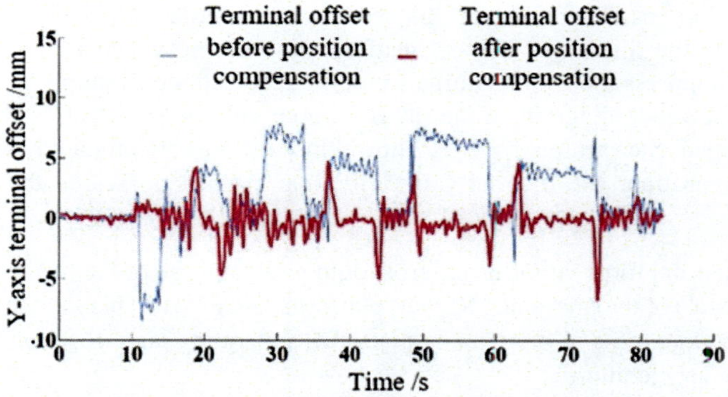

(b) Terminal position offset along the Y-axis.

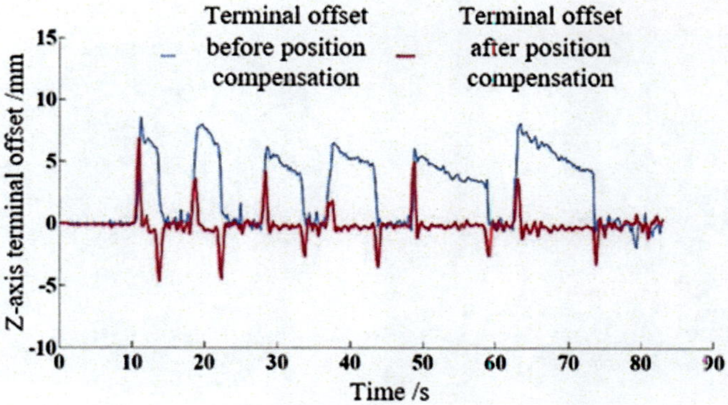

(c) Terminal position offset along the Z-axis

Fig. 8.18 Terminal position offsets along each coordinate axis in the vibration control vibration control, the internal force vibration control could fundamentally eliminate the negative influence of the reaction force of the rigid parallel robot. In Sect. 8.3, although the 1:15 scale model of the feed support system can meet the required terminal accuracy and the impact of the fine-tuned Stewart platform on the cable-driven parallel robot is weaken, the disturbance of internal force is not eliminated. As shown in Fig. 8.19, the mass magnitude and distribution of the moving platform of the 1:15 scale model and the prototype are of great difference. The mass of the moving platform of the 1:15 scale model is 2 kg, the total mass of the feed cabin is 200 kg, and the mass ratio of the moving platform to the feed cabin is 1:100. The moving platform of fine-tuned Stewart platform of the FAST prototype has the mass of 3 tons and the total mass of the feed cabin is 30 tons. The mass ratio of the moving platform to the feed cabin is 1:10. The heavy load characteristics of the fine-tuned Stewart platform in the prototype will significantly increase the reaction force under a slight acceleration, and the trajectory compensation vibration control will face serious challenges. Therefore, in this section, the internal force vibration control is applied to the fine-tuned Stewart platform, and the heavy-load characteristic is utilized to suppress the wind disturbance of the feed support system.

The logic block diagram of the internal force vibration control established for the fine-tuned Stewart platform is shown in Fig. 8.20. Compared with trajectory compensation, the internal force vibration control possesses the following advantages:

(1) The internal force vibration control could make full use of the heavy-load characteristics of the fine-tuned Stewart platform, and give full play to the advantage of the large mass of the moving platform, generating large reaction force with a little acceleration.

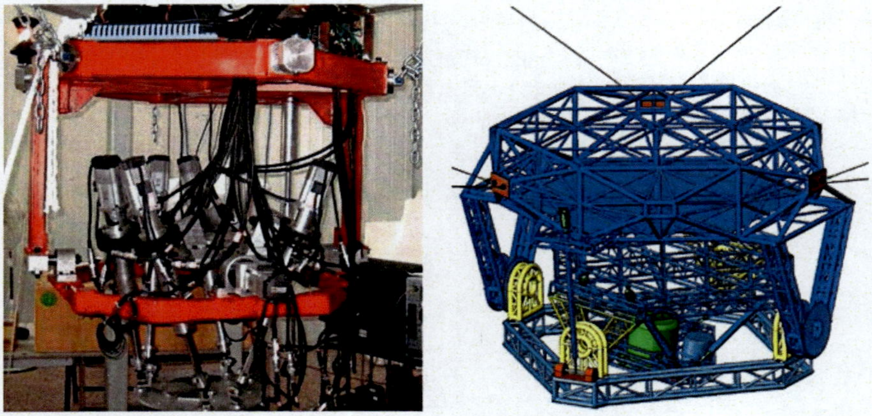

(a) 1:15 scale model. (b) 3D model of the prototype.

Fig. 8.19 Scale model and prototype of the feed cabin

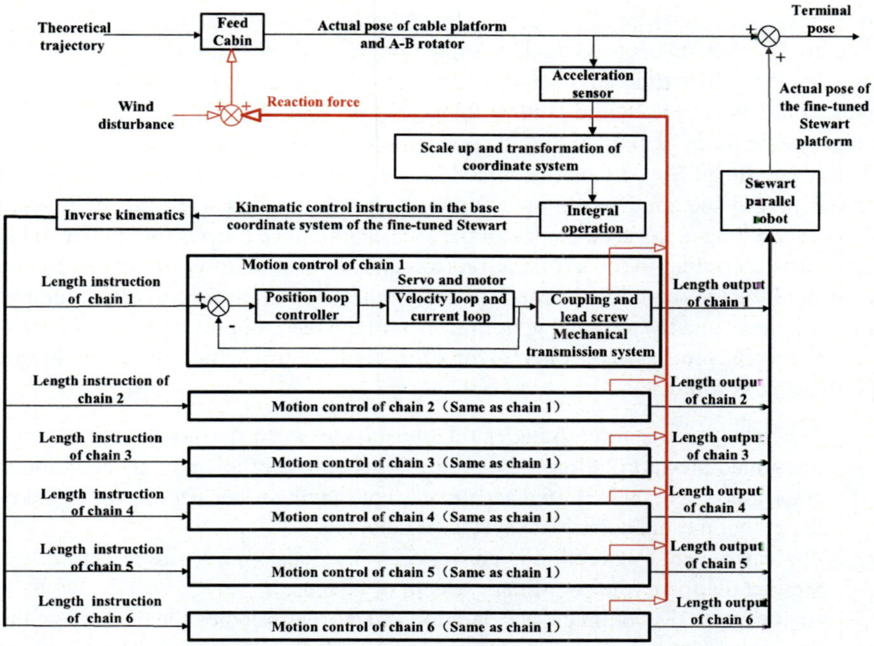

Fig. 8.20 Control block diagram of the internal force vibration control

(2) The internal force vibration control generates the reaction force and suppresses vibration through controlling the fine-tuned Stewart platform according to the acceleration of the feed cabin. It is a kind of advance compensation for the position error, which can generate effectively compensation with small position error, avoiding large following error.
(3) The feedback frequency of the acceleration sensor is higher than that of the non-contact position measurement, and the control frequency can be improved.

The acceleration sensors are installed in the feed cabin, and measured linear and angular accelerations of the feed cabin are described in the global coordinate system. The proportional control and coordinate system transform are introduced to ensure that the fine-tuned Stewart platform generates the opposite and appropriate reaction force against the disturbing force. The relationship between the measured acceleration of the cable platform and the command acceleration of the fine-tuned Stewart platform to generate the reaction force is shown in the following equation:

$$ {}^{\mathrm{B}}\ddot{\boldsymbol{t}}_{\mathrm{P}} = \lambda_{\mathrm{P}}\left({}^{\mathrm{G}}\boldsymbol{R}_{\mathrm{C}}{}^{\mathrm{C}}\boldsymbol{R}_{\mathrm{B}}\right)^{\mathrm{T}}{}^{\mathrm{G}}\ddot{\boldsymbol{t}}_{\mathrm{C}} \tag{8.12} $$

$$ {}^{\mathrm{B}}\ddot{\boldsymbol{\theta}}_{\mathrm{P}} = \lambda_{\mathrm{A}}\left({}^{\mathrm{G}}\boldsymbol{R}_{\mathrm{C}}{}^{\mathrm{C}}\boldsymbol{R}_{\mathrm{B}}\right)^{\mathrm{T}}{}^{\mathrm{G}}\ddot{\boldsymbol{\theta}}_{\mathrm{C}} \tag{8.13} $$

where, λ_P and λ_A are separately gains of linear acceleration and angular accelerations. The inverse dynamics of the flexibly supported parallel mechanism is complex and hard to solve in real time. In order to ensure the control frequency of the internal force vibration control, the hybrid control logic of force and position is adopted, which employs the force loop including the position closed loop. The target displacement of the fine-tuned Stewart platform in each cycle is deduced through the integral calculation of the command acceleration which are obtained with Eqs. (8.12) and (8.13) according to the acceleration of the feed cabin. Since the position command is generated according to the acceleration (force), the reaction force generated by the motion of the fine-tuned Stewart platform will directly inhibit the vibration trend of the feed cabin, and the terminal vibration error of the feed support system is reduced.

The specific process of the internal force vibration control of the fine-tuned Stewart platform is as follows:

(1) According to the force transfer relationship between the feed cabin and the fine-tuned Stewart platform, the proportional gain coefficient of acceleration is determined, and the required acceleration of the fine-tuned Stewart platform in the global inertial coordinate system is calculated.
(2) The calculated acceleration is converted to the local coordinate system of the Stewart platform (the coordinate system of the base).
(3) According to the command acceleration and current velocity, the feed rate of the fine-tuned Stewart platform is obtained through the integration. The length of each chain of the Stewart platform is obtained with inverse kinematics. The position controller directly controls the servo motors to complete the corresponding movement.

8.4.2 Vibration Control Simulation

This section focuses on improving the terminal positional accuracy of the feed support system, and we can define $\lambda_p = \lambda$, $\lambda_A = 0$. The wind load samples are exerted on the geometric center of the feed cabin along the positive direction of X-axis. The fine-tuned Stewart platform carries out the internal force vibration control. The relationship between the proportional coefficient of acceleration λ and the terminal RMS error along the X-axis is determined with the simulation. The relationship curve is shown in Fig. 8.21. When $\lambda = 2.6$, the terminal RMS error of the feed support system is the smallest, which is 4.1 mm. At this time, the terminal error of the feed support system is also the smallest, and the RMS error is 4.2 mm, which better meets the requirements of the terminal position accuracy.

The main direction of vibration control of the feed support system is along the X-axis of the global coordinate system. The relationship between the acceleration of the cable platform and the reaction force of the fine-tuned Stewart platform is illustrated along the X-axis, as shown in Fig. 8.22. In the process of the internal force vibration control, the reaction force of the Stewart platform on the cable platform and the acceleration of the cable platform, which are both described in the global

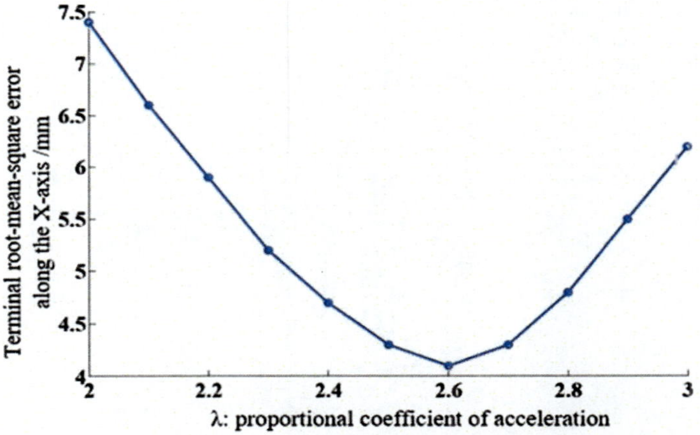

Fig. 8.21 Relationship between the proportional coefficient of acceleration and the terminal error

coordinate system, maintain opposite signs with the same trend. In other words, the reaction force of the Stewart platform on the cable platform counteracts the wind force well.

When $\lambda = 2.6$, the terminal trajectory error of the feed support system with the internal force vibration control is shown in Fig. 8.23. Since the wind disturbance acts on the feed cabin along the X-axis, the terminal error of the feed support system along the X-axis is large while the terminal errors along Y-axis and Z-axis are small. Compared with the terminal error curve without the internal force vibration control, it can be found that the terminal error of the feed support system has been significantly reduced. The maximum position error is reduced by about 85%, from 82.17 to 12 mm.

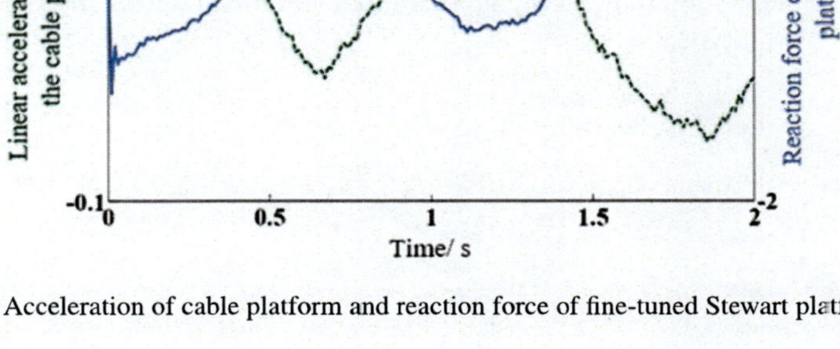

Fig. 8.22 Acceleration of cable platform and reaction force of fine-tuned Stewart platform

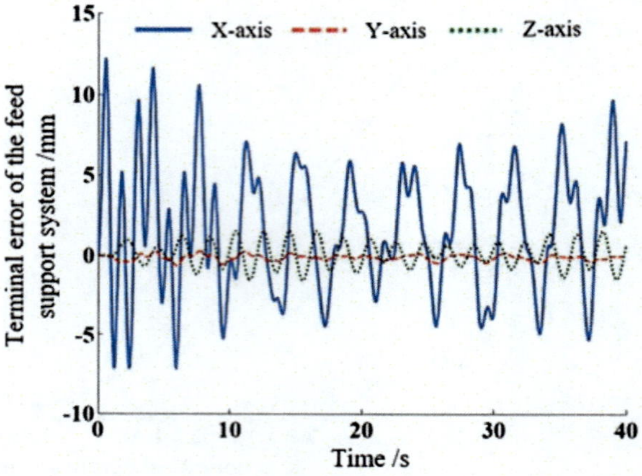

Fig. 8.23 Terminal error curve of the feed support system with the internal force vibration control

During the implementation of the internal force vibration control, the driving force of each telescopic chain of the Stewart platform are shown in Fig. 8.24. It can be seen that the driving forces change periodically near the initial value, and the driving forces of chains 1, 2, 5 and 6 change more intense than that of chains 3 and 4. In general, the maximum and minimum driving forces of telescopic chains of the Stewart platform are 8.1 kN and 6.8 kN respectively with limited fluctuation, and the chains are in good force condition.

The trajectory of the moving platform of the Stewart platform relative to the base under the local coordinate system is shown in Fig. 8.25. The maximum displacement offset of the moving platform appears along the x' axis of the local coordinate system, and the value is about 50 mm. The design workspace of the fine-tuned Stewart platform of FAST prototype is a sphere with the radius of 250 mm, and the motion of the fine-tuned Stewart platform is away from the workspace boundary during the internal force vibration control.

In summary, the internal force vibration control makes full use of the reaction force of the fine-tuned Stewart platform, effectively counteracts the wind disturbance acting on the feed cabin, reduces the vibration of the FAST feed support system, and significantly improves the terminal trajectory accuracy. From above simulation results, it can be seen that when the internal force vibration control is adopted, the telescopic chains of the fine-tuned Stewart platform are in a good condition of force with reasonable motion range. The internal force vibration control could be applied to the FAST prototype.

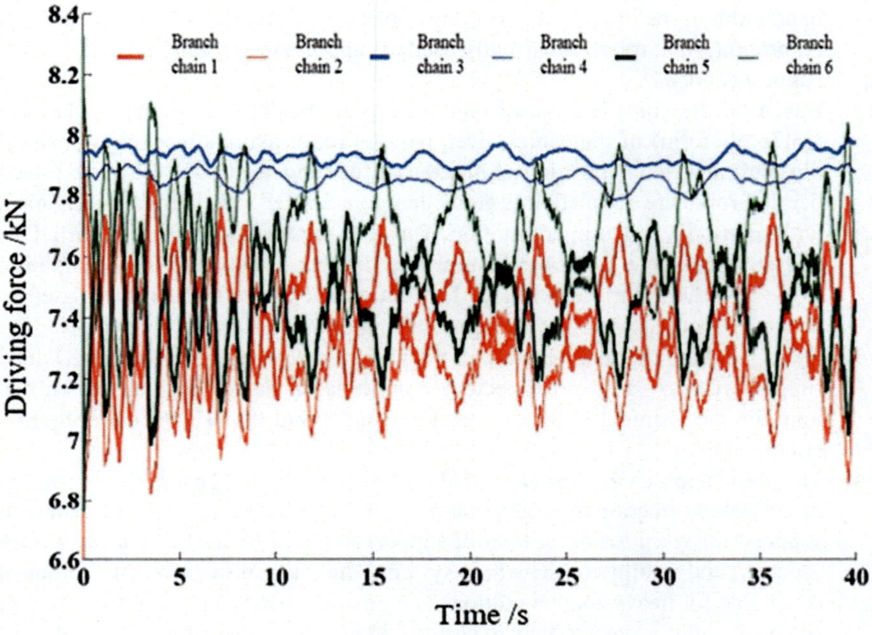

Fig. 8.24 Chain driving force of the fine-tuned Stewart platform

Fig. 8.25 Motion of the fine-tuned Stewart platform

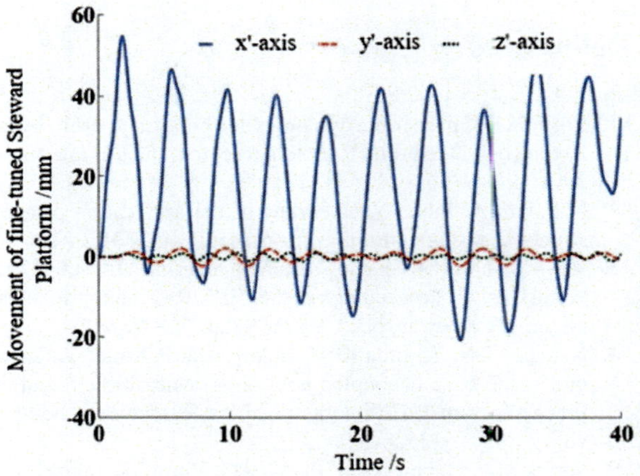

8.5 Conclusion

(1) This chapter discusses the existing vibration control methods of the flexibly supported robots, which can be divided into two categories: passive vibration control and active vibration control. Due to the low stiffness and limited control

bandwidth of the flexible support, the applicable vibration control is restricted. At present, rigid robots are usually used to suppress the vibration of the flexible support robots.

(2) Based on the consideration of engineering application, the moving platform (cable platform) of the cable-driven parallel robot maintains a certain natural dip angle at different positions. Since the attitude and position of the cable-driven parallel robot are coupled, it is equivalent to a 3-DOF translational mechanism with rotational concomitant motion. The A-B rotator has two rotational DOFs and is responsible for compensating the difference between the observation angle and the natural dip angle. The fine-tuned Stewart platform is used to improve terminal accuracy through carrying out the vibration control.

(3) By introducing the fuzzy PD logic, the vibration control method of FAST feed support system based on trajectory compensation is established, which can improve the terminal trajectory accuracy and avoid the self-excited vibration effectively.

(4) The internal force vibration control of the flexibly supported parallel robot makes full use of the reaction force generated by the motion of the rigid parallel robot to suppress the vibration of the flexible support and improve the terminal trajectory accuracy and stability of the whole system. The internal force vibration control is suitable for heavy-load condition. The position error can be compensated in advance, and it is easy to realize closed-loop control with high frequency.

References

1. Yoshida K. Space robot dynamics and control: to orbit, from orbit, and future. In: Hollerbach J, Koditschek D, editors. Robotics research, the 9th international symposium. Berlin: Springer; 1999. p. 449–56.
2. Nishitani A, Inoue Y. Overview of the application of active/semi-active control to building structures in Japan. Earthq Eng Struct Dyn. 2001;30:1565–74.
3. Van VJ, Sharf I. Frequency matching algorithm for aetive damping of macro-micro manipulator vibrations. In: Proceedings of the IEEE/RSJ international conference on intelligent robots and system, Piscataway, NJ, USA; 1998. p. 782–7.
4. Nenchev DN, Yoshida K, Vichitkulsawat P, Konno A, Uchiyama M. Experiments on reaction null-space base decoupled control of a flexible structure mounted manipulator system. In: Proceedings of IEEE international conference on robotics and automation, New York, USA, vol. 3; 1997. p. 2528–34.
5. Torres MA, Dubowsky S, Pisoni AC. Vibration control of deployment structures' long-reach space manipulators: the P-PED method. In: Proceedings of the IEEE international conference on robotics and automation, Minneapolis, MN, USA; 1996. p. 2498–504.
6. Lew JY, Trudnowski DJ. Vibration control of a micro/macro-manipulator system. IEEE Control System. 1996;16(1):26–31.
7. Yoshikawa T, Harada K, Matsumoto A. Hybrid position/force control of flexible-macro/rigid-micro manipulator systems. IEEE Trans Robot Autom. 1996;12(4):633–40.
8. Staffetti E, Bruyninckx H, Schutter J. On the invariance of manipulability indices. In: Lenarcic J, Thomas F, editors. Advances in robot kinematics: theory and applications. Dordrecht: Kluwer; 2002. p. 57–66.

9. Yoshikawa T, Hosoda K, Doi T, Murakami H. Quasi-static trajectory tracking control of flexible manipulator by macro-micro manipulator system. In: Proceedings of IEEE international conference on robotics and automation, Atlanta, USA; 1993. p. 210–4.
10. Cho C, Kang S, Kim M, Song JB. Macro-micro manipulation with visual tracking and its application to wheel assembly. Int J Control Autom Syst. 2005;3(3):461–8.
11. Nguyen LA, Wakler ID, Defigueiredo RJP. Dynamic control of flexible, kinematically redundant robot manipulator. IEEE Trans Rob Control. 1992;8(6):669–77.
12. Torres M A, Dubowsky S, Pisoni AC. Path-planning for elastically mounted space manipulators: experimental evaluation of the coupling map. In: Proceedings of IEEE international conference on robotics and automation, San Diego, USA; 1994. p. 2227–33.
13. Xie HP, Kalaycioglu S, Patel RV. Control of residual vibrations in the space shuttle remote manipulator system. Int J Robot Autom. 2000;15(2):68–77.
14. Lew J, Trudnowski D, Evans M, Bennett DW. Micro-manipulator motion control to suppress macro-manipulator structure vibrations. In: Proceedings of the IEEE international conference on robotics and automation, Nagoya, Japan; 1995. p. 3116–20.
15. Parsa K, Angeles J, Misra AK. Control of macro-micro manipulators revisited. Trans ASME. 2005;127(2):688–99.
16. Zhu B, Nie Y, Nan R, Peng B. The FAST/SKA site selection in Guizhou province. Phys Astron. 2001;278(1–2):213–8.
17. Taghirad HD, Nahon M. Kinematic analysis of a macro-micro redundantly actuated parallel manipulator. Adv Robot. 2008;22:657–87.
18. Taghirad HD, Nahon M. Dynamic analysis of a macro-micro redundantly actuated parallel manipulator. Adv Robot. 2008;24:949–81.
19. Su YX, Duan BY. The mechanical design and kinematics accuracy analysis of a fine tuning stable platform for the large spherical radio telescope. Mechatronics. 2000;10(7):819–34.
20. Zi B, Duan BY, Du JL, Qiu YY. Trajectory tacking sliding mode control of a cable parallel manipulator based on fuzzy logic. In: Proceedings of the world congress on intelligent control and automation, Dalian, China; 2006. p. 9203–07.
21. Zi B, Duan BY, Du JL, Bao H. Dynamic modeling and active control of a cable-suspended parallel robot. Mechatronics. 2008;18(1):1–12.
22. Su YX, Duan BY, Nan RD, et al. Development of a large parallel-cable manipulator for the feed-supporting system of a next-generation large radio telescope. J Robot Syst. 2001;18(11):633–43.
23. Duan XC, Qiu YY, Bao H, Du JL. Real-time motion planning based vibration control of a macro-micro parallel manipulator system for super antenna. J Vibroeng. 2014;16(2):694–703.
24. Zhou Q, Zhang H. Simulation and experimental analysis of the Stewart parallel mechanism for vibration control. In: IEEE international conference on systems, man & cybernetics, Washington D.C., USA, vol. 4; 2003. p. 3548–52.
25. Qi L, Zhang H, Duan GH. Task-space position/attitude tracking control of FAST fine tuning system. Front Mech Eng China. 2008;3(4):392–9.
26. Zhou Q, Qi L, Zhang H, Duan GH. Experiments on active large-amplitude vibration control of a coupled Stewart platform. China Mech Eng. 2004;16(18):1599–603.
27. Zhang H, Wang QM, Wang JS, Ren GX. Interpolation for active vibration control of a Stewart platform. J Tsinghua Univ (Sci Technol). 42(11):1473–76.
28. Tang XQ, Shao ZF. Trajectory generation and tracking control of a multi-level hybrid support manipulator in FAST. Mechatronics. 2013;23(8):1113–22.
29. Qu L, Tang XQ, Yao R, Chen X. Error analysis and compensation of cable driven parallel manipulators for the forty-meter aperture radio telescope. Chin High Technol Lett. 2010;20(3):303–8.
30. Shao ZF, Tang XQ, Wang LP, et al. Dynamic modeling and wind vibration control of the feed support system in FAST. Nonlinear Dyn. 2012;67(2):965–85.
31. Wang HO, Tanaka K, Griffin MF. An approach to fuzzy control of nonlinear systems: stability and design issue. IEEE Trans Fuzzy Syst. 2002;4(1):14–23.

32. Shao ZF, Tang XQ, Wang LP, You Z. A fuzzy PID approach for the vibration control of the FSPM. Int J Adv Rob Syst. 2013;10(1):59.

Chapter 9
Inertia Matching of Parallel Robots

In Chap. 8, it is found that inertia (mass distribution) affects the dynamic performance of the parallel mechanism significantly and is an important factor influencing the vibration control effect of the fine-tuning Stewart platform. In addition, inertia matching involves such engineering design issues as the driving motor and transmission chain selection. Therefore, this chapter investigates the inertia matching of the parallel robots.

Inertia matching means the ratio of the mechanism inertia converted to the motor shaft to the motor inertia is within the proper range, in order to give full play to the motor capacity and ensure the dynamic performance of the mechanism.

At present, there are many researches focusing on the inertia matching of the serial mechanism [1–3], especially the high-performance CNC machine tools, while researches on the parallel mechanism are limited. Parallel mechanisms, especially those with heavy load, high speed and attitude sensitive, must carry out the research on inertia matching. In the selection of motors and transmission chains for the parallel mechanism, the inertia matching relations must be considered on the basis of power, velocity and torque requirements. Otherwise, vibration, overshoot and even control system instability could appear. Although some motor manufacturers provide the recommended motor load requirements, they are only for the traditional serial mechanism, and the criteria given by different manufacturers vary greatly.

This chapter adopts the virtual work principle to derive the joint-space inertia matrix of the Stewart parallel mechanism, analyzes the coupling of the inertia matrix, and proposes the equivalent inertia index. Then, the distribution of the equivalent inertia is analyzed. Finally, the equivalent inertia is used to establish the inertia matching criterion for the parallel mechanism to ensure its dynamic performance. In this chapter, Sect. 9.1 briefly introduces inertia matching and presents the deduction of the joint-space inertia matrix; Sect. 9.2 derives the analytical equations of the inertia matrix with the virtual work principle; In Sect. 9.3, the joint-space inertia matrix of the Stewart parallel robot is investigated and the equivalent inertia index is proposed; Sect. 9.4 proposes the principle of inertia matching on the basis of the

© Tsinghua University Press 2024

X. Tang et al., *Research and Application of Cable-Driven and Rigid Parallel Robots*,
https://doi.org/10.1007/978-981-99-7452-8_9

comprehensive analysis on the resonance frequency, acceleration and deceleration performance as well as dynamic performance of limbs (branch chains).

The main contents of this chapter are:

(1) Inertia matching and joint-space inertia matrix of the parallel robot;
(2) Joint-space inertia matrix of the Stewart parallel mechanism;
(3) Equivalent inertia index of the parallel robot;
(4) Inertia matching guideline of the parallel robot.

9.1 Inertia Matching and Joint-Space Inertia Matrix of the Parallel Robot

There is a nonlinear kinematic mapping relation between the real axis (joint) space and the virtual axis space of the parallel mechanism. The motion trajectory of the parallel mechanism is defined in the virtual axis space. The uniform linear motion defined in the virtual axis space usually corresponds to the variable (acceleration/ deceleration) motions of limbs in the joint space. Therefore, for the parallel mechanism, the dynamic performance (inertia property) of the limbs affects not only the acceleration and deceleration performance of the moving platform, but also the trajectory accuracy under the uniform motion [4–6]. As the industrial applications of the parallel mechanism develop toward high acceleration, high velocity and heavy load, inertia matching begins to attract the attention of scholars. At present, inertia researches mainly include two aspects: the study of inertia matrix characteristics of the parallel mechanism and the study of inertia matching principles [7, 8].

Codourey [9] conducts dynamic modeling of the Delta parallel mechanism, proposes an estimation approach for the mass matrix and implement the computed torque control. Yang Haoquan [10] focuses on the effects of limb inertia in the dynamic modeling of the Stewart parallel mechanism. Ogbobe [11] analyzes the frequency coupling issue of a 6-DOF flight simulator based on the singular value decomposition of the joint-space inertia matrix. He Jingfeng [12] investigates the dynamic model of the joint space of a 6-DOF parallel robot with the Kane method, and reveals that there is a strong coupling relation among the limbs of the parallel robot, which affects the inertial load of the limbs. Yu [13] theoretically derives and proves the block diagonal dominance property of the joint-space inertia matrix. Professor Tian [14, 15] performs servomotors selection for the parallel mechanism with few DOFs (3-DOF translational parallel machine tools and 2-DOF high-speed parallel robots) considering the numerical distribution of the joint-space inertia matrix of the parallel mechanism. The inertia of the parallel mechanism is expressed in the matrix and changes in the workspace with the pose variation of the moving platform.

The parallel mechanism is driven by motors in limbs, and the inertia matching issue must be converted into the joint space. To achieve the conversion, the joint space inertia matrix needs to be deduced. The dynamic model of parallel mechanism is usually established in the virtual axis space, and the driving force of each limb is expressed as a function of the terminal motion parameters, structural dimensions and mass parameters, which can be expressed as:

$$\boldsymbol{M}_{\mathrm{P}}(\boldsymbol{X}_{\mathrm{P}})\,\ddot{\boldsymbol{X}}_{\mathrm{P}} + \boldsymbol{V}_{\mathrm{P}}\left(\dot{\boldsymbol{X}}_{\mathrm{P}}, \boldsymbol{X}_{\mathrm{P}}\right) + \boldsymbol{G}_{\mathrm{P}}(\boldsymbol{X}_{\mathrm{P}}) = \tau \tag{9.1}$$

where τ is the driving force vector of the limbs; $\boldsymbol{M}_P(\boldsymbol{X}_P)$ is the inertia matrix described in the virtual axis space; $\boldsymbol{V}_{\mathrm{P}}(\dot{\boldsymbol{X}}_{\mathrm{P}}, \boldsymbol{X}_{\mathrm{P}})$ is the vector of centripetal, Coriolis and friction forces; $\boldsymbol{G}_{\mathrm{P}}(\boldsymbol{X}_{\mathrm{P}})$ is the gravity vector; $\boldsymbol{X}_{\mathrm{P}}$ is the pose vector of the moving platform; $\dot{\boldsymbol{X}}_{\mathrm{P}}$ is the velocity vector of the moving platform; $\ddot{\boldsymbol{X}}_{\mathrm{P}}$ is the acceleration vector of the moving platform.

To obtain the joint space inertia matrix that is independent of the terminal coordinate system and reflects the inertia property of the parallel mechanism, the inverse kinematic equation ($\boldsymbol{J}$ is the inverse Jacobian matrix)

$$\ddot{\boldsymbol{L}} = \boldsymbol{J}\,\ddot{\boldsymbol{X}}_{\mathrm{P}} + \dot{\boldsymbol{J}}\,\dot{\boldsymbol{X}}_{\mathrm{P}} \tag{9.2}$$

is substituted into Eq. (9.1), and the joint space inertia matrix can be obtained as:

$$\boldsymbol{M}_{\mathrm{L}} = \boldsymbol{M}_{\mathrm{P}}(\boldsymbol{X}_{\mathrm{P}})\boldsymbol{J}^{-1} \tag{9.3}$$

The dynamic equations of the parallel mechanism have various deducing approaches. Although the dynamic model derived by the Newton–Euler method has clear physical meanings, it is difficult to be simplified to a standard form and to deduce the inertia matrix. Both the Lagrange formulation and the virtual work principle can obtain the standard form, and the deducing process of the virtual work principle is simpler and more convenient. The virtual work principle is adopted in Sect. 9.2 to deduce the joint space inertia matrix.

In the dynamic modeling with the virtual work principle, the following equation is used to facilitate the deduction of the limb Jacobian matrices:

$$\boldsymbol{a} \times \boldsymbol{b} = [\boldsymbol{a}\times]\boldsymbol{b} \tag{9.4}$$

where $[\boldsymbol{a}\times]$ is defined as the left cross product matrix and can be expressed as:

$$[\boldsymbol{a}\times] \triangleq \begin{bmatrix} 0 & -a_3 & a_2 \\ a_3 & 0 & -a_1 \\ -a_2 & a_1 & 0 \end{bmatrix}$$

where a_1, a_2, a_3 is the three elements of $\boldsymbol{a}$. Similarly, the right cross matrix can be defined as:

$$\boldsymbol{a} \times \boldsymbol{b} = [\times\boldsymbol{b}]\boldsymbol{a} \tag{9.5}$$

where $[\times\boldsymbol{b}]$ is the right cross product matrix and can be expressed as:

$$[\times \boldsymbol{b}] \triangleq \begin{bmatrix} 0 & b_3 & -b_2 \\ -b_3 & 0 & b_1 \\ b_2 & -b_1 & 0 \end{bmatrix}$$

where b_1, b_2, b_3 is the three elements of $\boldsymbol{b}$.

9.2 Joint Space Inertia Matrix of the Stewart Mechanism

A typical structure of the Stewart mechanism is shown in Fig. 9.1. The base platform is defined by the rotation centers ($B_1, B_2, B_3, B_4, B_5, B_6$) of the universal joints. The base coordinate system $\{\boldsymbol{B}\} : O - XYZ$ is fixed at the geometric center of the base platform, with the X-axis pointing to the midpoint of B_1B_2 and the Z-axis pointing vertically downward. The moving platform is defined by the rotation centers ($P_1, P_2, P_3, P_4, P_5, P_6$) of the spherical joints. The coordinate system $\{\boldsymbol{P}\} : o - xyz$ of the moving platform is fixed at the geometric center of the moving platform, with the x-axis pointing to the midpoint of P_1 P_6 and the z-axis perpendicular to the moving platform.

To simplify the deduction process, the virtual work of each limb is deduced under the corresponding local coordinate system. The local coordinate system $\{\boldsymbol{N}_i\} : O_i - X_iY_iZ_i$ is fixed at the rotation center of the universal joint of the i-th limb, with the X_i axis pointing to the geometric center of the base platform and the Z_i axis pointing vertically downward. The local coordinate system $\{\boldsymbol{K}_i\} : o_i - x_iy_iz_i$ is also fixed at the rotation center of the universal joint of the i-th limb, with the z_i axis pointing to the rotation center of the spherical joint along the limb and the y_i

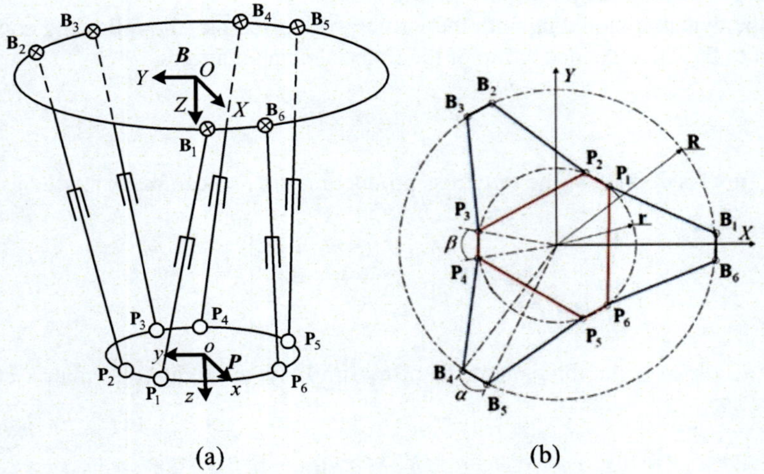

Fig. 9.1 Diagram of the Stewart mechanism. **a** Axonometric view; **b** upward view

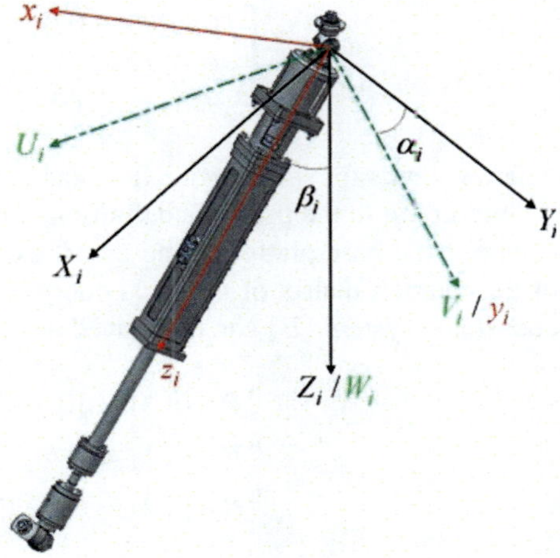

Fig. 9.2 Variation of the coordinate systems of the *i*-th limb

axis along the axis of the horizontal axis of the universal joint; the direction of the x_i axis can be determined by the cross product of the unit vectors of the z_i and y_i axes. Finally, to describe the lower part limb, a local coordinate system $\{\boldsymbol{M}_i\}$ (not marked in the figure) is established, with the origin fixed at the rotation center of the spherical joint and the direction of the coordinate axes parallel to the local coordinate system $\{\boldsymbol{K}_i\}$. The local coordinate system $\{\boldsymbol{N}_i\}$ is the link to establish the relations between $\{\boldsymbol{B}\}$ and $\{\boldsymbol{K}_i\}$. In the local coordinate system $\{\boldsymbol{K}_i\}$, such parameters as the mass center and inertia matrix of each limb are the same, which is convenient for the measurement and assignment of parameters. However, the rotation matrix of $\{\boldsymbol{K}_i\}$ with respect to $\{\boldsymbol{N}_i\}$ needs to be deduced in the following process. As shown in Fig. 9.2, the relations between $\{\boldsymbol{N}_i\}$ and $\{\boldsymbol{K}_i\}$ can be described after two successive rotations. The local coordinate system $\{\boldsymbol{N}_i\}$ is rotated around the Z_i axis with angle α_i to obtain $O_i - U_i V_i W_i$, and then rotated around V_i with angle β_i to obtain $\{\boldsymbol{K}_i\}$.

9.2.1 Pose Analysis

The pose of the moving platform of the Stewart parallel mechanism is represented by the vector $\boldsymbol{X} = [x, y, z, \psi, \theta, \varphi]^{\mathrm{T}}$, where x, y, z is the position coordinates of the origin of the coordinate system $\{\boldsymbol{P}\}$ of the moving platform under the base coordinate system. ψ, θ, φ is the Euler angle of the posture of the moving platform, which indicate that the coordinate system of the moving platform is rotated by three angles with respect to the base coordinate system around the fixed X, Y and Z axes in turn. The rotation matrix of the moving platform with respect to the base platform can be written as:

$$^{\mathrm{B}}\boldsymbol{R}_{\mathrm{P}} = \begin{bmatrix} c\phi c\theta & c\phi s\theta s\psi - s\phi c\psi & c\phi s\theta c\psi + s\phi s\psi \\ s\phi c\theta & s\phi s\theta s\psi + c\phi c\psi & s\phi s\theta c\psi - c\phi s\psi \\ -s\theta & c\theta s\psi & c\theta c\psi \end{bmatrix}$$

where s stands for sine operation; c stands for cosine operation.

According to the previous definition, the X_i axis of $\{\boldsymbol{N}_i\}$ points to the geometric center of the base platform, and the Z_i axis is vertically downward. The equations of the rotation matrix of the local coordinate system $\{\boldsymbol{N}_i\}$ with respect to the base coordinate system $\{\boldsymbol{B}\}$ can be derived as:

$$\begin{aligned} &^{\mathrm{B}}\boldsymbol{R}_{\mathrm{N}i}(:, 3) = [0, 0, 1]^{\mathrm{T}} \\ &^{\mathrm{B}}\boldsymbol{R}_{\mathrm{N}i}(:, 1) = -\boldsymbol{b}_i / \|\boldsymbol{b}_i\| \\ &^{\mathrm{B}}\boldsymbol{R}_{\mathrm{N}i}(:, 2) = {}^{\mathrm{B}}\boldsymbol{R}_{\mathrm{N}i}(:, 3) \times {}^{\mathrm{B}}\boldsymbol{R}_{\mathrm{N}i}(:, 1) \end{aligned}$$

As shown in Fig. 9.2, the rotation matrix $^{\mathrm{N}}\boldsymbol{R}_{\mathrm{K}i}$ of the coordinate system $\{\boldsymbol{K}_i\}$ with respect to $\{\boldsymbol{N}_i\}$ can be described by the rotation angles α_i and β_i as:

$$^{\mathrm{N}}\boldsymbol{R}_{\mathrm{K}i} = \begin{bmatrix} c\beta_i c\alpha_i & -s\beta_i & c\beta_i s\alpha_i \\ s\beta_i c\alpha_i & c\beta_i & s\beta_i s\alpha_i \\ -s\alpha_i & 0 & c\alpha_i \end{bmatrix} \tag{9.6}$$

The rotation matrix of the local coordinate system $\{\boldsymbol{K}_i\}$ of the i-th limb with respect to the base coordinate system $\{\boldsymbol{B}\}$ can be expressed as the product of the above two rotation matrices:

$$^{\mathrm{B}}\boldsymbol{R}_{\mathrm{K}i} = {}^{\mathrm{B}}\boldsymbol{R}_{\mathrm{N}i}{}^{\mathrm{N}}\boldsymbol{R}_{\mathrm{K}i} \tag{9.7}$$

where $^{\mathrm{B}}\boldsymbol{R}_{\mathrm{N}i}$ can be derived directly from the arrangement of the universuak joints of the base platform. In order to obtain $^{\mathrm{B}}\boldsymbol{R}_{\mathrm{K}i}$, we need to deduce $^{\mathrm{N}}\boldsymbol{R}_{\mathrm{K}i}$.

Thus, let's deduce the rotation angles α_i and β_i. The rotation center of the spherical joint of the moving platform can be expressed in the base coordinate system $\{\boldsymbol{B}\}$ as:

$$^{\mathrm{B}}\boldsymbol{P}_i = {}^{\mathrm{B}}\boldsymbol{R}_{\mathrm{P}}\boldsymbol{P}_i + \boldsymbol{t}$$

where $\boldsymbol{P}_i$ is the position vector of the rotation center of the spherical joint in the coordinate system of the moving platform; $\boldsymbol{t}$ is the position vector of the moving platform in the base coordinate system, i.e., $\boldsymbol{t} = [x, y, z]^{\mathrm{T}}$.

Based on the derivation in Chap. 6, the vector of the i-th limb can be obtained:

$$^{\mathrm{B}}\mathbf{S}_i = {}^{\mathrm{B}}\boldsymbol{R}_{\mathrm{P}}\boldsymbol{P}_i + \boldsymbol{t} - \boldsymbol{b}_i$$

Further, the unit limb vector can be deduced as:

$$^{\mathrm{B}}\mathbf{s}_i = {}^{\mathrm{B}}\mathbf{S}_i / L$$

where $L = \|{}^{\mathrm{B}}\mathbf{S}_{\mathrm{i}}\|$ is the length of the limb or branch chain.

According to the property of the rotation matrix:

$$^{\mathrm{N}}\boldsymbol{R}_{\mathrm{K}i}(:,3) = \begin{bmatrix} c\beta_i s\alpha_i \\ s\beta_i s\alpha_i \\ c\alpha_i \end{bmatrix} = {}^{\mathrm{N}}\boldsymbol{s}_i = {}^{\mathrm{B}}\boldsymbol{R}_{\mathrm{N}i}^{\mathrm{T}} \cdot {}^{\mathrm{B}}\boldsymbol{s}_i$$

The rotation angles α_i and β_i can be deduced as:

$$\alpha_i = -cos^{-1}\left({}^{\mathrm{N}}\boldsymbol{s}_{iz}\right)$$
$$\beta_i = tan^{-1}\left({}^{\mathrm{N}}\boldsymbol{s}_{iy} / {}^{\mathrm{N}}\boldsymbol{s}_{ix}\right)$$

where ${}^{\mathrm{N}}\boldsymbol{s}_{ix}$, ${}^{\mathrm{N}}\boldsymbol{s}_{iy}$ and ${}^{\mathrm{N}}\boldsymbol{s}_{iz}$ are the 1st, 2nd and 3rd elements of ${}^{\mathrm{N}}\boldsymbol{s}_i$. The obtained rotation angles are substituted into Eq. (9.6) to deduce ${}^{\mathrm{N}}\boldsymbol{R}_{\mathrm{K}i}$. The rotation matrix ${}^{\mathrm{N}}\boldsymbol{R}_{\mathrm{K}i}$ can be deduced with Eq. (9.7), determining the relations between the base coordinate system $\{\boldsymbol{B}\}$ and the local coordinate system $\{\boldsymbol{K}_i\}$.

9.2.2 Velocity Analysis and Limb Jacobian Matrices

The limb Jacobian matrices indicate the mapping from the limb velocity (including the angular velocity, the telescope velocity and linear velocities of the mass centers of the upper and lower parts) to the end-effector velocity of the moving platform. According to the definition of the Jacobian matrix, its equation is:

$$\boldsymbol{J}_{\mathrm{link}} = \left[\frac{\partial \boldsymbol{v}_{\mathrm{link}}}{\partial \dot{x}}, \frac{\partial \boldsymbol{v}_{\mathrm{link}}}{\partial \dot{y}}, \frac{\partial \boldsymbol{v}_{\mathrm{link}}}{\partial \dot{z}}, \frac{\partial \boldsymbol{v}_{\mathrm{link}}}{\partial \dot{\alpha}}, \frac{\partial \boldsymbol{v}_{\mathrm{link}}}{\partial \dot{\beta}}, \frac{\partial \boldsymbol{v}_{\mathrm{link}}}{\partial \dot{\gamma}}\right] \tag{9.8}$$

where $\boldsymbol{v}_{\mathrm{link}}$ is the limb velocity and $\dot{\boldsymbol{X}} = [\dot{x}, \dot{y}, \dot{z}, \dot{\psi}, \dot{\theta}, \dot{\varphi}]^{\mathrm{T}}$ is the end-effector velocity. The limb Jacobian matrices only contain the configuration, pose and inertia parameters of the mechanism.

The virtual work equation which is established in the local coordinate system and expressed by the limb velocity can be expressed with the end-effector velocity through the limb Jacobian matrices. This operation facilitates the acquisition of the standard form of the parallel mechanism dynamics. The kinematic analysis is performed to derive the limb Jacobian matrices.

The velocity at the rotation center of the spherical joint of the moving platform can be expressed as:

$${}^{\mathrm{B}}\dot{\boldsymbol{P}}_i = \boldsymbol{\omega} \times \boldsymbol{q}_i + \dot{\boldsymbol{t}} = \boldsymbol{J}_{\mathrm{P}i}\dot{\boldsymbol{X}} \tag{9.9}$$

where $\boldsymbol{q}_i = {}^{B}\boldsymbol{R}_{\mathrm{P}}\boldsymbol{P}_i$, and it can be derived from Eq. (9.5) that:

$$\mathbf{J}_{\mathrm{P}i}=\begin{bmatrix}1\,0\,0 & 0 & q_{iz} & -q_{iy}\\ 0\,1\,0 & -q_{iz} & 0 & q_{ix}\\ 0\,0\,1 & q_{iy} & -q_{ix} & 0\end{bmatrix}$$

Converting the above equation to the local coordinate system $\{\boldsymbol{K}_i\}$, it can be rewritten as:

$${}^{\mathrm{K}}\dot{\boldsymbol{P}}_i={}^{\mathrm{K}}\mathbf{R}_{\mathrm{B}i}{}^{\mathrm{B}}\dot{\mathbf{P}}_i={}^{\mathrm{K}}\mathbf{R}_{\mathrm{B}i}\mathbf{J}_{\mathrm{P}i}\dot{\mathbf{X}} \tag{9.10}$$

The velocity of the rotation center of the spherical joint of the moving platform can also be expressed by the telescopic velocity $\dot{L}_i$ of the limb and the sway velocity ${}^{\mathrm{K}}\boldsymbol{W}_i$:

$${}^{\mathrm{K}}\dot{\boldsymbol{P}}_i={}^{\mathrm{K}}\dot{\boldsymbol{S}}_i=\dot{L}_i{}^{\mathrm{K}}\boldsymbol{s}_i+L_i{}^{\mathrm{K}}\boldsymbol{W}_i\times{}^{\mathrm{K}}\boldsymbol{s}_i$$

Taking the dot product ${}^{\mathrm{K}}\boldsymbol{s}_i$ on each side of the above equation simultaneously obtains:

$$\dot{L}_i={}^{\mathrm{K}}\boldsymbol{s}_i\cdot{}^{\mathrm{K}}\dot{\boldsymbol{P}}_i={}^{\mathrm{K}}\dot{P}_{iz} \tag{9.11}$$

Taking the dot product of ${}^{\mathrm{K}}\boldsymbol{s}_i$ on both sides of the above equation, and bringing in Eq. (9.4), the equation of the sway angular velocity of the limb can be obtained as:

$${}^{\mathrm{K}}\boldsymbol{W}={}^{\mathrm{K}}\boldsymbol{s}\times{}^{\mathrm{K}}\dot{\boldsymbol{S}}/L=\frac{1}{L}\begin{bmatrix}0&-1&0\\1&0&0\\0&0&0\end{bmatrix}{}^{\mathrm{K}}\dot{\boldsymbol{S}}=\boldsymbol{J}_{\mathrm{W}i}\dot{\boldsymbol{X}} \tag{9.12}$$

where

$$\boldsymbol{J}_{\mathrm{W}i}=\frac{1}{L}\begin{bmatrix}0&-1&0\\1&0&0\\0&0&0\end{bmatrix}{}^{\mathrm{K}}\boldsymbol{R}_{\mathrm{B}i}\cdot\boldsymbol{J}_{\mathrm{P}i}$$

The velocity of the mass center of the upper part limb can be deduced from the sway angular velocity and the coordinates of the mass center, which can be expressed as:

$${}^{\mathrm{K}}\mathbf{v}_{\mathrm{U}i}={}^{\mathrm{K}}\mathbf{W}_i\times{}^{\mathrm{K}}\mathbf{r}_{\mathrm{U}i}$$

After simplifying with Eq. (9.5), we can obtain:

$${}^{\mathrm{K}}\boldsymbol{v}_{\mathrm{U}i}=\begin{bmatrix}0&{}^{\mathrm{K}}r_{\mathrm{U}iz}&-{}^{\mathrm{K}}r_{\mathrm{U}iy}\\-{}^{\mathrm{K}}r_{\mathrm{U}iz}&0&{}^{\mathrm{K}}r_{\mathrm{U}iy}\\{}^{\mathrm{K}}r_{\mathrm{U}iy}&-{}^{\mathrm{K}}r_{\mathrm{U}ix}&0\end{bmatrix}{}^{\mathrm{K}}\boldsymbol{W}_i=\boldsymbol{J}_{\mathrm{U}i}\dot{\boldsymbol{X}} \tag{9.13}$$

where

$$\boldsymbol{J}_{\mathrm{U}i} = \frac{1}{L_i}\begin{bmatrix} r_{\mathrm{U}iz} & 0 & 0 \\ 0 & r_{\mathrm{U}iz} & 0 \\ -r_{\mathrm{U}ix} & -r_{\mathrm{U}iy} & 0 \end{bmatrix} {}^{\mathrm{K}}\boldsymbol{R}_{\mathrm{B}i}\,\boldsymbol{J}_{\mathrm{P}i}$$

The coordinate of the mass center of the upper part limb in the local coordinate system $\{\boldsymbol{K}_i\}$ is ${}^{\mathrm{K}}\boldsymbol{r}_{\mathrm{U}i} = [r_{\mathrm{U}ix}, r_{\mathrm{U}iy}, r_{\mathrm{U}iz}]^{\mathrm{T}}$.

The velocity of the mass center of the lower part limb can be expressed as the sum of the cross product of the sway angular velocity and the coordinates of the mass center with the telescopic velocity. Similarly, Eq. (9.5) can be written as:

$${}^{\mathrm{K}}\boldsymbol{v}_{\mathrm{L}i} = {}^{\mathrm{K}}\boldsymbol{W}_i \times \left(L_i\,{}^{\mathrm{K}}\boldsymbol{s}_i + \boldsymbol{r}_{\mathrm{L}i}\right) + \dot{L}_i\,{}^{\mathrm{K}}\boldsymbol{s}_i = \boldsymbol{J}_{\mathrm{L}i}\dot{\boldsymbol{X}} \tag{9.14}$$

where

$$\boldsymbol{J}_{\mathrm{L}i} = \frac{1}{L_i}\begin{bmatrix} L_{\mathrm{i}} + r_{\mathrm{L}iz} & 0 & 0 \\ 0 & L_{\mathrm{i}} + r_{\mathrm{L}iz} & 0 \\ -r_{\mathrm{L}ix} & -r_{\mathrm{L}iy} & 1 \end{bmatrix} {}^{\mathrm{K}}\boldsymbol{R}_{\mathrm{B}i}\,\boldsymbol{J}_{\mathrm{P}i}$$

The coordinate of the mass center of the upper part limb in the local coordinate system $\{\boldsymbol{M}_i\}$ is ${}^{\mathrm{M}}\boldsymbol{r}_{\mathrm{L}i} = [r_{\mathrm{L}ix}, r_{\mathrm{L}iy}, r_{\mathrm{L}iz}]^{\mathrm{T}}$.

The deduction of the inverse Jacobian matrix for the Stewart parallel mechanism is carried out below. The two velocity equations for the rotation center of the spherical joint of the moving platform are combined as:

$$\dot{\boldsymbol{t}} + \boldsymbol{\omega} \times \left({}^{\mathrm{B}}\boldsymbol{R}_{\mathrm{P}}\boldsymbol{p}_i\right) = L_i\,{}^{\mathrm{B}}\boldsymbol{W}_i \times {}^{\mathrm{B}}\boldsymbol{s}_i + \dot{L}_i\,{}^{\mathrm{B}}\boldsymbol{s}_i$$

Taking the dot product of ${}^{\mathrm{B}}\boldsymbol{s}_i$ on both sides simultaneously and the vector product formula ((a, b, c) = (b, c, a) = (c, a, b)) is adopted to organize it into a matrix form as:

$$\boldsymbol{J}\dot{\boldsymbol{X}} = \dot{\boldsymbol{L}} \tag{9.15}$$

where the inverse Jacobian matrix of the Stewart parallel mechanism is:

$$\boldsymbol{J} = \begin{bmatrix} {}^{\mathrm{B}}\boldsymbol{s}_1^{\mathrm{T}} \left({}^{\mathrm{B}}\boldsymbol{R}_{\mathrm{P}}\boldsymbol{P}_1 \times {}^{\mathrm{B}}\boldsymbol{s}_1\right)^{\mathrm{T}} \\ {}^{\mathrm{B}}\boldsymbol{s}_2^{\mathrm{T}} \left({}^{\mathrm{B}}\boldsymbol{R}_{\mathrm{P}}\boldsymbol{P}_2 \times {}^{\mathrm{B}}\boldsymbol{s}_2\right)^{\mathrm{T}} \\ {}^{\mathrm{B}}\boldsymbol{s}_3^{\mathrm{T}} \left({}^{\mathrm{B}}\boldsymbol{R}_{\mathrm{P}}\boldsymbol{P}_3 \times {}^{\mathrm{B}}\boldsymbol{s}_3\right)^{\mathrm{T}} \\ {}^{\mathrm{B}}\boldsymbol{s}_4^{\mathrm{T}} \left({}^{\mathrm{B}}\boldsymbol{R}_{\mathrm{P}}\boldsymbol{P}_4 \times {}^{\mathrm{B}}\boldsymbol{s}_4\right)^{\mathrm{T}} \\ {}^{\mathrm{B}}\boldsymbol{s}_5^{\mathrm{T}} \left({}^{\mathrm{B}}\boldsymbol{R}_{\mathrm{P}}\boldsymbol{P}_5 \times {}^{\mathrm{B}}\boldsymbol{s}_5\right)^{\mathrm{T}} \\ {}^{\mathrm{B}}\boldsymbol{s}_6^{\mathrm{T}} \left({}^{\mathrm{B}}\boldsymbol{R}_{\mathrm{P}}\boldsymbol{P}_6 \times {}^{\mathrm{B}}\boldsymbol{s}_6\right)^{\mathrm{T}} \end{bmatrix}$$

9.2.3 Acceleration Analysis

The deduction process of the acceleration parameters is similar to the process of velocity parameters for the Stewart parallel mechanism. The acceleration of the rotation center of the spherical joint of the moving platform can be expressed as:

$$^{\mathrm{B}}\ddot{\boldsymbol{P}}_i = \ddot{\boldsymbol{t}} + \boldsymbol{\varepsilon} \times \boldsymbol{q}_i + \boldsymbol{\omega} \times (\boldsymbol{\omega} \times \boldsymbol{q}_i) = \boldsymbol{J}_{\mathrm{P}i}\ddot{\boldsymbol{X}} + \boldsymbol{\omega} \times (\boldsymbol{\omega} \times \boldsymbol{q}_i) \tag{9.16}$$

The sway angular acceleration can be expressed as:

$$^{\mathrm{K}}\boldsymbol{A} = \left(^{\mathrm{K}}\boldsymbol{s} \times {}^{\mathrm{K}}\ddot{\boldsymbol{S}} - 2\dot{L}\,^{\mathrm{K}}\boldsymbol{W}\right)\Big/ L = \boldsymbol{J}_{\mathrm{A}}\ddot{\boldsymbol{X}} + {}^{\mathrm{K}}\boldsymbol{A}' \tag{9.17}$$

where

$$^{\mathrm{K}}A_{\mathrm{i}}' = \begin{bmatrix} 0 & -1 & 0 \\ 1 & 0 & 0 \\ 0 & 0 & 0 \end{bmatrix} {}^{\mathrm{K}}R_{\mathrm{B}i}\left[\omega \times (\omega \times \boldsymbol{q}_i)\right]$$

The telescopic acceleration of the limb is:

$$\ddot{L}_{\mathrm{i}} = L_i\,^{\mathrm{K}}\boldsymbol{W}_i \cdot {}^{\mathrm{K}}\boldsymbol{W}_i + {}^{\mathrm{K}}\boldsymbol{s}_i \cdot {}^{\mathrm{K}}\ddot{\boldsymbol{S}}_i = {}^{\mathrm{K}}\ddot{S}_{iz} + L_i\,^{\mathrm{K}}\boldsymbol{W}_i \cdot {}^{\mathrm{K}}\boldsymbol{W}_i \tag{9.18}$$

Further, the acceleration of the mass center of the upper and lower parts limb can be derived as:

$$^{\mathrm{K}}\boldsymbol{a}_{\mathrm{U}i} = \boldsymbol{J}_{\mathrm{U}i}\ddot{\boldsymbol{X}} + {}^{\mathrm{K}}\boldsymbol{a}'_{\mathrm{U}i} \tag{9.19}$$

$$^{\mathrm{K}}\boldsymbol{a}_{\mathrm{L}i} = \boldsymbol{J}_{\mathrm{L}i}\ddot{\boldsymbol{X}} + {}^{\mathrm{K}}\boldsymbol{a}'_{\mathrm{L}i} \tag{9.20}$$

where

$$\begin{aligned}
^{\mathrm{K}}\boldsymbol{a}'_{\mathrm{U}i} &= \frac{1}{L_i}\begin{bmatrix} r_{\mathrm{U}iz} & 0 & 0 \\ 0 & r_{\mathrm{U}iz} & 0 \\ -r_{\mathrm{U}ix} & -r_{\mathrm{U}iy} & 0 \end{bmatrix} {}^{\mathrm{K}}\boldsymbol{R}_{\mathrm{B}i}\left[\boldsymbol{\omega} \times (\boldsymbol{\omega} \times \boldsymbol{q})\right] + \frac{2\dot{L}_i\,^{\mathrm{K}}\boldsymbol{W}_i \times \boldsymbol{r}_{\mathrm{U}i}}{L_i} \\
&\quad + {}^{\mathrm{K}}\boldsymbol{W}_i \times \left(^{\mathrm{K}}\boldsymbol{W}_i \times \boldsymbol{r}_{\mathrm{U}i}\right) \\
^{\mathrm{K}}\boldsymbol{a}'_{\mathrm{L}i} &= \frac{2\dot{L}_i\,^{\mathrm{K}}\boldsymbol{W}_i \times \left(^{\mathrm{K}}\boldsymbol{s}_i + r_{\mathrm{L}}\right)}{L_i} + L_i\,^{\mathrm{K}}\boldsymbol{W}_i^{2\mathrm{K}}\boldsymbol{s}_i + {}^{\mathrm{K}}\boldsymbol{W}_i \times \left[^{\mathrm{K}}\boldsymbol{W}_i \times \left(^{\mathrm{K}}\boldsymbol{S}_i + \boldsymbol{r}_{\mathrm{L}i}\right)\right] \\
&\quad + 2\dot{L}_i\,^{\mathrm{K}}\boldsymbol{W}_i \times {}^{\mathrm{K}}\boldsymbol{s}_i
\end{aligned}$$

The limb Jacobian matrices of the upper and lower parts and the inverse Jacobian matrix of the Stewart parallel mechanism have been deduced. The motion parameters of each component of the mechanism can be deduced with the motion of the moving platform.

9.2.4 Force System Analysis

Before establishing the dynamic equations with the virtual work principle, the force state of each component needs to be analyzed. With good lubrication, the friction at the joint can be neglected, and is not considered in the force analysis below. The inertia forces of the moving platform are deduced in the base coordinate system $\{\boldsymbol{B}\}$, while inertia forces of the limbs are deduced in their local coordinate system $\{\boldsymbol{K}_i\}$ respectively.

The forces and moments of the moving platform at the geometric center are expressed as:

$$\boldsymbol{F}_{\mathrm{P}} = \begin{bmatrix} \boldsymbol{f}_{\mathrm{E}} + m_{\mathrm{P}}\boldsymbol{g} - m_{\mathrm{P}}\ddot{\boldsymbol{t}} \\ \boldsymbol{n}_{\mathrm{E}} + m_{\mathrm{P}}\boldsymbol{e} \times \boldsymbol{g} - {}^{\mathrm{B}}\boldsymbol{I}_{\mathrm{P}}\boldsymbol{\varepsilon} - \boldsymbol{\omega} \times \left({}^{\mathrm{B}}\boldsymbol{I}_{\mathrm{P}}\boldsymbol{\omega}\right) \end{bmatrix} \tag{9.21}$$

where $\boldsymbol{f}_E$ and $\boldsymbol{n}_E$ are the external force and moment acting on the moving platform; $\boldsymbol{g} = [0, 0, 9.8]^{\mathrm{T}}$ m/s^2 is the gravity acceleration; m_{P} is the mass of the moving platform; $\mathbf{e} = [e_x, e_y, e_z]^{\mathrm{T}}$ is the coordinate of the mass center in the coordinate system of the moving platform; ${}^{B}\boldsymbol{I}_{\mathrm{P}}$ is the inertia matrix of the moving platform with respect to the geometric center in the global coordinate system $\{\boldsymbol{B}\}$, which can be deduced based on the parallel axis theorem:

$${}^{\mathrm{B}}\boldsymbol{I}_{\mathrm{P}} = {}^{\mathrm{B}}\boldsymbol{R}_{\mathrm{P}}\left\{\boldsymbol{I}_{\mathrm{PC}} + m_{\mathrm{P}} \cdot \mathrm{diag}\left(\left[\sqrt{e_z^2 + e_y^2}, \sqrt{e_z^2 + e_x^2}, \sqrt{e_x^2 + e_y^2}\right]\right)\right\}{}^{\mathrm{B}}\boldsymbol{R}_{\mathrm{P}}^{\mathrm{T}} \tag{9.22}$$

where $\boldsymbol{I}_{\mathrm{PC}}$ is the inertia matrix of the moving platform with respect to the mass center in the coordinate system $\{\boldsymbol{P}\}$ of the moving platform.

According to the standard form of the dynamic equation (9.1), $\boldsymbol{F}_P$ can be decomposed into four parts: the inertial force, the gravity, the external force as well as the Coriolis and the centripetal forces, which is expressed as:

$$\begin{aligned} \boldsymbol{F}_{\mathrm{P}} &= \begin{bmatrix} -m_{\mathrm{P}}\ddot{\boldsymbol{t}} \\ -{}^{\mathrm{B}}\boldsymbol{I}_{\mathrm{P}}\boldsymbol{\varepsilon} \end{bmatrix} + \begin{bmatrix} m_{\mathrm{P}}\boldsymbol{g} \\ m_{\mathrm{P}}\boldsymbol{e} \times \boldsymbol{g} \end{bmatrix} + \begin{bmatrix} \boldsymbol{f}_{\mathrm{E}} \\ \boldsymbol{n}_{\mathrm{E}} \end{bmatrix} + \begin{bmatrix} 0 \\ -\boldsymbol{\omega} \times \left({}^{\mathrm{B}}\boldsymbol{I}_{\mathrm{P}}\boldsymbol{\omega}\right) \end{bmatrix} \\ &= \boldsymbol{F}_{\mathrm{PA}} + \boldsymbol{F}_{\mathrm{PG}} + \boldsymbol{F}_{\mathrm{PE}} + \boldsymbol{F}'_{\mathrm{P}} \end{aligned}$$

The force and moment exerted to the upper part limb at B_i in the local coordinate system $\{\boldsymbol{K}_i\}$ are:

$$\boldsymbol{F}_{\mathrm{U}i} = \begin{bmatrix} m_{\mathrm{U}}{}^{\mathrm{K}}\boldsymbol{R}_{\mathrm{B}i}\boldsymbol{g} - m_{\mathrm{U}}{}^{\mathrm{K}}\boldsymbol{a}_{\mathrm{U}i} \\ m_{\mathrm{U}}{}^{\mathrm{K}}\boldsymbol{r}_{\mathrm{U}i} \times \left({}^{\mathrm{K}}\boldsymbol{R}_{\mathrm{B}i}\boldsymbol{g}\right) - \boldsymbol{I}_{\mathrm{U}}{}^{\mathrm{K}}\boldsymbol{A}_i - {}^{\mathrm{K}}\boldsymbol{W}_i \times \left(\boldsymbol{I}_{\mathrm{U}}{}^{\mathrm{K}}\boldsymbol{W}_i\right) \end{bmatrix} \tag{9.23}$$

where m_U is the mass of the upper part limb; $\boldsymbol{I}_U$ is the rotational inertia matrix of the upper part limb with respect to B_i in $\{\boldsymbol{K}_i\}$.

According to the definition of Eq. (9.1), $\boldsymbol{F}_{Ui}$ can be decomposed into three parts: the inertial force, the gravity, and the Coriolis and centripetal forces. And the equation is shown as:

$$\begin{aligned}\boldsymbol{F}_{\mathrm{U}i} &= \begin{bmatrix} -m_{\mathrm{U}}\boldsymbol{J}_{\mathrm{U}i}\ddot{\boldsymbol{X}} \\ -\boldsymbol{I}_{\mathrm{U}}\left(\boldsymbol{J}_{\mathrm{W}i}\ddot{\boldsymbol{X}}\right) \end{bmatrix} + \begin{bmatrix} m_{\mathrm{U}}{}^{\mathrm{K}}\boldsymbol{R}_{\mathrm{B}i}\boldsymbol{g} \\ m_{\mathrm{U}}{}^{\mathrm{K}}\boldsymbol{r}_{\mathrm{U}i} \times \left({}^{\mathrm{K}}\boldsymbol{R}_{\mathrm{B}i}\boldsymbol{g}\right) \end{bmatrix} + \begin{bmatrix} -m_{\mathrm{U}}{}^{\mathrm{K}}\boldsymbol{a'}_{\mathrm{U}i} \\ -\boldsymbol{I}_{\mathrm{U}}{}^{\mathrm{K}}\boldsymbol{A'}_{\mathrm{i}} - {}^{\mathrm{K}}\boldsymbol{W}_{\mathrm{i}} \times \left(\boldsymbol{I}_{\mathrm{U}}{}^{\mathrm{K}}\boldsymbol{W}_{\mathrm{i}}\right) \end{bmatrix} \\ &= \boldsymbol{F}_{\mathrm{UA}i} + \boldsymbol{F}_{\mathrm{UG}i} + \boldsymbol{F'}_{\mathrm{U}i}\end{aligned}$$

where $\boldsymbol{I}_{\mathrm{U}}$ can be obtained based on the parallel axis theorem:

$$\boldsymbol{I}_{\mathrm{U}} = \boldsymbol{I}_{\mathrm{UC}} + m_{\mathrm{U}} \cdot \mathrm{diag}\left(\left[\sqrt{r_{\mathrm{U}iz}^2 + r_{\mathrm{U}iy}^2}, \sqrt{r_{\mathrm{U}iz}^2 + r_{\mathrm{U}ix}^2}, \sqrt{r_{\mathrm{U}ix}^2 + r_{\mathrm{U}iy}^2}\right]\right) \tag{9.24}$$

where $\boldsymbol{I}_{\mathrm{UC}}$ is the inertia matrix of the upper part limb with respect to its mass center in.

In the local coordinate system $\{\boldsymbol{K}_i\}$ of the i-th limb, the force and the moment of the lower part limb at B_i can be expressed as:

$$\boldsymbol{F}_{\mathrm{L}i} = \begin{bmatrix} m_{\mathrm{L}}{}^{\mathrm{K}}\boldsymbol{R}_{\mathrm{B}i}\boldsymbol{g} - m_{\mathrm{L}}{}^{\mathrm{K}}\boldsymbol{a}_{\mathrm{L}i} \\ {}^{\mathrm{K}}\boldsymbol{r}_{\mathrm{L}i} \times \left({}^{\mathrm{K}}\boldsymbol{R}_{\mathrm{B}i}\boldsymbol{g}\right) - \boldsymbol{I}_{\mathrm{L}}{}^{\mathrm{K}}\boldsymbol{A}_{\mathrm{i}} - {}^{\mathrm{K}}\boldsymbol{W}_{\mathrm{i}} \times \left({}^{\mathrm{K}}\boldsymbol{I}_{\mathrm{L}}{}^{\mathrm{K}}\boldsymbol{W}_i\right) \end{bmatrix} \tag{9.25}$$

where m_{L} is the mass of the lower part limb; $\boldsymbol{I}_{\mathrm{L}}$ is the rotational inertia matrix of the lower part limb with respect to B_i in the local coordinate system $\{\boldsymbol{K}_i\}$, which can be deduced by the following equation:

$$\boldsymbol{I}_{\mathrm{L}} = \boldsymbol{I}_{\mathrm{LC}} + m_{\mathrm{L}} \cdot \mathrm{diag}\left(\left[\sqrt{(L + r_{\mathrm{L}iz})^2 + r_{\mathrm{U}iy}^2}, \sqrt{(L + r_{\mathrm{L}iz})^2 + r_{\mathrm{L}ix}^2}, \sqrt{r_{\mathrm{L}ix}^2 + r_{\mathrm{L}iy}^2}\right]\right) \tag{9.26}$$

where

$\boldsymbol{I}_{\mathrm{LC}}$ is the inertia matrix of the lower part limb with respect to its mass center in the local coordinate system $\{\boldsymbol{M}_i\}$.

According to the definition of Eq. (9.1), $\boldsymbol{F}_{\mathrm{L}i}$ is decomposed into three parts: the inertial force, the gravity, and the Coriolis and centripetal forces, and written as:

$$\begin{aligned}\boldsymbol{F}_{\mathrm{L}i} &= \begin{bmatrix} -m_{\mathrm{L}}\boldsymbol{J}_{\mathrm{L}i}\ddot{\boldsymbol{X}} \\ -\boldsymbol{I}_{\mathrm{L}}\boldsymbol{J}_{\mathrm{A}i}\ddot{\boldsymbol{X}} \end{bmatrix} + \begin{bmatrix} m_{\mathrm{L}}{}^{\mathrm{K}}\boldsymbol{R}_{\mathrm{B}i}\boldsymbol{g} \\ {}^{\mathrm{K}}\boldsymbol{r}_{\mathrm{L}i} \times \left({}^{\mathrm{K}}\boldsymbol{R}_{\mathrm{B}i}\boldsymbol{g}\right) \end{bmatrix} + \begin{bmatrix} -m_{\mathrm{L}}{}^{\mathrm{K}}\boldsymbol{a'}_{\mathrm{L}i} \\ -\boldsymbol{I}_{\mathrm{L}}{}^{\mathrm{K}}\boldsymbol{A'}_i - {}^{\mathrm{K}}\boldsymbol{W}_i \times \left(\boldsymbol{I}_{\mathrm{L}}{}^{\mathrm{K}}\boldsymbol{W}_i\right) \end{bmatrix} \\ &= \boldsymbol{F}_{\mathrm{LAi}} + \boldsymbol{F}_{\mathrm{LGi}} + \boldsymbol{F'}_{\mathrm{Li}}\end{aligned}$$

9.2.5 Joint Space Inertia Matrix

In this section, the joint space inertia matrix of the Stewart parallel mechanism is deduced with the virtual work principle. It is assumed that the terminal of the Stewart mechanism produces a virtual displacement $\delta \boldsymbol{X}$ that is within the workspace and does not disturb its motion law. According to the virtual work principle, the dynamic equation of the Stewart mechanism is established as:

$$\sum_{i=1}^{6}\left(\delta \boldsymbol{t}_{\mathrm{L}i}^{\mathrm{T}} \boldsymbol{F}_{\mathrm{L}i} + \delta \boldsymbol{t}_{\mathrm{U}i}^{\mathrm{T}} \boldsymbol{F}_{\mathrm{U}i}\right) + \delta \boldsymbol{L}^{\mathrm{T}} \boldsymbol{F} + \delta \boldsymbol{X}^{\mathrm{T}} \boldsymbol{F}_{\mathrm{P}} = 0 \tag{9.27}$$

where $\boldsymbol{F}$ is the driving forces of the limbs. The following equations can be obtained according to the derivation in Sect. 9.2.2:

$$\begin{aligned}
\delta \boldsymbol{t}_{\mathrm{U}i} &= \left[{}^{\mathrm{K}}\boldsymbol{v}_{\mathrm{U}i};\, {}^{\mathrm{K}}\boldsymbol{W}\right] = [\boldsymbol{J}_{\mathrm{U}i};\, \boldsymbol{J}_{\mathrm{W}i}]\delta \boldsymbol{X} \\
\delta \boldsymbol{t}_{\mathrm{L}i} &= \left[{}^{\mathrm{K}}\boldsymbol{v}_{\mathrm{L}i};\, {}^{\mathrm{K}}\boldsymbol{W}\right] = [\boldsymbol{J}_{\mathrm{L}i};\, \boldsymbol{J}_{\mathrm{W}i}]\delta \boldsymbol{X} \\
\delta \boldsymbol{L} &= \boldsymbol{J} \cdot \delta \boldsymbol{X}
\end{aligned}$$

Substitute Eqs. (9.21), (9.23), and (9.25) into Eq. (9.27) and convert the dynamic equation into a standard form as:

$$\begin{aligned}
\boldsymbol{F} = &-\boldsymbol{J}^{-\mathrm{T}}\left[\sum_{i=1}^{6}\left(\boldsymbol{J}'^{\mathrm{T}}_{\mathrm{L}i} \boldsymbol{F}_{\mathrm{LA}i} + \boldsymbol{J}'^{\mathrm{T}}_{\mathrm{U}i} \boldsymbol{F}_{\mathrm{UA}i}\right) + \boldsymbol{F}_{\mathrm{PA}}\right] \\
&- \boldsymbol{J}^{-\mathrm{T}}\left[\sum_{i=1}^{6}\left(\boldsymbol{J}'^{\mathrm{T}}_{\mathrm{L}i} \boldsymbol{F}_{\mathrm{LG}i} + \boldsymbol{J}'^{\mathrm{T}}_{\mathrm{U}i} \boldsymbol{F}_{\mathrm{UG}i}\right) + \boldsymbol{F}_{\mathrm{PG}}\right] \\
&- \boldsymbol{J}^{-\mathrm{T}}\left[\sum_{i=1}^{6}\left(\boldsymbol{J}'^{\mathrm{T}}_{\mathrm{L}i} \boldsymbol{F}'_{\mathrm{L}i} + \boldsymbol{J}'^{\mathrm{T}}_{\mathrm{U}i} \boldsymbol{F}'_{\mathrm{U}i}\right) + \boldsymbol{F}'_{\mathrm{P}}\right] - \boldsymbol{J}^{-\mathrm{T}} \boldsymbol{F}_{\mathrm{PE}}
\end{aligned} \tag{9.28}$$

where

$$\boldsymbol{J}'_{\mathrm{U}i} = [\boldsymbol{J}_{\mathrm{U}i};\, \boldsymbol{J}_{\mathrm{W}i}],\quad \boldsymbol{J}'_{\mathrm{L}i} = [\boldsymbol{J}_{\mathrm{L}i};\, \boldsymbol{J}_{\mathrm{W}i}]$$

This chapter mainly studies the joint space inertia matrix of the Stewart mechanism, and we focus on the first term on the right side of Eq. (9.28), which is the acceleration-related term. After expanding, we obtains:

$$\boldsymbol{F}_{\mathrm{A}} = -\boldsymbol{J}^{-\mathrm{T}}\left[\sum_{i=1}^{6}\left(\boldsymbol{J}_{\mathrm{Li}}^{\mathrm{T}} \boldsymbol{F}_{\mathrm{LAi}} + \boldsymbol{J}_{\mathrm{Ui}}^{\mathrm{T}} \boldsymbol{F}_{\mathrm{UAi}}\right) + \boldsymbol{F}_{\mathrm{PA}}\right]$$

$$= \boldsymbol{J}^{-\mathrm{T}}\left[\sum_{i=1}^{6}\left(\boldsymbol{J}'^{\mathrm{T}}_{\mathrm{Li}}\begin{bmatrix} m_{\mathrm{L}}\boldsymbol{J}_{\mathrm{Li}} \\ \boldsymbol{I}_{\mathrm{L}}\boldsymbol{J}_{\mathrm{Wi}} \end{bmatrix}\ddot{\boldsymbol{X}} + \boldsymbol{J}'^{\mathrm{T}}_{\mathrm{Ui}}\begin{bmatrix} m_{\mathrm{U}}\boldsymbol{J}_{\mathrm{Ui}} \\ \boldsymbol{I}_{\mathrm{U}}\boldsymbol{J}_{\mathrm{Wi}} \end{bmatrix}\ddot{\boldsymbol{X}}\right) + \begin{bmatrix} m_{\mathrm{P}}\boldsymbol{E} & 0_{3\times3} \\ 0_{3\times3} & {}^{\mathrm{B}}\boldsymbol{I}_{\mathrm{P}} \end{bmatrix}\ddot{\boldsymbol{X}}\right]$$

The joint space inertia matrix reflects the relations between the motor torque of the limb and the limb acceleration and can deduced as:

$$\boldsymbol{I} = (\eta\boldsymbol{J})^{-\mathrm{T}}\left[\sum_{i=1}^{6}\left(\boldsymbol{J}'^{\mathrm{T}}_{\mathrm{Li}}\begin{bmatrix} m_{\mathrm{L}}\boldsymbol{E} & 0_{3\times3} \\ 0_{3\times3} & \boldsymbol{I}_{\mathrm{L}} \end{bmatrix}\boldsymbol{J}'_{\mathrm{Li}} + \boldsymbol{J}'^{\mathrm{T}}_{\mathrm{Ui}}\begin{bmatrix} m_{\mathrm{U}}\boldsymbol{E} & 0_{3\times3} \\ 0_{3\times3} & \boldsymbol{I}_{\mathrm{U}} \end{bmatrix}\boldsymbol{J}'_{\mathrm{Ui}}\right) + \begin{bmatrix} m_{\mathrm{P}}\boldsymbol{E} & 0_{3\times3} \\ 0_{3\times3} & {}^{\mathrm{B}}\boldsymbol{I}_{\mathrm{P}} \end{bmatrix}\right](\eta\boldsymbol{J})^{-1}$$

where $\eta = 2n\pi/\xi$, and ξ is the screw lead, n is the reduction ratio of the reducer, and $\boldsymbol{E}$ is the 3×3 unit diagonal matrix.

Now, the derivation of the joint space inertia matrix of the Stewart mechanism is illustrated. Next, its characteristics will be studied, and the inertia index will be proposed.

9.3 Equivalent Inertia of the Parallel Robot

The study object of the inertia matching experiment is shown in Fig. 9.3, which is a 1:10 scale model of the fine-tuning Stewart platform. To ensure the unification of the theoretical simulations with the physical experiments, the mechanism's inertia and the dimensional parameters obtained in the dynamic verification experiment of Chap. 6 are used:

$$\boldsymbol{e} = [0, 0, 0.0089]^{\mathrm{T}}\ \mathrm{m},$$

$$\boldsymbol{I}_{\mathrm{PC}} = \begin{bmatrix} 0.0865 & 0 & 0 \\ 0 & 0.0865 & 0 \\ 0 & 0 & 0.1494 \end{bmatrix}\ \mathrm{kg\,m^2}$$

Coordinates of the center of mass of the i-th upper part limb in $\{\boldsymbol{K}_i\}$ and the inertia matrix about its center of mass are:

$${}^{\mathrm{K}}\boldsymbol{r}_{\mathrm{U}i} = [0.0051, 0.0002, 0.2538]^{\mathrm{T}}\ \mathrm{m},$$

$$\boldsymbol{I}_{\mathrm{UC}} = \begin{bmatrix} 0.2167 & 0 & 0.0057 \\ 0 & 0.2175 & -0.0001 \\ 0.0057 & -0.0001 & 0.0075 \end{bmatrix}\ \mathrm{kg\,m^2}$$

Coordinates of the center of mass of the i-th lower part limb in $\{\boldsymbol{M}_i\}$ and the inertia matrix about its center of mass are:

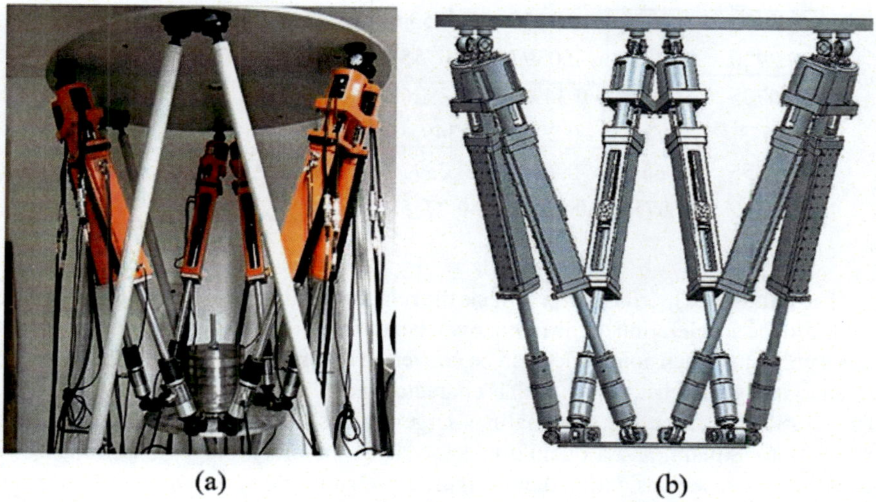

Fig. 9.3 Experimental model of the Stewart mechanism. **a** Physical model, **b** 3D virtual model

$$ {}^{\mathrm{M}}\boldsymbol{r}_{\mathrm{L}i} = [-0.0007, 0.0001, -0.2752]^{\mathrm{T}}\ \mathrm{m}, $$

$$ \boldsymbol{I}_{\mathrm{LC}} = \begin{bmatrix} 0.1444 & 0 & 0.0001 \\ 0 & 0.1446 & 0.0001 \\ 0.0001 & 0.0001 & 0.0011 \end{bmatrix} \mathrm{kg\ m^2} $$

Other parameters of the experimental object are shown in Table 9.1.

With above parameters, the joint space inertia matrix of the 1:10 scale model of the fine-tuning Stewart platform can be deduced for arbitrary pose. When the moving platform is at a given pose ($\boldsymbol{X} = [0\ \mathrm{m}, 0\ \mathrm{m}, 0.967\ \mathrm{m}, 0°, 0°, 0°]^{\mathrm{T}}$) in the workspace, the deduced joint space inertia matrix is:

Table 9.1 Parameters of the 1:10 scale model of the fine-tuning platform

Parameters	Values
Radius of the joint distribution circle of the moving platform r_a (mm)	225
Radius of the joint distribution circle of the base platform r_b (mm)	450
Central angle of adjacent joints of the moving platform β (°)	25
Central angle of adjacent joints of the base platform α (°)	10
Mass of the moving platform m_{P} (kg)	6.513
Mass of the upper part limb m_{U} (kg)	7.470
Mass of the lower part limb m_{L} (kg)	3.355

$$I = \begin{bmatrix} 0.3492 & -0.2926 & -0.0755 & 0.1653 & -0.0755 & -0.0397 \\ -0.2926 & 0.3492 & -0.0397 & -0.0755 & 0.1653 & -0.0755 \\ -0.0755 & -0.0397 & 0.3492 & -0.2926 & -0.0755 & 0.1653 \\ 0.1653 & -0.0755 & -0.2926 & 0.3492 & -0.0397 & -0.0755 \\ -0.0755 & 0.1653 & -0.0755 & -0.0397 & 0.3492 & -0.2926 \\ -0.0397 & -0.0755 & 0.1653 & -0.0755 & -0.2926 & 0.3492 \end{bmatrix} \times 10^{-4}\left(\text{kg}\,.\,\text{m}^2\right)$$

The joint space inertia matrix reflects the relations between the motor torque of the limb and the acceleration of the motor, which has three significant characteristics: ① it has a uniform dimension, which can be converted into standard inertia units (kg m^2); ② it is a real symmetric matrix; ③ it is characterized by the block diagonal dominance (the 2×2 matrix on the diagonal has larger values). The first two characteristics are easy to explain, and a qualitative analysis of the block diagonal dominance is carried out below, considering the configuration characteristics of the Stewart parallel mechanism.

As shown in Fig. 9.1b, limbs 1 and 2, 3 and 4, as well as 5 and 6 are three groups of adjacent limbs on the moving platform of the Stewart mechanism. Each group of adjacent limbs can be equivalently simplified to a planar 2-R$\underline{\text{P}}$R parallel mechanism (R represents the rotational joint, P represents the prismatic joint, and the underline indicates the active joint) for coupling analysis. Since the structure of the Stewart mechanism is circumferentially symmetrical, limbs 1 and 2 are selected as examples for analysis, and the structure diagram of the 2-R$\underline{\text{P}}$R mechanism composed of two limbs is shown in Fig. 9.4. It is assumed that an acceleration of $[1, 0]^{\text{T}}$ is generated in the joint space from a stationary and initial left–right symmetric position (the extension direction of the limb is defined as the positive direction), the limb 1 motor of the 2-R$\underline{\text{P}}$R generates a positive torque and a driving force along the ς_1 direction. When the angle between the two limbs is acute, limb 2 will generate a constraint force along the ς_2 direction to keep its stationary state, and finally the moving platform will generate an acceleration along the ς_3 direction (perpendicular to the direction of limb 2). The limb 2 motor generates a reverse constraint torque, i.e., the direction of the constraint force is along the shortening direction of the limb. At this point, the constraint effects of limb 2 on the motion of limb 1 are equivalent to a circular-arc surface. The constraint (or coupling) effects are manifested in the numerical distribution of the joint space inertia matrix of the 2-R$\underline{\text{P}}$R parallel mechanism, which results in negative nonzero elements outside the principal diagonal:

$$\boldsymbol{I}_{\text{U}} = \begin{bmatrix} A & -B \\ -B & A \end{bmatrix}$$

where both A and B are positive real numbers. The diagonal element A reflects the inertia property of the limb in this configuration, and the non-diagonal element B reflects the coupling interaction between limbs.

In addition, the two limbs mounted far apart on the moving platform can be equivalently simplified to an RPR-RPR mechanism (e.g., limbs 1 and 3), as shown in Fig. 9.5. The mechanism is under-constrained, and the distance between the connecting points of the moving platform plays a coordination role in reducing the constraint relations between the two limbs, and the acceleration motion of limb 1 is less constrained by limb 3. The effects reflected in the inertia matrix are: compared to the 2-RPR mechanism, the ratio of the absolute value of the non-diagonal elements to the diagonal elements of the joint space inertia matrix of the RPR-RPR mechanism decreases, reflecting the reduction of the coupling relations.

Through the above analysis, it is found that: ① the diagonal elements of the joint space inertia matrix of the parallel mechanism reflect the inertia properties of the corresponding limbs, and the non-diagonal elements reflect the coupling relations

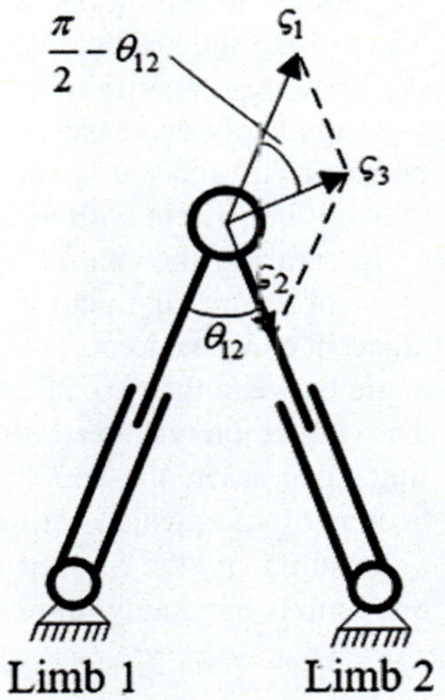

Fig. 9.4 Structure diagram of the equivalent 2-RPR mechanism

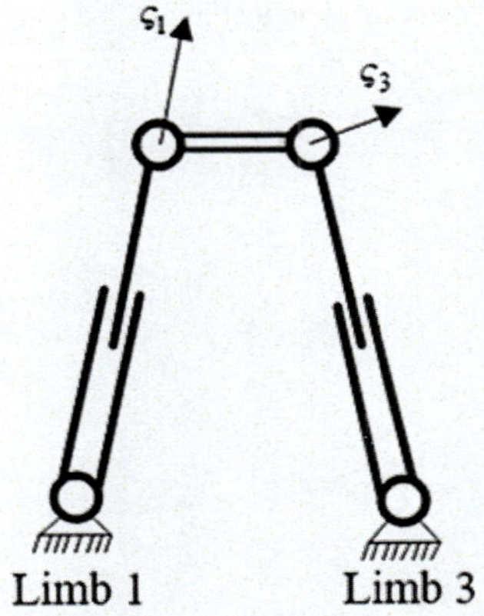

Fig. 9.5 Structure diagram of the equivalent RPR-RPR mechanism

between the limbs; ② adjacent limbs on the moving platform of the Stewart mechanism have strong coupling relations, which determines that its joint space inertia matrix is characterized by the block diagonal dominance.

The following numerical simulation verifies the relations between the coupling of the adjacent limbs on the moving platform and the configuration of the Stewart mechanism. The previous qualitative analysis finds that: when the angle between the two adjacent limbs decreases, the coupling increases and the value of B/A of the small matrix on the diagonal increases; when the distance between the spherical joint of the two adjacent limbs increases, the mutual motion constraint and the value of B/A decrease.

The central angle corresponding to the adjacent universal joints of the base platform is set as α, and the central angle of adjacent spherical joints of the moving platform is set as β. Let the moving platform be in the center of the workspace and the attitude is kept horizontal, $\alpha = 10°$, and β gradually increases from 5° to 50°. The ratio of the absolute value (B/A) of the small 2×2 matrix on the diagonal of the joint space inertia matrix can be obtained. As shown in Fig. 9.6, the value of B/A gradually decreases with the increase of the central angle β, i.e., the distance between the adjacent spherical joints of the moving platform gradually increases, which is consistent with the qualitative analysis results.

In contrast, the central angle $\beta = 10°$ corresponding to the adjacent spherical joints of the moving platform is fixed, and the central angle α corresponding to the adjacent universal joints of the base platform gradually increases from 5° to 100°. The angle between the two adjacent limbs on the moving platform decreases gradually. The simulation results are shown in Fig. 9.7. As the angle between the two adjacent limbs decreases, the coupling is enhanced, and the value of B/A increases as α becomes larger, which verifies the correctness of the previous conclusion.

In summary, the Stewart mechanism has six limbs, and the motion tendency of one limb is inevitably limited by other five limbs. The magnitude of the interaction

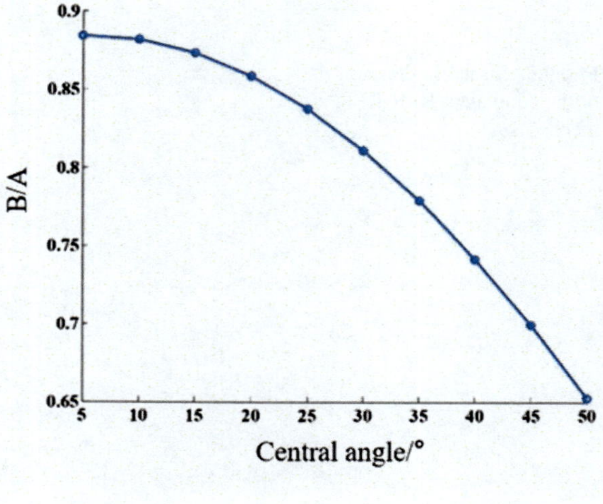

Fig. 9.6 B/A change with the central angle β of adjacent spherical joints

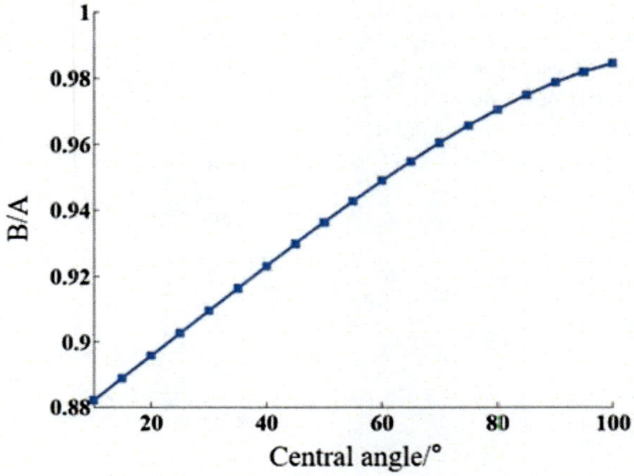

Fig. 9.7 Variation of B/A with the central angle α of adjacent universal joints

between the limbs depends on the relative positions of the action points on the moving platform and the angular relations of their axial vectors. At the same time, the principal diagonal elements of the joint space inertia matrix reflect the inertia property of each limb, while the non-diagonal elements reflect the mutual coupling relations between limbs.

The distribution of the principal diagonal elements of the joint space inertia matrix within the maximum horizontal section of the workspace is presented below, as shown in Fig. 9.8. The distribution of the principal diagonal elements is approximated as a set of concentric arc curves. Combined with the structure of the Stewart mechanism (see Fig. 9.1), it can be found that the contour lines of diagonal elements of the inertia matrix approximate the length gradient curves of the limb. Taking I_{11} as an example, I_{11} becomes larger and smaller with the extension and shortening of limb 1. The principal diagonal element I_{ii} of the joint space inertia matrix becomes larger and smaller with the extension and shortening of the *i*-th limb. When the limb extends, the overall mass center of the limb and the moving platform move away from the rotation center of the universal joint of the base platform, and the rotational inertia of the limb around the universal joint increases. It also proves that the principal diagonal elements of the joint space inertia matrix reflect the inertia property of the corresponding limb.

Based on the above analysis of the joint space inertia matrix, the mechanism's inertia index—the equivalent inertia $\boldsymbol{I}_{\mathrm{E}}$ is established.

Definition 9.1 Based on the joint-space inertia matrix $\boldsymbol{I} = \left(a_{ij}\right)$ of the parallel mechanism, the mean value of its eigenvalues is defined as the equivalent inertia $\boldsymbol{I}_{\mathrm{E}}$.

Since the trace of the joint space inertia matrix is equal to the sum of its eigenvalues, the equivalent inertia equals the mean value of the diagonal elements. The equivalent inertia of the parallel mechanism is a quantitative description of the inertia property in the given pose, which can be expressed as:

Fig. 9.8 Distribution of the principal diagonal elements of the joint space inertia matrix of the Stewart mechanism (kg m^2). **a** I_{11}; **b** I_{22}; **c** I_{33}; **d** I_{44}; **e** I_{55}; **f** I_{66}

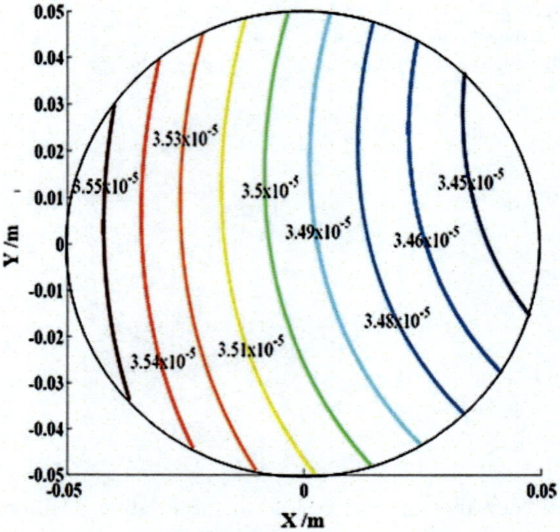

(a)

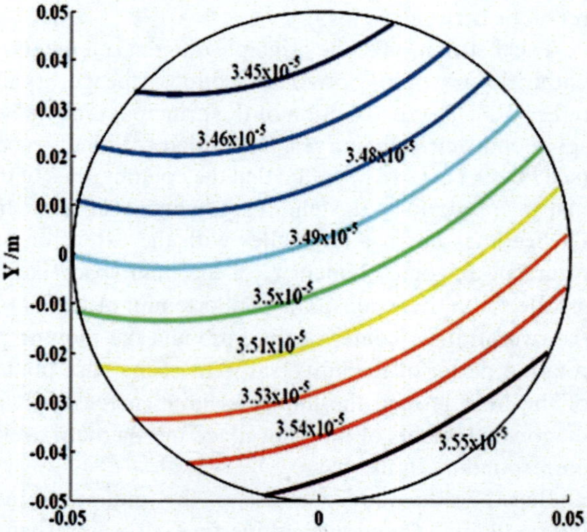

(b)

Fig. 9.8 (continued)

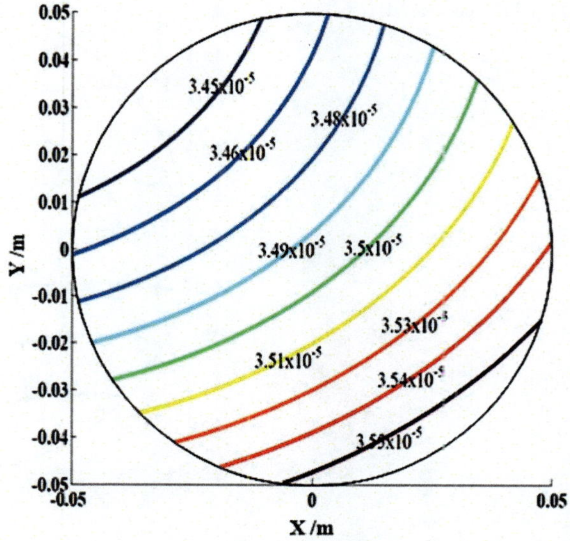

(c)

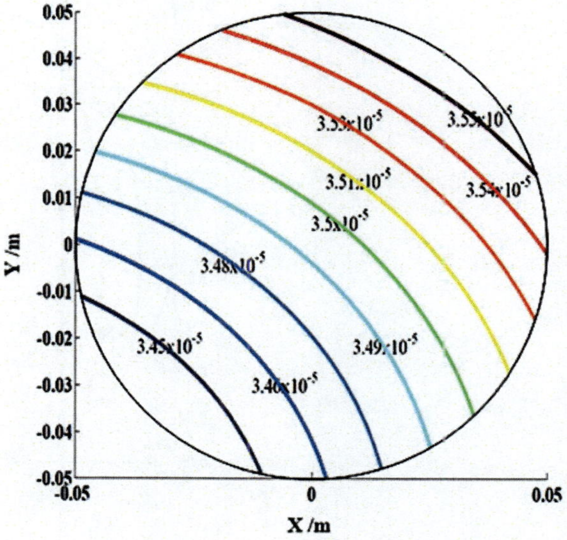

(d)

Fig. 9.8 (continued)

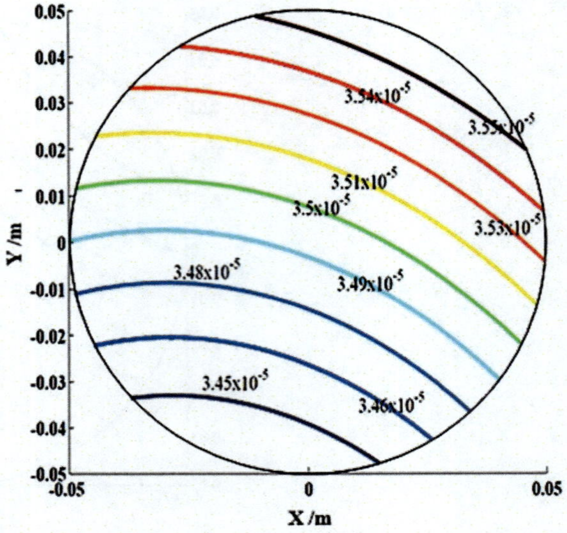

(e)

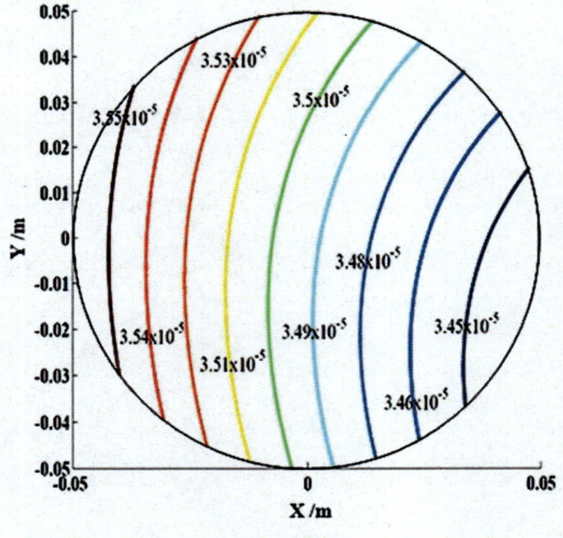

(f)

$$\boldsymbol{I}_{\mathrm{E}} = \frac{1}{6}\sum_{i=1}^{6}(\lambda_i) = \frac{1}{6}\sum_{i=1}^{6}(a_{ii}) \tag{9.29}$$

where λ_i and a_{ii} are the eigenvalue and the principal diagonal element of the joint space inertia matrix, respectively.

9.4 Inertia Matching Guideline of the Parallel Robot

The joint space inertia matrix represents the inertia property of the parallel mechanism. From the previous analysis, it is clear that the non-diagonal elements of the joint space inertia matrix reflect the coupling relations between the limbs, while the diagonal elements reflect the inertia property of the corresponding limbs. The equivalent inertia of the parallel mechanism is equal to the mean value of the diagonal elements of the joint space inertia matrix. The equivalent inertia index of the parallel mechanism is a comprehensive reflection of the inertia property of all limbs. The inertia matching issue of the parallel mechanism can be solved by analyzing the inertia matching of the single limb. To establish the inertia matching principle, the study of inertia matching of the single limb is carried out. The following three aspects are considered: mechanical resonance frequency, acceleration torque and dynamic performance [5].

Firstly, the mechanical resonance frequency is studied. The structure of the mechanical transmission system of the limb of the Stewart mechanism is shown in Fig. 9.9. It can be simplified to a torsional vibration model consisting of springs and concentrated masses. The motor and the coupling are parts with lower stiffness. After neglecting the system damping, the limbs can be simplified to the torsional vibration model, as shown in Fig. 9.10.

k_1 and k_2 denote the stiffness of the motor and coupling (or belt drive), respectively; I_{M} and I_{L} are the motor inertia and the inertial load of the mechanism converted to the motor shaft $I_{\mathrm{L}} = I_{\mathrm{Load}}/\mu^2$; μ is the transmission ratio; $\theta_{\mathrm{i}}(t)$ is the input angle of the motor (input of the system); $\theta_{\mathrm{o}}(t)$ is the output angle of the transmission system (output of the system). The rotary motion of the motor is transformed into the linear

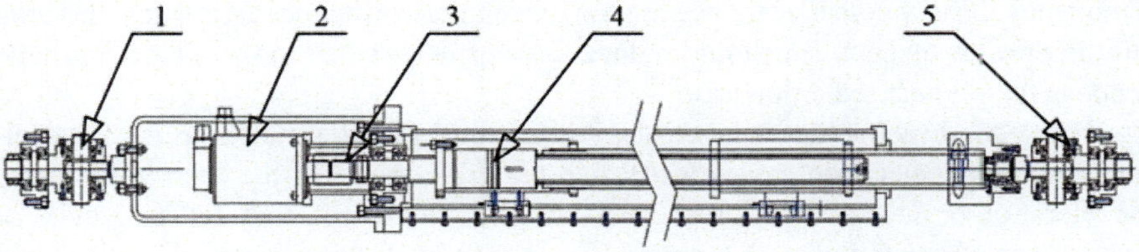

1: universal joint, 2: motor, 3: coupling, 4: ball screw, 5: spherical joint

Fig. 9.9 Mechanical transmission structure of the limb of the Stewart mechanism

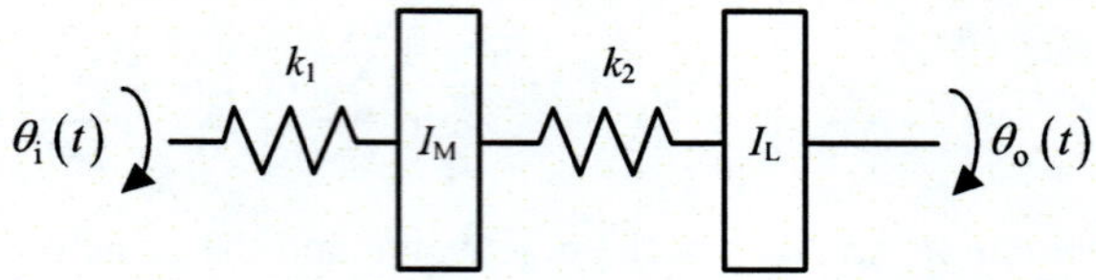

Fig. 9.10 Torsional vibration model of the limb of the Stewart mechanism

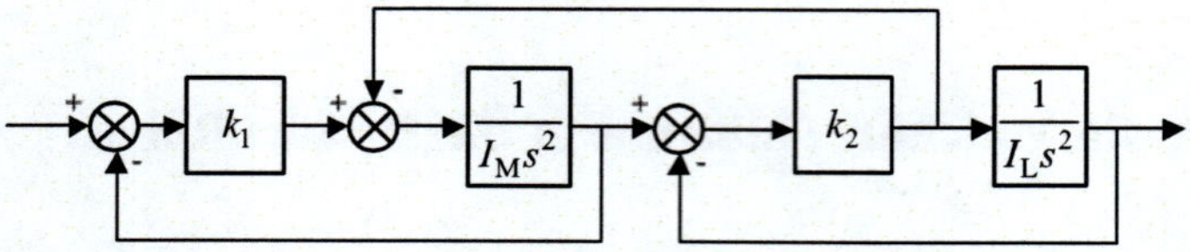

Fig. 9.11 Block diagram of the torsional vibration model of the limb

feed of the limb with a screw. The block diagram of the torsional vibration model of the limb is shown in Fig. 9.11.

Further, the transfer function of the mechanical transmission system can be obtained:

$$TF = \frac{\frac{k_1 k_2}{I_M I_L}}{s^4 + \left(\frac{k_1+k_2}{I_M} + \frac{k_2}{I_L}\right)s^2}$$

When the transfer function achieves a maximum value, i.e., the denominator is zero, the mechanical system resonates, and the frequency is the mechanical resonance frequency of the transmission system, which can be expressed as a function of the inertia and stiffness of the mechanism:

$$f = \frac{1}{2\pi}\sqrt{\frac{k_1 + k_2}{I_M} + \frac{k_2}{I_L}}$$

The resonance frequency of the mechanical system should be much higher than the maximum control frequency, at least 5–10 times of the control frequency. From the analysis, it can be concluded that increasing the reduction ratio of the transmission system (reducing I_L) is an effective method to increase the inherent mechanical frequency. If only consider increasing the mechanical resonance frequency, the limb of the Stewart mechanism should reduce the inertia reflected to the motor shaft and choose the highest reduction ratio.

The second factor that needs to be considered in inertia matching is the acceleration and deceleration capacities. When the drive motor of the limb implements acceleration or deceleration, it needs to overcome the force generated by gravity as well as forces generated by the motion (Coriolis force, friction force, etc.). Newton's second law equation of the limb can be expressed as:

$$T_{acc} = (T_{Motor} - T_G - T_V) = I \cdot \ddot{\theta}_{Motor}$$

where

$$I = I_{\mathrm{M}} + \frac{I_{\mathrm{Load}}}{\mu^2}$$
$$\ddot{\theta}_{\mathrm{M}} = \mu \ddot{\theta}_{\mathrm{Load}}$$

I_{Load} is the load inertia of the mechanism converted to the reducer. By combining the above three equations, the equation for the load acceleration (terminal acceleration) can be derived as:

$$\ddot{\theta}_{\mathrm{Load}} = \frac{T_{\mathrm{acc}}}{\mu I_{\mathrm{M}} + I_{\mathrm{Load}}/\mu} \tag{9.30}$$

When the acceleration torque T_{acc} is given, to obtain the maximum acceleration, the derivative of the denominator of the above equation is deduced with respect to μ. The ideal deceleration ratio that enables the terminal maximum acceleration is obtained as:

$$\mu_{\mathrm{ideal}} = \sqrt{\frac{I_{\mathrm{Load}}}{I_M}}$$

The optimal reduction ratio should ensure that the load inertia converted to the motor shaft I_{Load}/μ^2 and the motor inertia I_{M} are equal, when only consider the acceleration and deceleration capacities.

To obtain an optimal range of reduction ratio, the variation relations of the acceleration torque T_{acc} with the reduction ratio μ is analyzed. When the motor rotor inertia, the load inertia and the load acceleration are given, the relations between the required acceleration torque and reduction ratio can be derived according to Eq. (9.30), as shown in Fig. 9.12. When the actual reduction ratio μ is close to the ideal reduction ratio μ_{ideal}, the required acceleration torque is small. Conversely, as μ deviates from μ_{ideal}, the required acceleration torque gradually increases. Therefore, if only consider to make the limb have good acceleration and deceleration capacities, the load inertia reflected to the motor shaft should be 0.5–2 times of the motor inertia.

The inertia is also an important parameter that affects the dynamic performance of the limb. The Simulink software is applied to simulate the position control system of a limb to analyze the effects of the inertia on its response characteristics. Firstly, a three-loop (position loop, velocity loop and current loop) position control system model is established in Simulink. The control block diagram is shown in Fig. 9.13, and the simulation parameters are listed in Table 9.2. To reflect the effects of the inertia on the dynamic performance of the control system, a disturbance torque, which is the product of the output acceleration and the load inertia converted to the motor shaft, reacts to the control system.

The simulation experiment of the position step response is conducted for the established control system, and the results are shown in Fig. 9.14. When the ratio

Fig. 9.12 Variation curve of acceleration torque with deceleration ratio

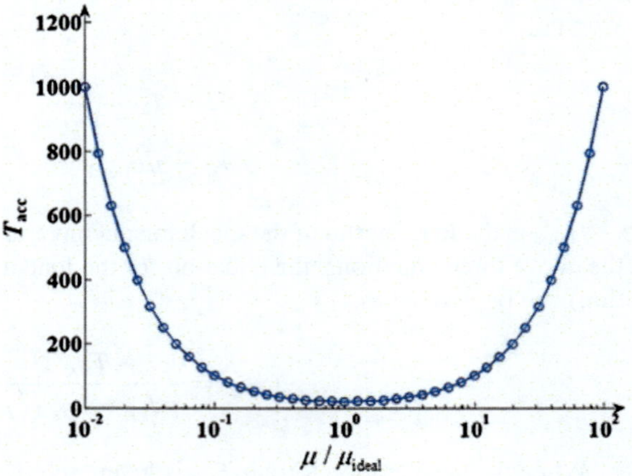

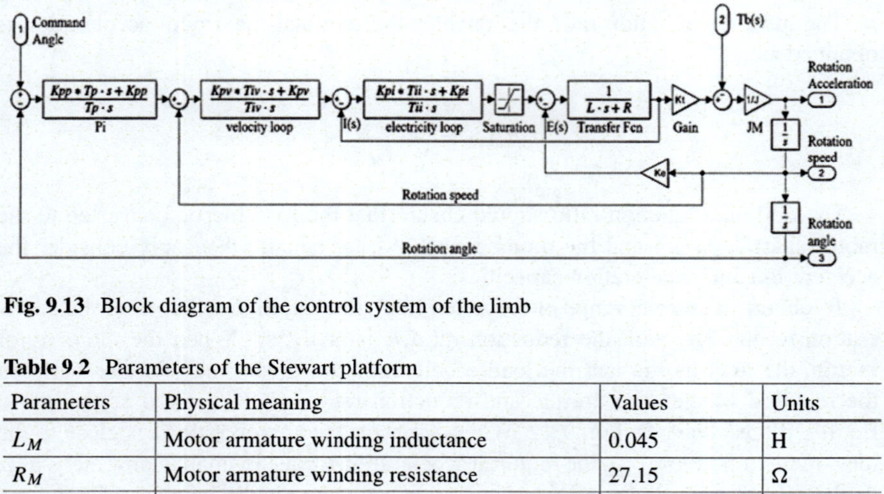

Fig. 9.13 Block diagram of the control system of the limb

Table 9.2 Parameters of the Stewart platform

Parameters	Physical meaning	Values	Units
L_M	Motor armature winding inductance	0.045	H
R_M	Motor armature winding resistance	27.15	Ω
J_{M}	Motor rotor inertia	2.82×10^{-5}	kg m^2
K_{e}	Motor counter-electromotive force constant	0.33	V s/rad
K_{t}	Motor torque constant	0.57	N m/A
K_{pi}	Current loop scaling factor	200	V/A
T_{ii}	Current loop integration time constant	0.0017	s
K_{pv}	Velocity loop scaling factor	80	A s/rad
T_{iv}	Velocity loop integration time constant	0.5	s
K_{pp}	Position loop scaling factor	35	1/s
T_{p}	Position loop integration time constant	0.1	s

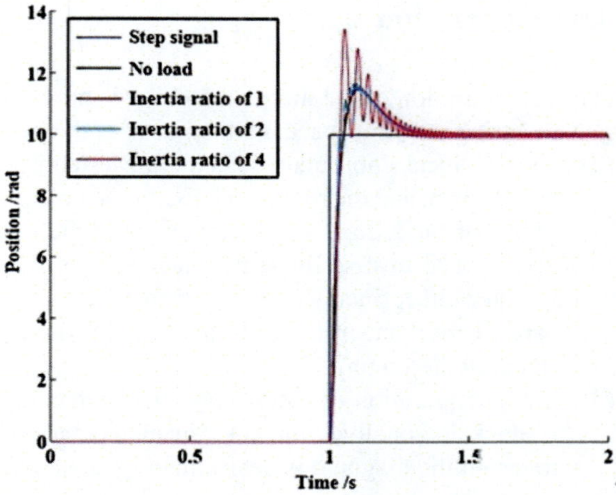

Fig. 9.14 Position step response curve of the control system

of the load inertia to the motor rotor inertia reaches 4:1, the stabilization time and overshoot of the step response of the limb increase significantly, indicating that the dynamic performance of the control system are seriously affected. When the ratio of inertia reflected to the motor shaft to motor inertia is not larger than 2:1, the dynamic performance of the system is almost the same as when the motor is unloaded, i.e., the system has good dynamic performance.

With a comprehensive analysis of the previous three factors discussed earlier, Table 9.3 can be obtained. The parallel mechanism is a closed-loop structure with high stiffness, and the mechanical resonance frequency is not the principal factor affecting its performance. The acceleration capacity and dynamic performance of the Stewart mechanism are the main factors to determine the inertia matching. Considering the efficiency of the motor, the inertia matching principle of the Stewart mechanism can be obtained as: the ratio of the equivalent inertia I_E of the parallel mechanism to the motor inertia I_M should be within the range of 1:1–2:1. The ratio of the equivalent inertia to motor inertia should be within this range in the whole workspace of the parallel mechanism. Since the equivalent inertia changes with the pose of the parallel robot in the workspace, the change law and the extreme value of the equivalent inertia in the workspace should be analyzed to complete the inertia matching of the parallel mechanism.

Table 9.3 Inertia matching requirements

	Mechanical resonance frequency	Acceleration capacity	Dynamic performance
Inertia matching requirements	The smaller I_E, the better	$I_E/I_M \in [0.5, 2]$	$I_E/I_M \in [1, 2]$

9.5 Conclusion

(1) The definition of the inertia matching is presented, and a general approach for deducing the joint space inertia matrix of the parallel mechanism is established.
(2) Unified local coordinate systems are established in the limb of the Stewart mechanism, and the rotation center of the universal joint is chosen as the pivotal point for modeling, which can simplify the deduction of the partial velocity matrix. The inverse dynamic equation in a simple form is obtained with the virtual work principle. The derived joint space inertia matrix of the Stewart parallel mechanism have a unified dimension, which can be converted to the standard inertia unit.
(3) The configuration of the Stewart mechanism determines that its inertia matrix is block diagonally dominant. The elementary coupling relations of the submatrices on the diagonal is determined by the configuration of the parallel mechanism. The degree of coupling of submatrices is high when two adjacent limbs on the moving platform are mounted close with small angle; and vice versa. Based on the analysis of the coupling characteristics of the joint space inertia matrix, the mean value of its eigenvalues is defined as the equivalent inertia of the parallel mechanism, which is characterized by unified dimension and clear physical meaning.
(4) Through comprehensive consideration of mechanical vibration frequency, acceleration and deceleration capacities and dynamic performance of the limb of the Stewart mechanism, the inertia matching principle is determined, and the ratio of the equivalent inertia to motor inertia is best to be ensured within the range of 1:1–2:1.

References

1. Zhu D, Liu H. The role of inertia matching in improving the performance of CNC gantry boring and milling machines. Des Res. 2007;2:61–3.
2. Zhang GG, Furusho JJ. Speed control of two-inertia system by PI/PID control. In: Proceedings of the IEEE international conference on power electronics and drive systems, Hong Kong; 1999. p. 567–72.
3. Cetinkunt S. Optimal design issues in high-speed high-precision motion servo systems. Mechatronics. 1991;1(2):187–201.
4. Mo J, Shao ZF, Guan L, Xie F, Tang X. Dynamic performance analysis of the X4 high-speed pick-and-place parallel robot. Robot Comput-Integr Manuf. 2017;46:48–57.
5. Tang X, Liu Z, Shao Z, Wang L. Self-excited vibration analysis for the feed support system in FAST. Int J Adv Robot Syst. 2014;11(4):63.
6. Shao Z, Tang X, Wang L. Optimum design of 3–3 Stewart platform considering inertia property. Adv Mech Eng. 2015;5:249121.
7. Shao Z, Tang X, Chen X, Wang L. Inertia match research of reconfigurable planar 3-RRR parallel mechanism. Chin J Mech Eng. 2009;22(6):791–9.
8. Shao Z, Tang X, Chen X, Wang L. Research on the inertia matching of the Stewart parallel mechanism. Robot Comput-Integ Manuf. 2012;28(6):649–59.

9. Codourey A. Dynamic modeling and mass matrix evaluation of the DELTA parallel robot for axes decoupling control. In: Proceedings of the IEEE/RSJ international conference on intelligent robots and systems, Osaka, Japan, vol. 3; 1996. p. 1211–8.
10. Yang H, Wu S, Cao J, et al. Study of Stewart platform dynamics considering the influence of drive branch inertia. China Mech Eng. 2002;13(12):1009–12.
11. Ogbobe P, Jang HZ, He JF, Yang CF, Han JW. Analysis of coupling effects on hydraulic controlled 6 degrees of freedom parallel mechanism using joint space inverse mass matrix. In: The second international conferences on intelligent computation technology and automation, Changsha, China; 2009. p. 845–8.
12. He J, Ye Z, Jiang H, et al. Coupling analysis of parallel robots based on joint space model. J Mech Eng. 2006;42(6):161–5.
13. Yao Y, Fu S, Han L. Diagonal dominance analysis and discrimination of spatial inertia matrix blocks of Stewart platform joints. J Mech Eng. 2008;44(6):101–6.
14. Huang T, Zhao XY, Wang Y, Mei JP, Ni YB. Determination of servomotor parameters of a tripod-based parallel kinematic machine. Progr Nat Sci. 2001;11(8):612–21.
15. Huang T, Mei JP, Li ZX, Zhao XM. A method for estimating servomotor parameters of a parallel robot for rapid pick-and-place operations. Trans ASME. 2005;127(4):596–601.

Chapter 10
Final Scale Model of the Feed Support System

The previous chapters focus on the theoretical study and numerical simulation of the FAST feed support system. However, considering the complexity of the practical problems, a complete functional model is required for comprehensive validation of theories and methods. The dimension of the FAST prototype is huge (the diameter is 520 m), and the span of the feed support system reaches 600 m. It is costly and time-consuming to build an equivalent-scale experimental prototype, and the experimental study is carried out with a small-scale model. The final scale model with a span of 40-meter and a size ratio of 1:15 is built at the Miyun Observing Station of NAOC in Beijing, which verifies the research contents of this book, realizes the astronomical observation function and guides the engineering construction of the FAST prototype.

This chapter first presents the mechanical structure and control system of the FAST final scale model. Then, experimental research on control methods is implemented on this scale model. Finally, astronomical observation experiments are conducted to systematically verify the research contents of this book.

The main contents of this chapter are:

(1) Mechanical structure of the 40-meter scale model;
(2) Development and implementation of the control system;
(3) Control experiments of the CDPR;
(4) Astronomical observation experiments of the 40-meter scale model.

10.1 Mechanical Structure of the 40-meter Scale Model

The completed 1:15 scale model of the FAST is built at the Miyun Observing Station of NAOC, as shown in Fig. 10.1. The reflector is a network structure composed of articulated triangular modules, and the distributed drive units are used to drag and control the articulation joints to implement the parabolic fitting of the reflector. The feed cabin is attached to the moving platform of the cable-driven parallel robot,

© Tsinghua University Press 2024

X. Tang et al., *Research and Application of Cable-Driven and Rigid Parallel Robots*,
https://doi.org/10.1007/978-981-99-7452-8_10

Fig. 10.1 1:15 scale model of the FAST

as shown in Fig. 10.2. The feed cabin can be divided into the upper and lower parts. As shown in Fig. 10.2a, the control and communication systems (including an industrial PC, eight AC servo drivers and a fiber connector, etc.) are installed in the upper part. As shown in Fig. 10.2b, the A-B rotator and the fine-tuning Stewart platform are mounted in the lower part. To reduce the mass of the feed cabin, the fine-tuning Stewart platform uses aluminum alloy extensively. Three total station targets are installed on the outside of the feed cabin and uniformly distributed in a circle for real-time measurement of the attitude. Communication optical fibers and power cables are connected from the ground to the feed cabin along two symmetrical cables to realize data communication and power distribution.

10.1.1 Fine-Tuning Stewart Platform

Before dimension design of the fine-tuning Stewart platform, the workspace requirements should be determined and the kinematic modeling should be established.

1. **Workspace requirements of the Stewart platform**

(1) Through the analysis of the cable-driven parallel mechanism, its terminal accuracy reaches 10 mm, and the workspace of the Stewart platform (including the rotation angle of the attitude) is a sphere with a radius larger than 10 mm.

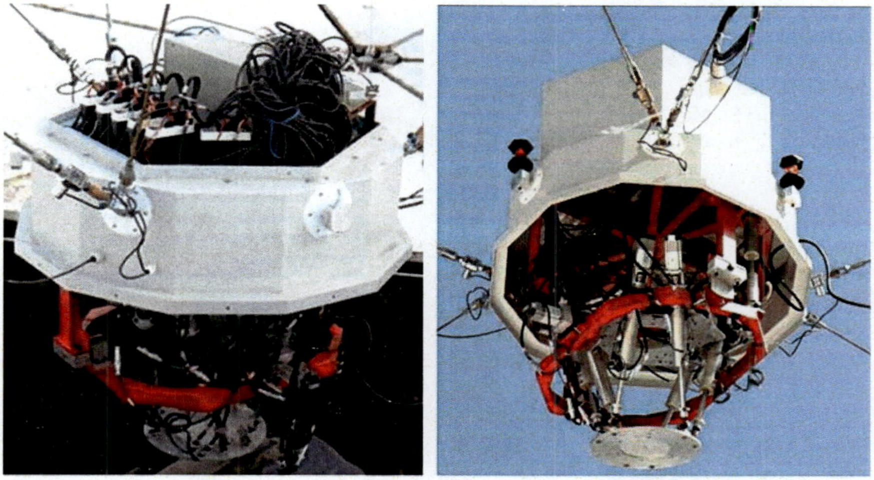

(a) (b)

Fig. 10.2 Scale model of the feed cabin. **a** Upper part of the feed cabin; **b** lower part of the feed cabin

(2) The Stewart platform need the posture adjustment ability of ±5° pitch, roll and yaw angle to improve the terminal posture accuracy.
(3) According to the preliminary design of the FAST prototype, the dimensions of the base and the moving platform of the small-scale model are obtained through the similarity ratio calculation. The diameter of the base is 550 mm, and that of the moving platform is 350 mm.

Considering the above requirements of (1), (2) and (3), workspace of the Stewart platform (including the rotation angle of the attitude) is given as a sphere with a radius of 25 mm, and the adjustment range of the moving platform is −5° to 5°.

2. **Kinematic modeling of the Stewart platform**

As shown in Fig. 10.3, two coordinate systems are established for the Stewart platform: one is the inertial coordinate system $\Re : O - XYZ$, the origin of which is located at the center of the base platform with the Z-axis upward; the other is the moving coordinate system $\Re' : O' - X'Y'Z'$, the origin of which is located at the center of the moving platform with the Z'-axis upward along the normal of the moving platform. $B_i (i = 1, 2, \ldots, m)$ are the rotation centers of the universal joints on the base platform and $T_i (i = 1, 2, \ldots, 6)$ are the rotation centers of the spherical joints on the moving platform.

The symbols are defined as follows. $\boldsymbol{O}'^{\Re}$ is the position vector of the origin of the moving coordinate system O' in the inertial coordinate system. $\boldsymbol{B}_i^{\Re}$ is the position vector of B_i in the inertial coordinate system. $\boldsymbol{T}_i^{\Re}$ is the position vector of T_i in the inertial coordinate system. $\boldsymbol{T}_i^{\Re'}$ is the position vector of T_i in the moving coordinate system. r_b is the radius of the base platform, and r_a is the radius of the moving

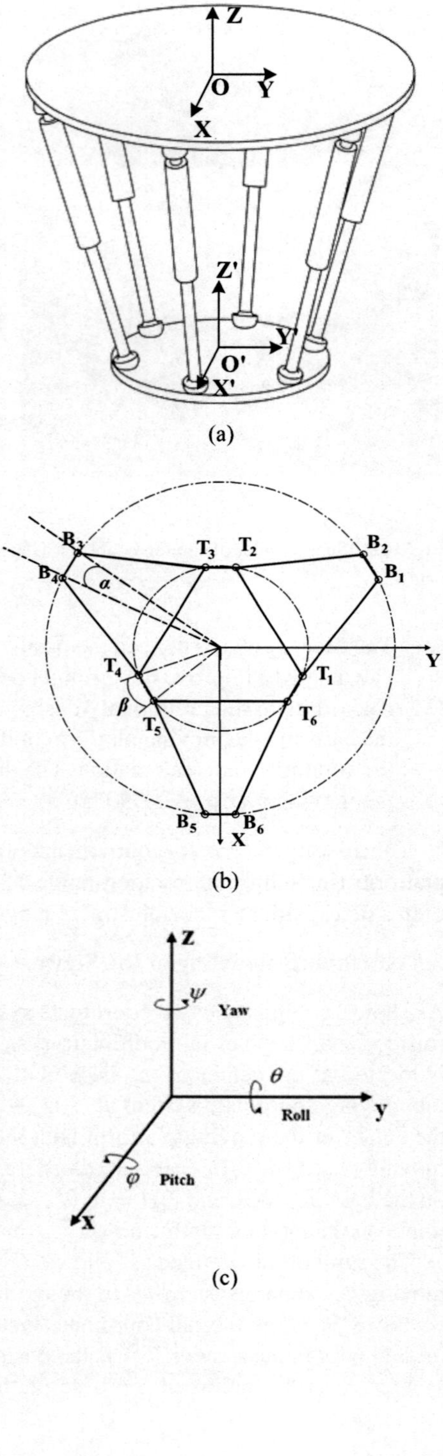

Fig. 10.3 Modeling of the fine-tuning platform. **a** 3D model of the Stewart platform; **b** kinematic diagram of the Stewart platform; **c** diagram of pitch, roll and yaw angles

platform. α is the distribution angle of the spherical joints of the moving platform; β is the distribution angle of the universal joints of the base platform.

According to Fig. 10.3, the position vector of the spherical joints of the Stewart platform is deduced as:

$$\boldsymbol{T}_i^{\Re} = \boldsymbol{R} \cdot \boldsymbol{T}_i^{\Re'} + \boldsymbol{O}'^{\Re}$$

where $\boldsymbol{R}$ is the rotation matrix of the moving coordinate system with respect to the inertial coordinate system.

$$\boldsymbol{R} = \begin{bmatrix} \cos\theta\cos\psi + \sin\varphi\sin\theta\sin\psi & -\cos\theta\sin\psi + \sin\varphi\sin\theta\cos\psi & \cos\varphi\sin\theta \\ \cos\varphi\sin\psi & \cos\varphi\cos\psi & -\sin\varphi \\ -\sin\theta\cos\psi + \sin\varphi\cos\theta\sin\psi & \sin\theta\sin\psi + \sin\varphi\cos\theta\cos\psi & \cos\varphi\cos\theta \end{bmatrix}$$

where φ is the pitch angle around the X-axis, θ is the roll angle around the Y-axis, and ψ is the yaw angle around the Z-axis. c denotes the cosine operation while s denotes the sine operation.

The vector of the limb can be obtained as:

$$\boldsymbol{l}_i = \boldsymbol{B}_i^{\Re}\boldsymbol{T}_i^{\Re}$$

The length of the limb can be deduced as:

$$l_i = \boldsymbol{l}_i$$

3. **Dimension design of the Stewart platform**

According to the above dimensional parameters and workspace requirements of the Stewart platform, the optimized results can be obtained as:

(1) Diameter of the base platform: 550 mm;
(2) Diameter of the moving platform: 350 mm;
(3) Distribution angle of the universals joints of the base: $\alpha = 30°$;
(4) Distribution angle of the spherical joints of the moving platform: $\beta = 15°$;
(5) The shortest length of the limb: $l_{\mathrm{min}} = 360$ mm;
(6) The longest length of the limb: $l_{\mathrm{max}} = 480$ mm;
(7) Initial location: $z_{\mathrm{m}} = -380$ mm.

The workspace of the Stewart platform is determined as shown in Fig. 10.4. The inner circle is the required workspace, i.e., a sphere with a radius of 25 mm. From Fig. 10.4, it can be known that the optimized dimensions of the Stewart platform well meet the workspace requirements.

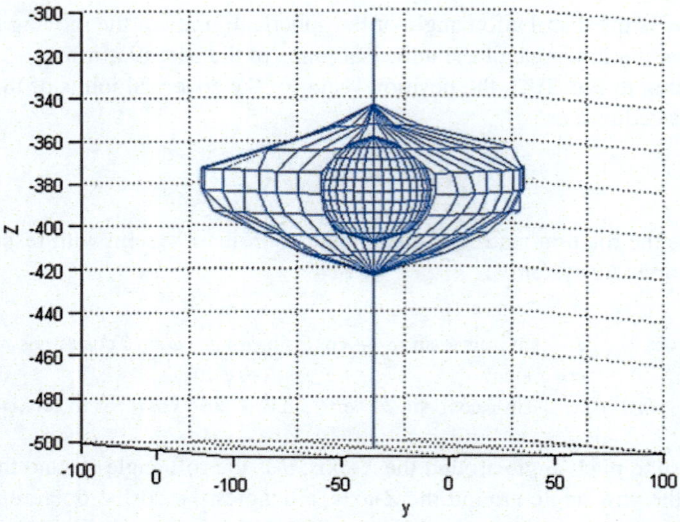

Fig. 10.4 Workspace of the scale fine-tuning Stewart platform

10.1.2 A-B Rotator

The dimensions of the A-B rotator is determined to meet the rotation angle of ±25°. The rotation angle α of the *A*-axis is controlled with the telescopic rod; the rotation angle β of the *B*-axis is controlled with the servo motor and reducer.

Figure 10.5 illustrate the design of the upper and lower platforms of the A-B rotator. The upper platform is a hexagon while the lower platform is an octagon.

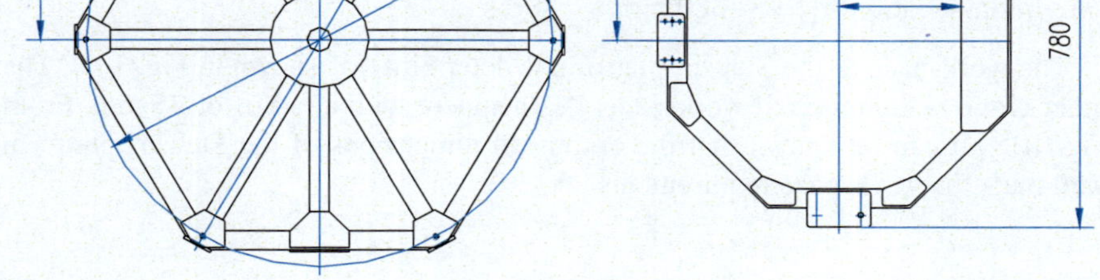

Fig. 10.5 Upper and lower platforms of the A-B rotator

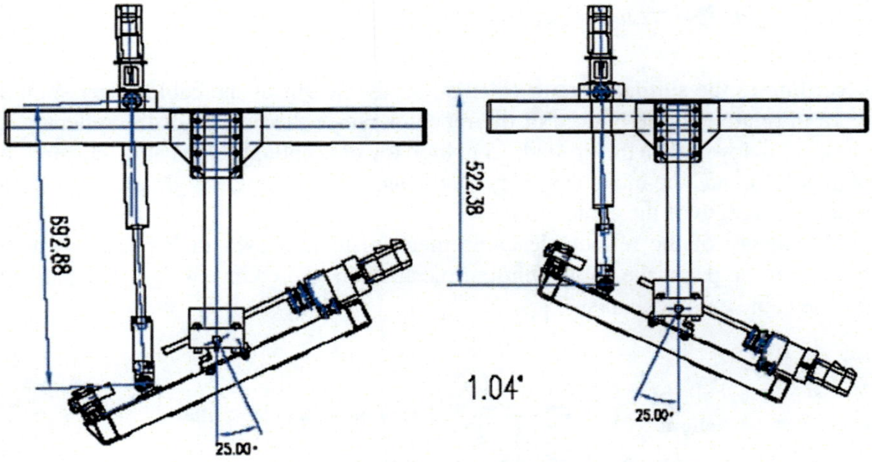

Fig. 10.6 Design of the extreme positions of the A-B rotator

Figure 10.6 shows the design of the A-B rotator at the two extreme positions of the *A*-axis to determine the length of the telescopic rod.

The motion range of the telescopic rod is obtained as 692.88 mm − 522.38 mm = 170.5 mm.

The control accuracy of the A-B rotator is analyzed. The adopted Panasonic servo motor feedbacks 10,000 counts per rotation, and the lead screw is 4 mm. The maximum error of the motor dynamic control is no more than 100 counts. The distance between the action point of the *A*-axis telescopic rod and the rotation shaft is 350 mm, and the angle error of the *A*-axis can be obtained as:

$$\Delta = \frac{100}{10000} \times 4 = 0.04 \text{ mm}$$

$$\Delta \text{rad} = \text{arctg}\left(\frac{0.04}{350}\right) = 0.006548^{\circ} = 5.714 \times 10^{-5} \text{ rad}$$

There is a reducer on *B*-axis, and the reduction ratio is 1:240. Under the same condition as above, the angle error of the *B*-axis can be obtained as:

$$\Delta \text{rad} = \frac{100}{10000} \times 360^{\circ} \times \frac{1}{240} = 0.015^{\circ} = 1.309 \times 10^{-4} \text{ rad}$$

The above accuracy completely meets the requirement of the rotation angle error of the A-B rotator.

The feed cabin is a hexagon, the diameter of its circumscribed circle is 1 m, and the height is 450 mm. Six mounting interfaces are set on the six top corners of the hexagon.

10.1.3 Cable-Driven Parallel Robot

According to the similarity ratio calculation, the height of the cable tower should be 18 m relative to the bottom of the reflector. Since the bottom of the reflector is about 4.7 m lower than the ground, 15 m high tower is enough. The cable will stretch 12 m is 12 m, and the movable pulley is adopted in the counterweight mechanism to double the length of the cable.

The design of the retractable cable mechanism is shown in Fig. 10.7, which includes three parts: the cable winding mechanism, the counterweight mechanism and the cable exit mechanism.

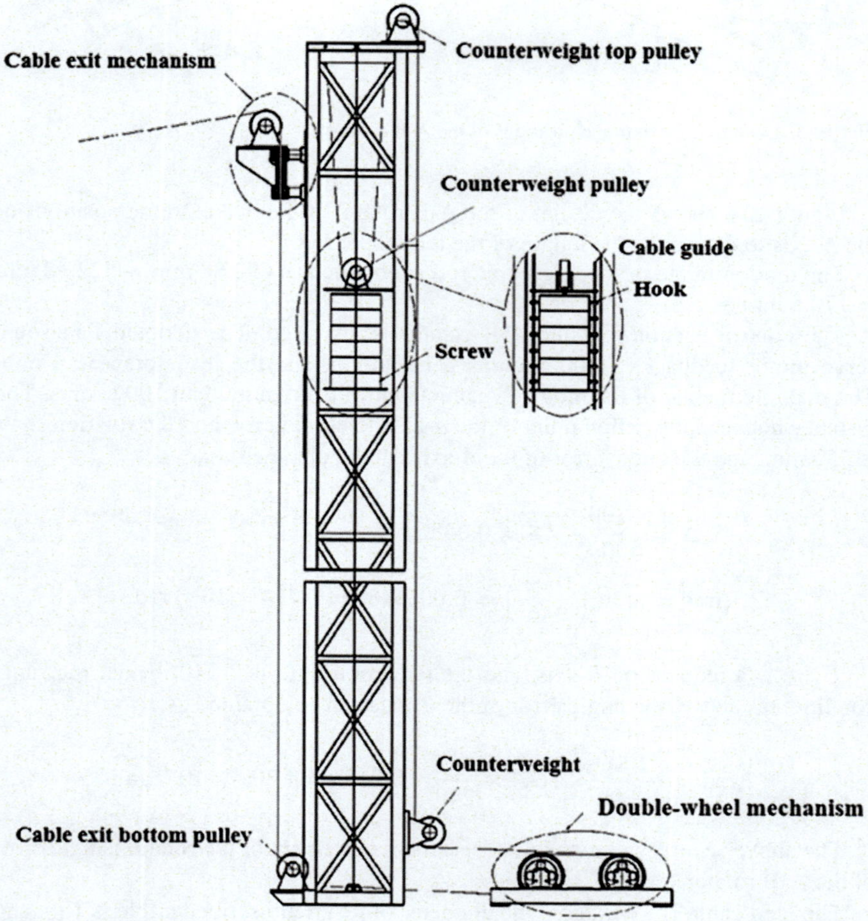

Fig. 10.7 Structure of the retractable cable mechanism

For the cable-driven parallel mechanism, the cable drum is generally used directly for cable winding, which will cause a large load on the motor, and the spiral rolling device could cause the position change of the cable exit point, damaging the accuracy. In this book, a double-wheel mechanism with a counterweight structure is proposed to ensure the pretension of the cable and the accuracy of the cable length.

The double-wheel mechanism is shown in Fig. 10.8, which is composed of the driving wheel and the passive wheel. The driving wheel is driven by a motor. The groove spacing of the driving wheel and the passive wheel is the same, but the installation is staggered by half-groove spacing. The cable is wound in the groove of the pulley, overlapping or slipping of the spiral cable are avoided by adjusting the staggered distance. The diameter of the wheel is 240 mm and the cable diameter is 8 mm.

The diameters of the counterweight and guide pulleys are 240 mm. To reduce the motor load, the counterweight mechanism is used for auxiliary control, and its weight W_0 is determined by referring to the elevator counterweight as

$$W_0 = \frac{\sigma_{\max} + \sigma_{\min}}{2} = \frac{2500 + 500}{2} = 1500 \text{ N}$$

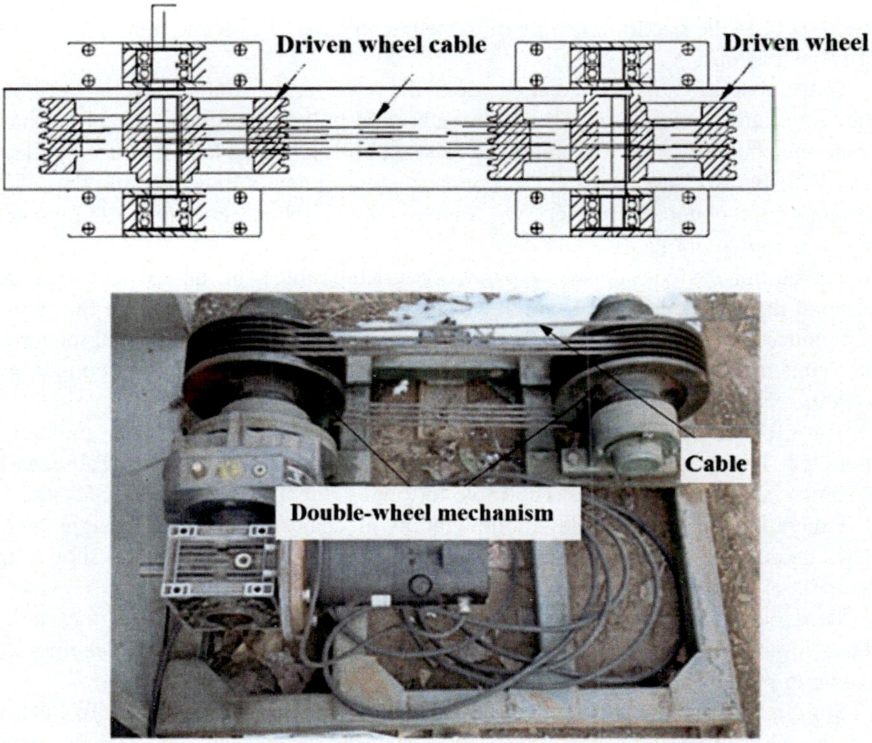

Fig. 10.8 Double-wheel mechanism

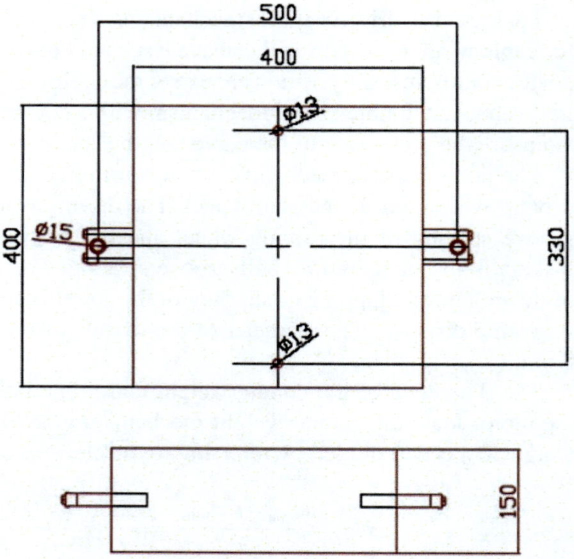

Fig. 10.9 Design of the counterweight

where σ_{max} is the maximum value of cable tension, and σ_{min} is the minimum value of cable tension.

One end of the cable is fixed at the top of the tower, and the other end connects the moving platform sequentially through the bottom pulley, the double-wheel mechanism, the counterweight top pulley, the bottom pulley, the cable exit bottom pulley and the cable exit mechanism. The counterweight pulley is movable, and the total weight of the counterweight is $2W_0$. Five counterweights are used in each tower, which is shown in Fig. 10.9.

The motion the feed support system will cause the cable to be inconsistent with the plane of the cable exit pulley, which will increase the friction and wear of the cable and reduce the control accuracy. To solve this problem, a passive rotational tracking mechanism is adopted, which makes the plane of the cable exit pulley consistent with the cable in real time.

The cable exit mechanism adopts a swing bracket, where a cable exit pulley is mounted. The swing bracket can rotate according to the direction of the cable, and is shown in Fig. 10.10a and b. The cable following effect is shown in Fig. 10.10c.

Except for the double-pulley in the winding mechanism that is self-designed, all other guide pulleys are standard parts, and the diameter is 240 mm, as shown in Fig. 10.11.

To realize the power supply and communication to the feed cabin, the cable retracting mechanism is designed by adopting a curtain-like following scheme, as shown in Fig. 10.12.

Sliding hooks in the cable retracting mechanism are divided into pairs. To ensure that the cable will not be bent sharply or broken due to the great change of the distance of the sliding hooks, a plastic bar is installed under each pair of sliding hooks and

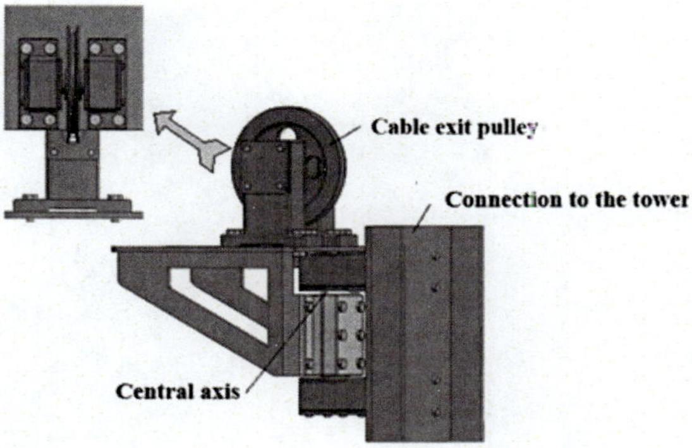

(a)

(b)

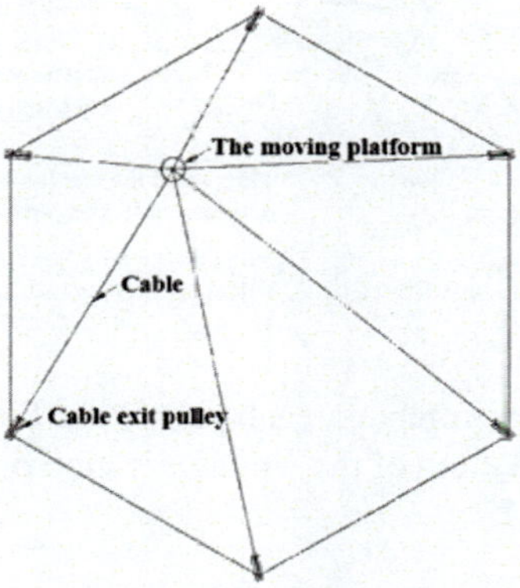

(c)

Fig. 10.10 Cable exit mechanism. **a** Structure of the swing bracket; **b** photo of the swing bracket; **c** the cable following effect

Fig. 10.11 Standard pulley

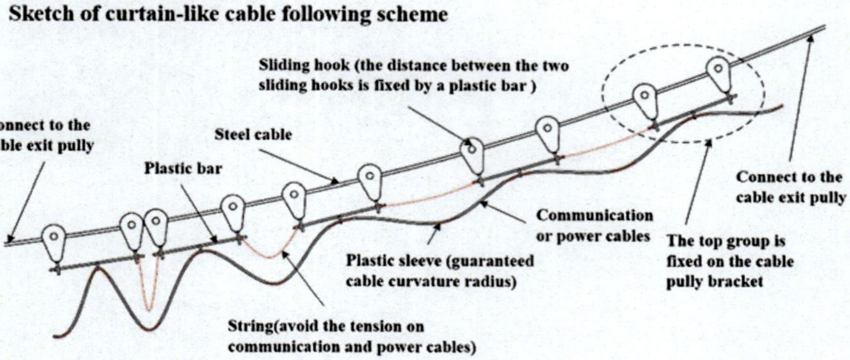

Fig. 10.12 Diagram of the cable retracting mechanism

there is a grommet on each plastic bar for connecting the string. Plastic sleeves are set at the bottom of the communication or power cable. The final structure is shown in Fig. 10.13.

Fig. 10.13 Photos of the cable retracting mechanism

10.1.4 Cable Towers

According to the similarity ratio calculation, the height of the cable tower should be 18 m. Since the bottom of the reflector is −4.7 m from the ground, the height of the cable tower only needs to be 15 m from the ground to meet the design requirements, and the pulleys are installed at 15 m and 12 m. The six towers are mounted uniformly around the circumference. The radius of the circle is 20 m, and the dimensions of the tower pedestal are 1.2 m × 1.2 m × 2.0 m. The towel location is illustrated in Fig. 10.14.

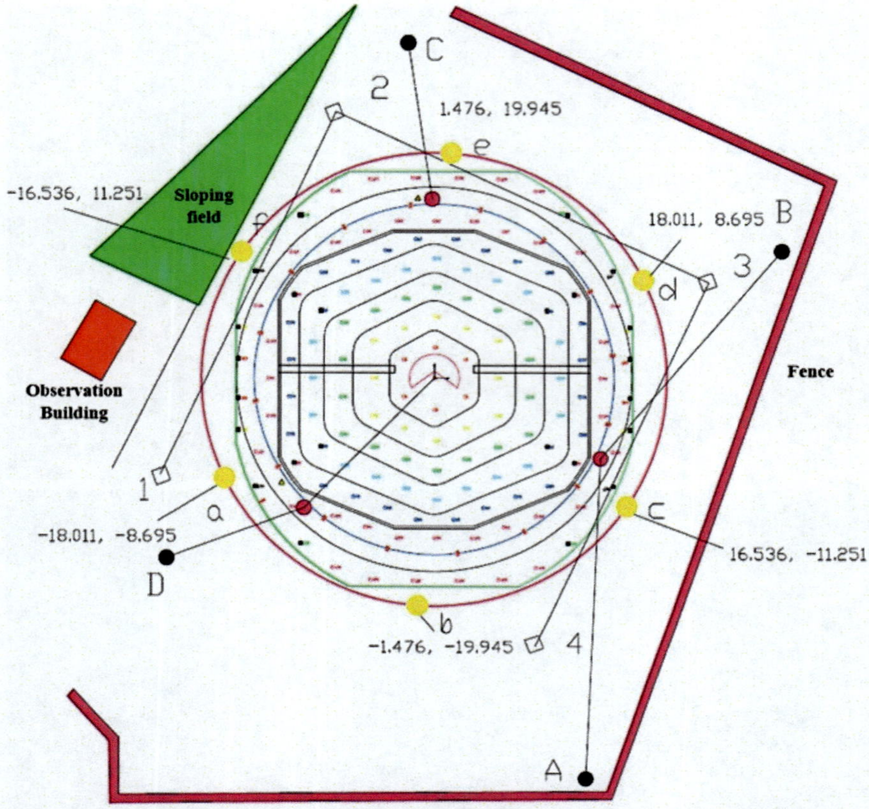

Fig. 10.14 Coordinates of the tower locations

10.1.5 Calibration of the CDPR

The final scale model of the FAST adopts the established reflector and measurement system in Miyun Observing Station, it is necessary to determine the location information of the feed support system with respect to the reflector and measurement system to ensure the matching of the feed support system with the existing system. The coordinates of the cable-driven parallel robot are established on the coordinate of the measuring system.

Figure 10.15 shows the cable driven parallel mechanism in FAST. Firstly, two coordinate systems are established: one is the inertial coordinate system $\Re : O - XYZ$, and the origin is located at the center of the base platform with the Z-axis upward. The other is the moving coordinate system $\Re' : O' - X'Y'Z'$, and the origin is located at the center of the moving platform with the Z'-axis upward along the normal of the moving platform. $B_i (i = 1, 2, \ldots, m)$ is the cable exit point of the

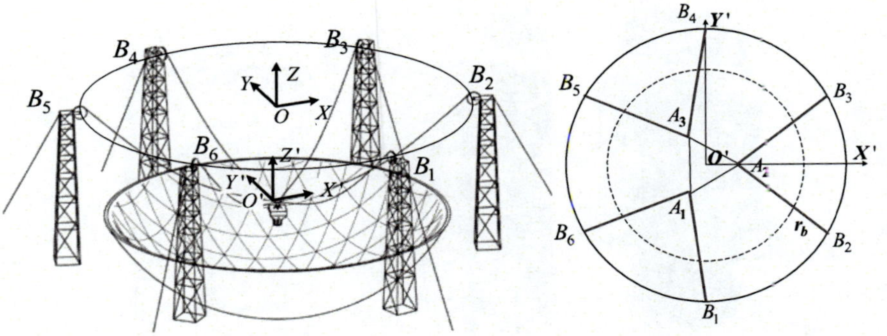

Fig. 10.15 Coordinates establishment of final scale model of the cable driven parallel robot

cable on the base platform and $A_j(j = 1, 2, 3)$ is the cable anchor point of the cable on the moving platform.

The symbols are defined as follows: $\boldsymbol{O}'^{\Re}$ is the position vector of the origin of the moving coordinate system O' in the inertial coordinate system. $\boldsymbol{B}_i^{\Re}$ is the position vector of B_i in the inertial coordinate system. $\boldsymbol{A}_j^{\Re}$ is the position vector of A_j in the inertial coordinate system. $\boldsymbol{A}_j^{\Re'}$ is the position vector of A_j in the moving coordinate system.r_b is the radius of the base platform, that is, the radius of the distribution circle of the cable tower. r_a is the radius of the moving platform.

According to Fig. 10.15, vectors are deduced as:

$$\begin{aligned}
&\boldsymbol{B}_1^{\Re} = [-1.476, -19.945, 0]^{\mathrm{T}};\ \boldsymbol{B}_2^{\Re} = [16.536, -11.251, 0]^{\mathrm{T}} \\
&\boldsymbol{B}_3^{\Re} = [18.011, 8.695, 0]^{\mathrm{T}};\ \boldsymbol{B}_4^{\Re} = [1.476, 19.945, 0]^{\mathrm{T}} \\
&\boldsymbol{B}_5^{\Re} = [-16.536, 11.251, 0]^{\mathrm{T}};\ \boldsymbol{B}_6^{\Re} = [-18.011, -8.695, 0]^{\mathrm{T}}
\end{aligned}$$

Unit: m.

$$\begin{aligned}
&\boldsymbol{A}_j^{\Re'} = \left[r_a \cos[(j-2)\times 120^\circ],\ r_a \sin[(j-2)\times 120^\circ], 0\right]^{\mathrm{T}} \quad (j = 1, 2, 3) \\
&\boldsymbol{A}_j^{\Re} = \boldsymbol{R}\cdot \boldsymbol{A}_j^{\Re'} + \boldsymbol{O}'^{\Re}
\end{aligned}$$

where $\boldsymbol{R}$ is the rotation matrix from the moving to the inertial coordinate system.

Accurate control of the system cannot be achieved without measurement, and a high-precision control algorithm is meaningful only if high-precision feedback data are obtained. In this system, non-contact total stations (see Fig. 10.16) are employed to complete the pose measurement of the cable platform.

Errors are inevitable during the construction of the cable tower, and the cable exit point needs to be calibrated before conducting the experiment. Since the swing bracket is adopted, the rotation shaft should be figured out to determine the position of the cable exit point. As shown in Fig. 10.17, target 1 and target 2 are mounted on the roller support. The projection distance between the center of the rotation shaft

Fig. 10.16 Leica total station TC2003

on the roller support and the mounting plate is h, and that between the center of the rotation shaft and the cable exit point is r.

The following equations can be obtained:

$$x_3 = \frac{x_1 + x_2}{2};\ y_3 = \frac{y_1 + y_2}{2}$$

x_4, y_4 satisfy the equations:

$$y_4 - y_3 = -\frac{x_1 - x_2}{y_1 - y_2}(x_4 - x_3)$$
$$(y_4 - y_3)^2 + (x_4 - x_3)^2 = h^2$$

Therefore, x_4, y_4 can be obtained.

The calibration points x_b, y_b satisfy the equations:

$$(y_b - y_4)^2 + (x_b - x_4)^2 = r^2$$

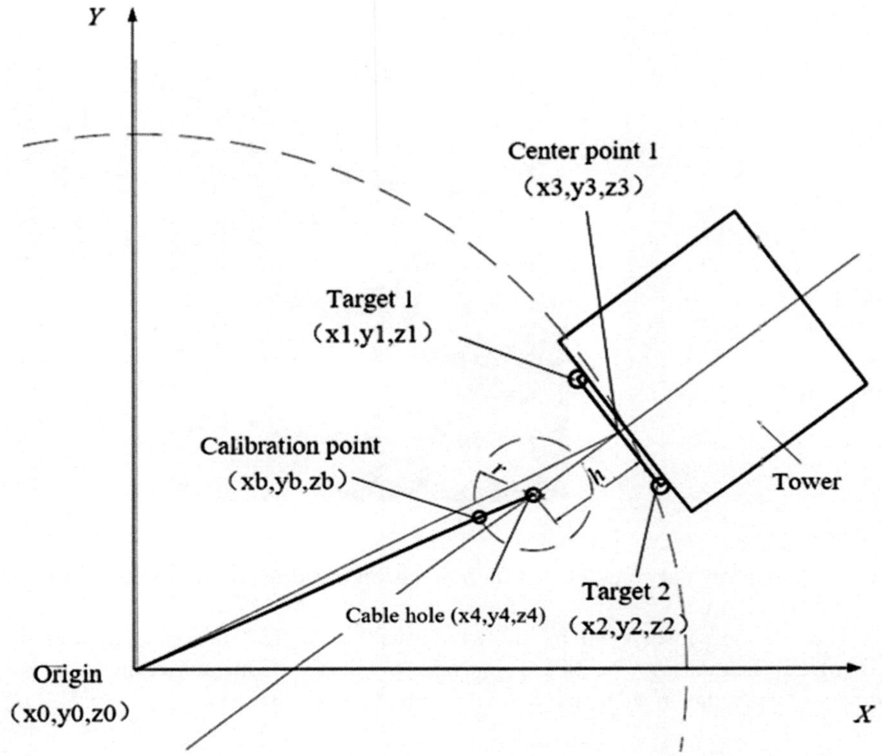

Fig. 10.17 Calibration scheme

$$\frac{x_4 - x_0}{y_4 - y_0} = \frac{x_b - x_0}{y_b - y_0}$$

Therefore, x_b, y_b can be deduced.

With the measurement of the total station, the coordinates of the cable exit point of the cable-driven parallel mechanism in the inertial coordinate system are obtained as:

$$\begin{aligned}
&\boldsymbol{B}_1^{\Re} = [-1.5913, -19.3618, 0.2777]^{\mathrm{T}};\ \boldsymbol{B}_2^{\Re} = [16.1397, -10.8135, 0.2937]^{\mathrm{T}} \\
&\boldsymbol{B}_3^{\Re} = [17.5657, 8.4683, 0.3037]^{\mathrm{T}};\ \boldsymbol{B}_4^{\Re} = [1.5974, 19.3185, 0.3697]^{\mathrm{T}} \\
&\boldsymbol{B}_5^{\Re} = [-16.1203, 10.9188, 0.3307]^{\mathrm{T}};\ \boldsymbol{B}_6^{\Re} = [-17.5047, -8.508, 0.3327]^{\mathrm{T}}
\end{aligned}$$

To avoid the sagging problem of the first-level support system, and to ensure the balance of the cable tension in the initial state, the cable tensions of the six cables are monitored in real time, and the cable tension compensation is implemented. The cables are equipped with high-precision tension sensors, which monitor the variations of the cable tension in real time, and implement the force/position control to keep

Fig. 10.18 Force sensor

cables in tension and complete the high-precision tracking control of the first-level support platform.

The tension sensor used in this experiment is MCL-Z designed by Beijing Zhengkai Technology Co., Ltd. The working voltage of this sensor is 24 V, the range is 10 kN, the impedance is 650 Ω, and the output voltage is 0–5 V (Fig. 10.18).

10.2 Control System of the Scale Model

The structure of the control system of the final scale model of FAST is shown in Fig. 10.19. Its control system is divided into five modules, including the master-control computer, the feed cabin control cabinet, the cable control cabinets (including I and II), the master-control cabinet and the total stations (measurement module). The distribution of electrical energy among modules is indicated by hollow arrows, and the signal transmission is denoted by solid arrows. The cable-driven parallel mechanism is driven by a low-speed, high-torque DC servo motor, and the motor speed feedback relies on the tachogenerator. The pose of the cable platform is measured by the total station in real time and fed back to the master-control computer through the network. With the dynamic model established in Chap. 6, the rated power and torque of the limb motor of the fine-tuning Stewart platform is calculated, and limb drive motor and transmission chain (a Panasonic AC servo motor MHMD022P1V with 200 W and a single-start ball screw with a lead of 4 mm) are determined with the inertia matching principle proposed in Chap. 9. The speed feedback of the A-B rotator and fine-tuning Stewart platform is completed by the incremental encoder in drive motors, and the semi-closed loop control is adopted by the industrial computer in the feed cabin and the PMAC card. The cable tension is measured by the tension sensor connected in series in each cable, and the analog signal returned from the tension sensor is read by the industrial computer through the Advantech PCI-1711L A/D card.

The master-control cabinet is responsible for the electrical power distribution and electrical protection of the feed support system. According to the power calculation, the three-phase AC power is divided into three two-phase approximately equal AC powers for each design power of 5 kW to supply the cable control cabinet I, cable control cabinet II, and other modules (the control cabinet in the feed cabin, the master-control computer, and the total station). The master-control cabinet is equipped with fuses, air switches, and relays, which can realize separate power on and off operation of different modules as well as over-current, over-voltage, and leakage protection functions. The high-voltage AC power is controlled through low-voltage DC power to ensure operation safety.

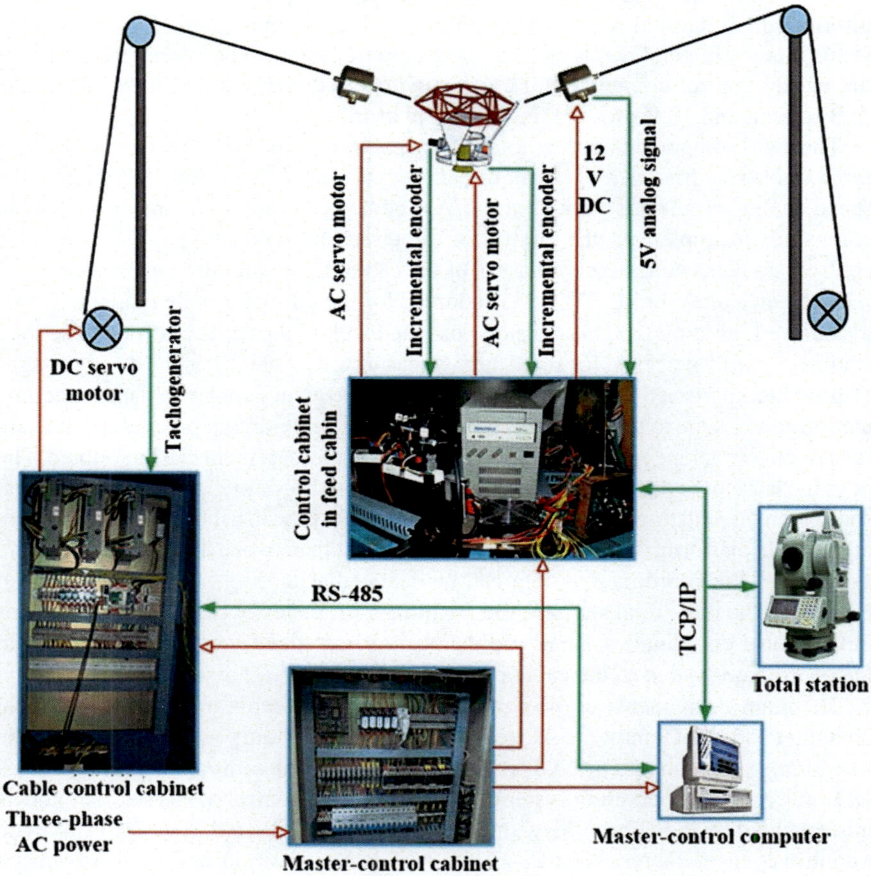

Fig. 10.19 Control system of the 40-meter scale model of FAST

The master-control computer adopts the Windows operating system, which mainly completes the functions of interface display, trajectory planning, status monitoring, etc. Based on the feedback from the total stations, the master-control computer calculates the actual pose of the cable platform and the terminal pose of the feed support system and implements the trajectory planning (coarse interpolation) to determine the motion trajectory of the cable-driven parallel mechanism, the A-B rotator and the fine-tuning Stewart platform according to the given astronomical observation trajectory. The motion commands of the cable-driven parallel mechanism are transmitted to the Mitsubishi PLC in the cable control cabinet through the RS-485 serial communication protocol, and that of the A-B rotator and the fine-tuning platform are transmitted to the industrial computer in the control cabinet through the UDP protocol. At the same time, the Mitsubishi PLCs in the cable control cabinets feedback the operation status information of the cable-driven parallel mechanism to the master-control computer, and the industrial computer feedbacks the cable tension data, the movement status of the fine-tuning Stewart platform and the movement information of the A-B rotator to the master-control computer, which completes the status monitoring and display of the feed support system. The coarse frequency of the master-control computer is 1 Hz for the cable-driven parallel mechanism and A-B rotator, and 10 Hz for the fine-tuning platform.

The main components in the cable control cabinet include the Mitsubishi PLC and three DC servo drives. In each interpolation cycle, the Mitsubishi PLC outputs the speed analog commands through the D/A module according to the received motion commands to implement the control of the three DC servo drives. The two cable control cabinets control the six cables of the cable-driven parallel mechanism.

The Leica total station TC2003 is adopted for the 40-meter scale model of FAST to achieve long-range, high-precision pose feedback of the cable platform. The total station is mainly designed for static target measurement, which has the disadvantages of time lag, increased errors, and low sampling frequency under dynamic tracking measurement. Due to the low terminal speed of the cable-driven parallel mechanism (observation tracking speed is 3 mm/s), total station measurement can be adopted. The pose measurement principle of the cable platform is shown in Fig. 10.20. Three total stations are used to track three targets which are uniformly distributed around the edge of the cable platform, and the relative pose relationship between the global coordinate system and the coordinate system can be deduced according to the distances feed backed by the three total stations, the position coordinates of the three total stations in the global coordinate system, and the position coordinates of the three targets in the coordinate system of the cable platform.

The main components of the control cabinet in feed cabin include the industrial computer and eight Panasonic AC servo drives (corresponding servo motors drive the fine-tuning platform and the A-B rotator). The industrial computer adopts Linux + RTLinux real-time operating system, and the real-time control of AC servo motors is implemented by the Tubro PMAC motion control card plugged in the PCI slot, with a control cycle of 2.2 ms. The PCI-1711L A/D data acquisition card from Advantech is mounted in the industrial computer, which collects cable tensions and feeds the information back to the master-control computer in real-time.

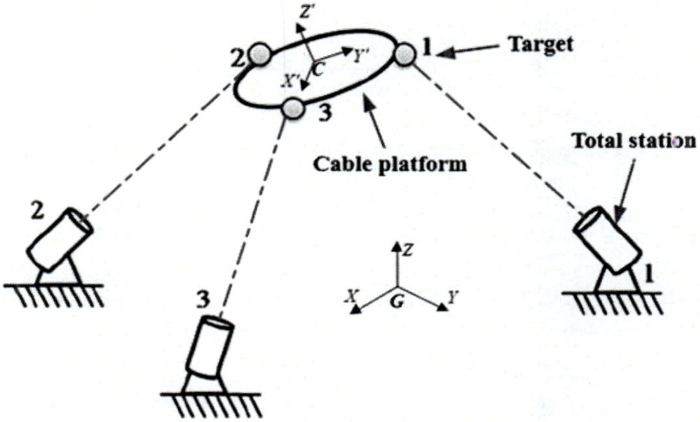

Fig. 10.20 Pose measurement principle of the cable platform

The control software of the feed support system for the 40-meter FAST scale model consists of three parts: the master-control software, the feed cabin software and the PLC program. The core function of the master-control software is to deduce the motion trajectory of the cable-driven parallel mechanism, A-B rotator and fine-tuning Stewart platform in the next interpolation cycle based on the feedback from the total stations and the theoretical trajectory, and send the commands to the Mitsubishi PLCs and the industrial computer through the RS-485 serial communication protocol and UDP network communication protocol respectively. The Mitsubishi PLC program mainly implements simple manual control logic and sends the received speed commands signal to the DC servo drives through the D/A module. This section focuses on the feed cabin software, which realizes the connection between the non-real-time speed commands of the master-control computer and the real-time interpolation calculation inside the Turbo PMAC card, and is the key to realizing the vibration control of the fine-tuning platform of the 40-meter FAST scale model.

The feed cabin software adopts Linux + RTLinux as the system platform, which consists of three parts: the real-time program under the RTLinux, the non-real-time program under the Linux kernel and the open servo program inside the Turbo PMAC card. The program of the RTLinux real-time kernel is shown in Fig. 10.21. Firstly, the initialization of the motion control card is performed, including: implementing the port and shared memory read/write functions of the PMAC card, and configuring the important *I* and *M* variables. Then, insert the RTLinux real-time kernel, start the real-time interrupt thread, and establish a software interrupt with a period of 2.2 ms. Subsequently, the RT FIFO function is set up to call the PMAC card to complete the corresponding functions according to the commands in FIFO (stop control, return to mechanical zero point, return to control zero point and vibration control, etc.), while updating the status flags and data in the shared memory. The data exchange between the PMAC card and the industrial computer is realized through the dual-port memory during each real-time interrupt processing.

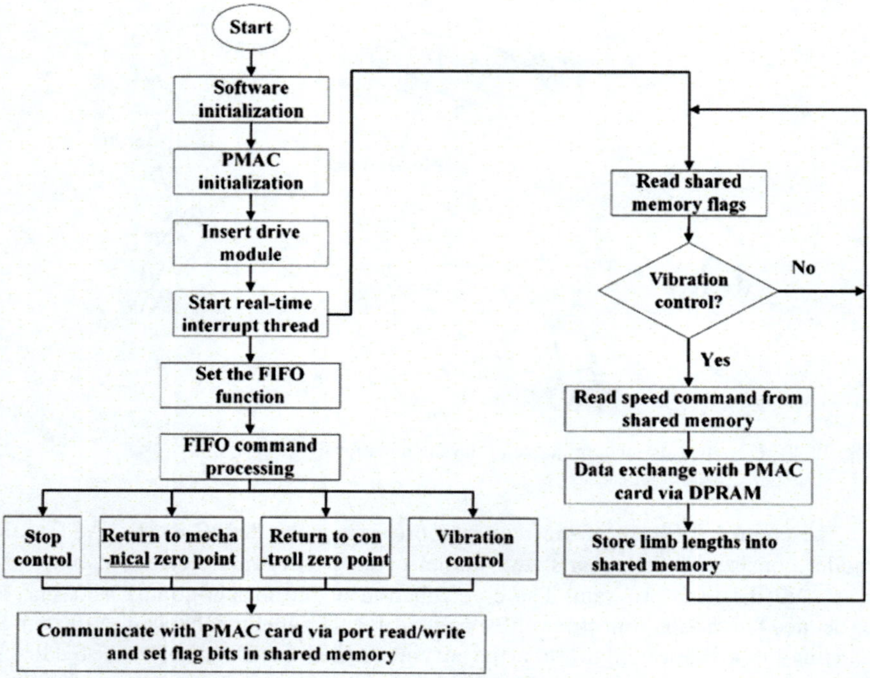

Fig. 10.21 Flow of the real-time kernel software of the industrial computer

The non-real-time program is mainly used to establish UDP network communication, receives commands from the master-control computer, and feedbacks the motion status of the A-B rotator and fine-tuning Stewart platform as well as the data of the cable tension sensors to the master-control computer. The specific flow is shown in Fig. 10.22. Firstly, the A-B rotator and the fine-tuning Stewart platform are controlled to return the zero point. The commands to return to mechanical zero point and control zero point are sent to the real-time software through FIFO. Then, set up the socket server and initialize the A/D card. Next, the data sending and receiving cycle starts, receives commands from the master-control computer, extracts the control command and data, exchanges information with the real-time software through FIFO and shared memory respectively, and sends the obtained status information back to the master-control computer.

The open servo function of the Turbo PMAC card allows users to write servo control logic in a high-level language. During each servo control cycle, each control axis executes a servo control program once. The servo control logic written by users is compiled into a machine language that is directly executable by the DSP microprocessor and downloaded into the dynamic memory in the Turbo PMAC motion control card. Once the appropriate *I*-variable is configured, the Turbo PMAC card will use the user-written program instead of the built-in servo program in the card. In the 40 m

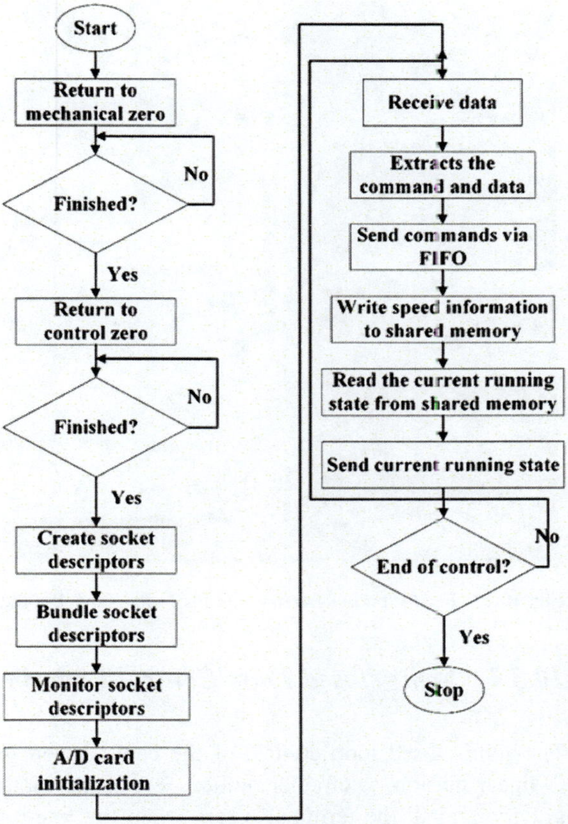

Fig. 10.22 Non-real-time kernel software flow of the industrial computer

FAST scale model, the PMAC motion control card reads the target position and target velocity commands from the shared memory every 5 servo cycles (2.2 ms), which is used to calculate the following errors of each limb. By continuously updating the target length of each limb of the fine-tuning Stewart platform, the following errors are generated to realize the tracking control of the fine-tuning Stewart platform.

10.3 Control Experiments of the CDPR

The feed support system of the 40-meter aperture radio telescope built at the Miyun Observing Station of NAOC in Beijing is shown in Fig. 10.23.

For the cable-driven parallel mechanism, two types of experiments are conducted: (1) open-loop control experiment of the cable-driven parallel mechanism without total stations; (2) closed-loop control experiment with total stations.

Fig. 10.23 Feed support system of the final scale model of the FAST

10.3.1 Semi-closed Loop Control Experiments

The semi-closed loop control of the cable-driven parallel robot is completed by ① linear motion, ② circular motion, ③ circular motion with the variable center of gravity, and ④ the terminal errors when the fine-tuning Stewart platform carries out square wave motion and the CDPR is controlled at a fixed point. The feedback of servo motors is adopted to control the CDPR. In the movement processes, the cable tension variations of six cables are measured to verify the kinematic and static analysis. The cable tensions are analyzed when the center of gravity of the CDPR changes or with the motion of the Stewart platform.

The experiments are designed as follows:

Experiment 1: The cable-driven parallel robot implements the linear motion along the Z-axis direction from (0, 0, −14 m) to (0, 0, −9 m);

Experiment 2: The cable-driven parallel robot carries out the linear motion from (0, 0, −9 m) back to (0, 0, −14 m);

Experiment 3: The cable-driven parallel robot moves along a circle with a diameter of 2 m and its center at the point (0, 0, −14 m);

Experiment 4: Adjust the attitude through the A-B rotator, and change the center of gravity of the cable-driven parallel robot, and the CDPR repeats the circular motion of experiment 3;

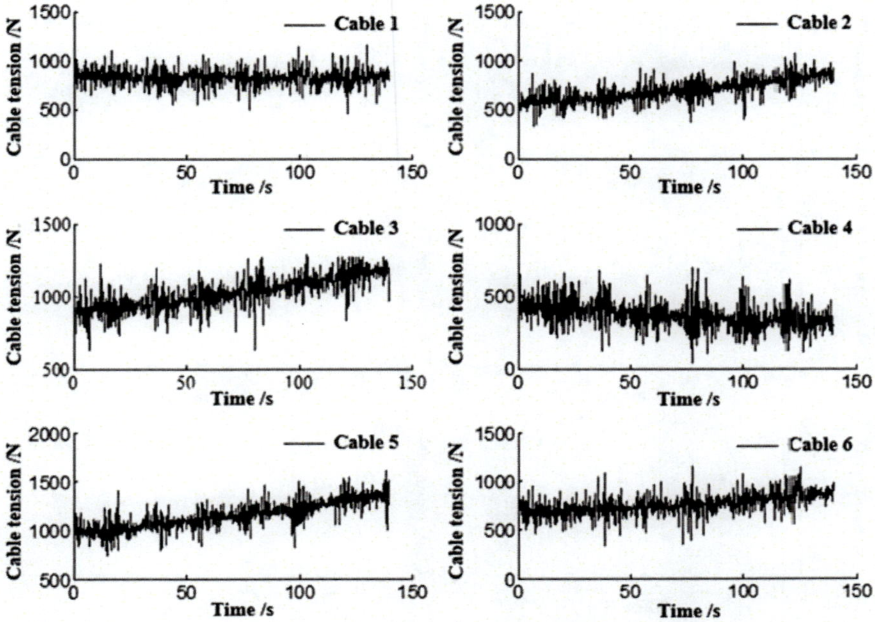

Fig. 10.24 Experiment 1: Cable tension curves of upward motion

Experiment 5: The cable-driven parallel robot is controlled at a fixed point, and the fine-tuning Stewart platform carries out the square wave motion with an amplitude of 20 mm and frequency of 5 Hz.

The variations of the cable tension of Experiments 1 to 5 are shown in Figs. 10.24, 10.25, 10.26, 10.27 and 10.28.

It can be deduced from the cable tension variations of Experiment 4 that when the A-B rotator no longer maintains the initial state, the center of gravity and the mechanical state of the CDPR will change. Compared with Experiment 3, there are a few changes on cable tensions.

From the cable tension variations of Experiment 5, we can tell that the motion of the fine-tuning Stewart platform will affect the CDPR a little. Since the vibration suppression frequency of the fine-tuning Stewart platform is low and its weight is light, the disturbance of reaction force of the Stewart platform is weak. The scale model can ignore the reaction effects of the fine-tuning Stewart platform on the cable support system.

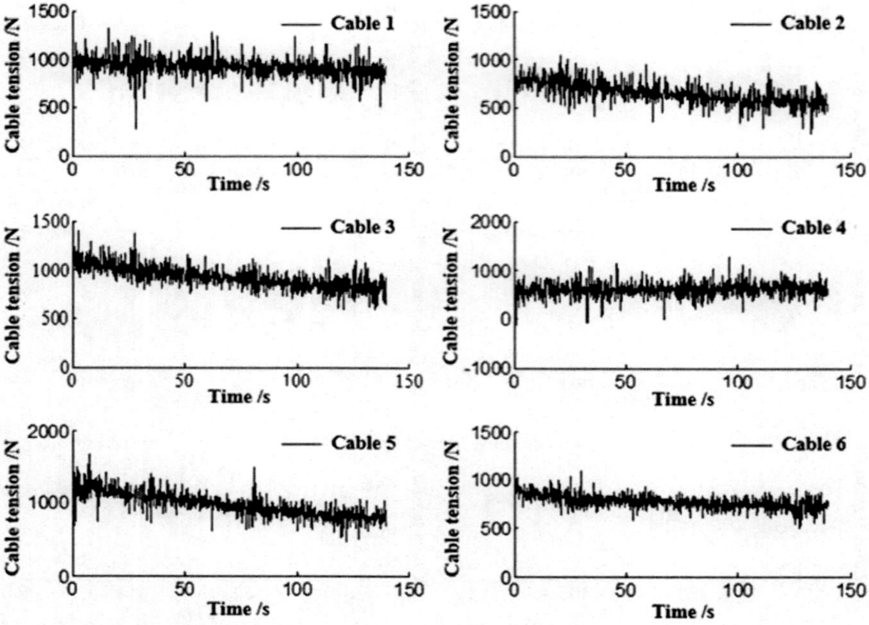

Fig. 10.25 Experiment 2: Cable tension curves of downward motion

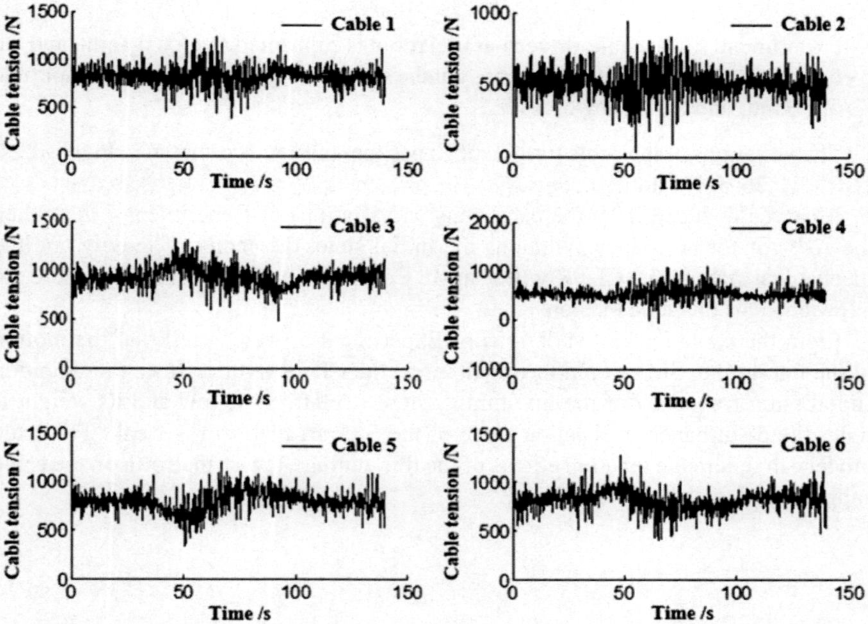

Fig. 10.26 Experiment 3: Cable tension curves of the circumferential motion

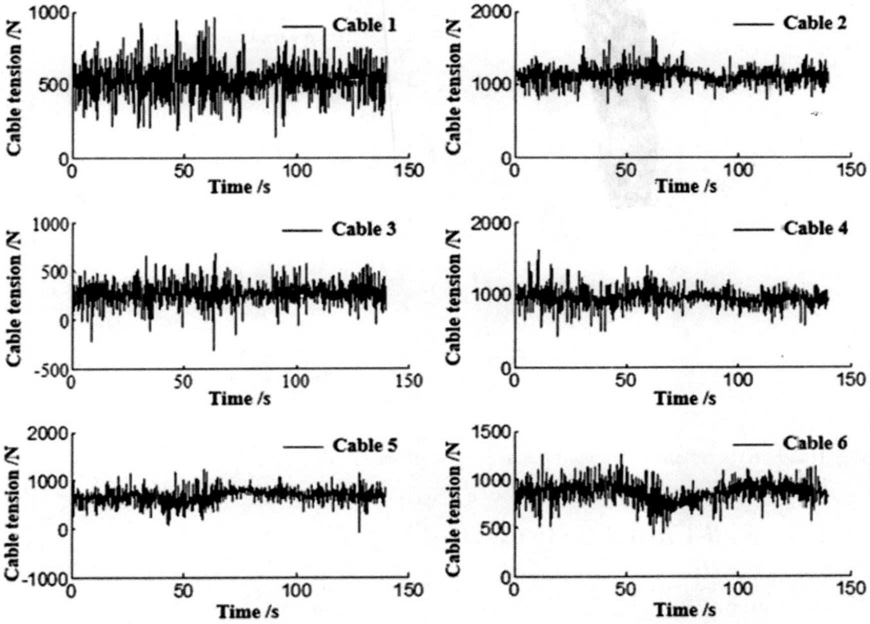

Fig. 10.27 Experiment 4: Influence of the center of gravity change on cable tensions

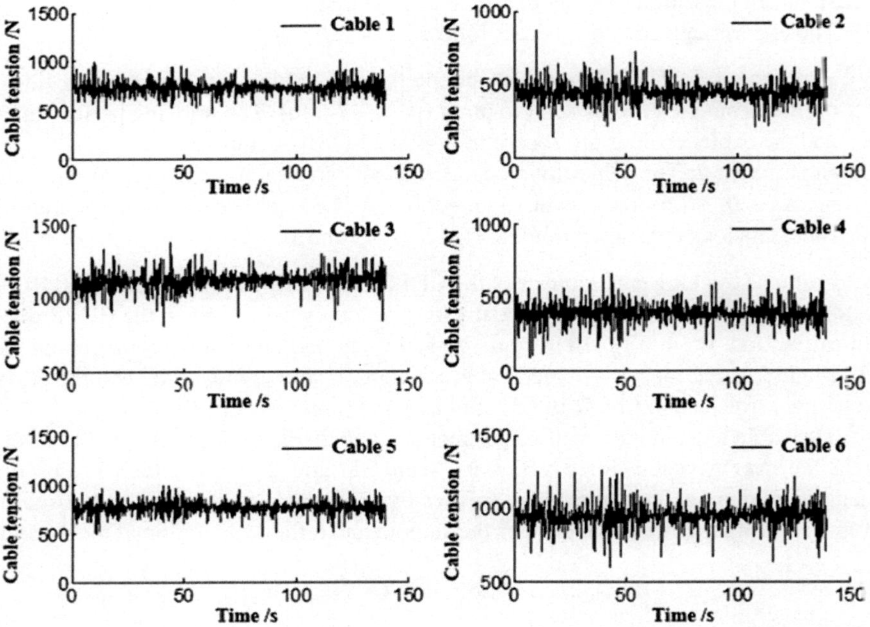

Fig. 10.28 Experiment 5: Effects of motion of the fine-tuning system on cable tensions

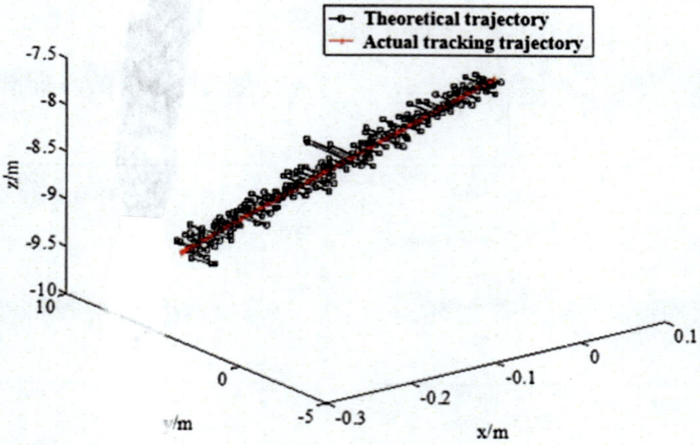

Fig. 10.29 Experiment 1: Upward linear motion of the CDPR

10.3.2 Full-Closed Loop Control Experiments

The measurement signal of the total stations is transmitted to the master-control computer as the feedback and the full-closed loop control is implemented. Linear and circular motions are carried out, and the terminal accuracy of the system and cable tensions are measured to study the force characteristics and the control accuracy.

The experiments are designed as follows:

Experiment 1: The cable-driven parallel robot implements the upward linear motion from the current point to point (0, 0, −7.6 m). The tracking performance and the cable tensions are shown in Figs. 10.29, 10.30 and 10.31.

Experiment 2: The cable-driven parallel robot carries out the linear motion from point (0, 0, −7.6 m) to point (0, 0, −9.6 m). The tracking performance and the cable tensions are shown in Figs. 10.32, 10.33 and 10.34.

From the tracking performance of the CDPR of Experiments 1e and 2 (Figs. 10.29 and 10.32), it can be known that the tracking accuracy is high when the CDPR does linear motion. From Figs. 10.30 and 10.33, we can find out that the errors along the *X*, *Y* and *Z* directions of the feed cabin are numerically comparable, and the RMS tracking errors of the CDPR is 3.37 mm and 3.59 mm respectively.

The cable tension curves of Experiments 1 and 2 (Figs. 10.31 and 10.34) reveal that the average cable force varies between 500 and 2500 N, which avoids the sagging phenomenon of the cable, verifies the effectiveness of the hybrid position/force control algorithm, and ensures the smooth and reliable operation of the system.

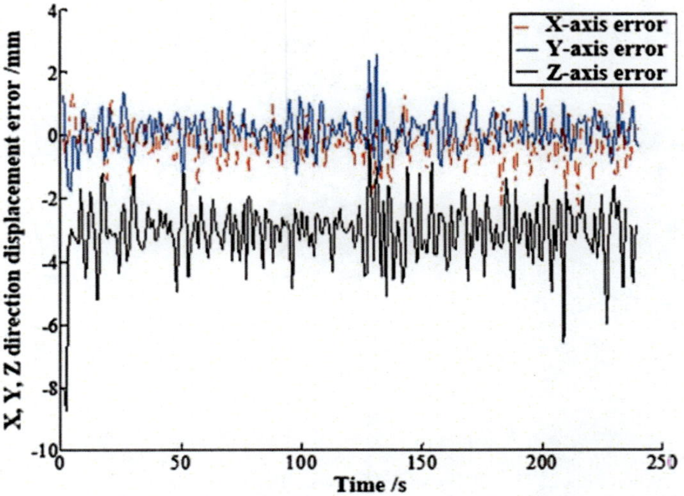

Fig. 10.30 Experiment 1: Error of the upward linear motion of the CDPR along the *X*, *Y* and *Z* directions

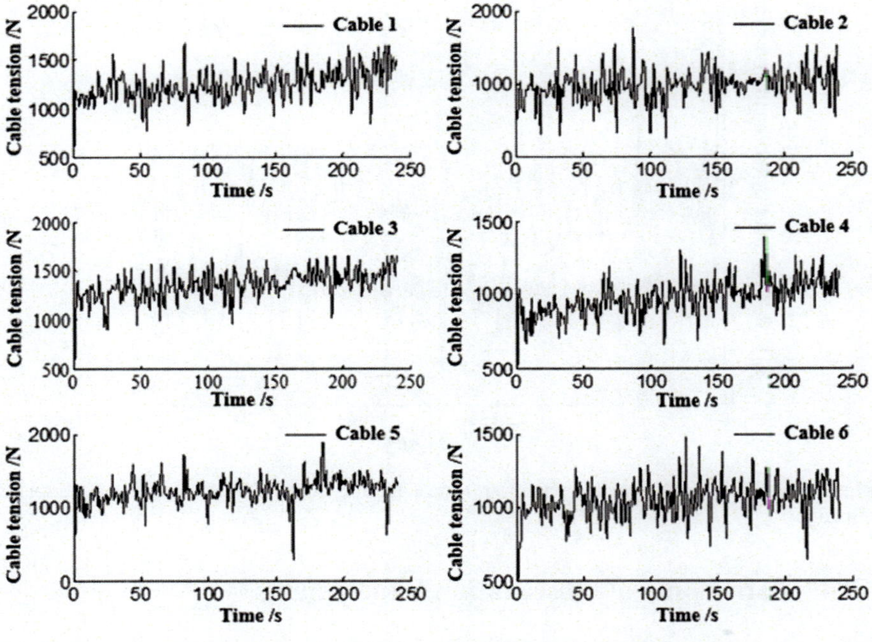

Fig. 10.31 Experiment 1: Cable tension curves of with the upward linear motion

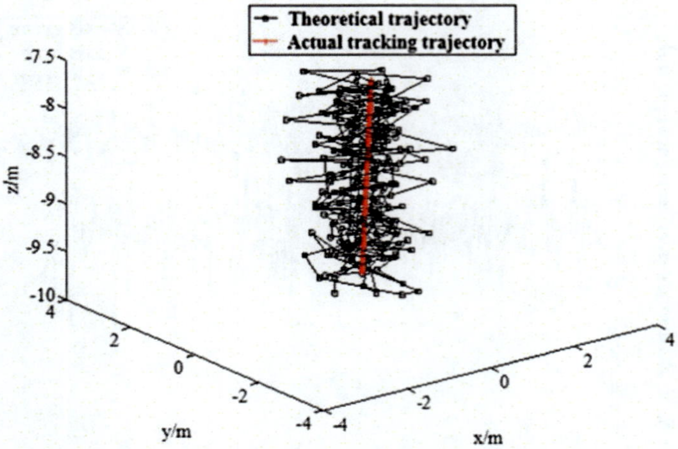

Fig. 10.32 Experiment 2: The CDPR with downward linear motion

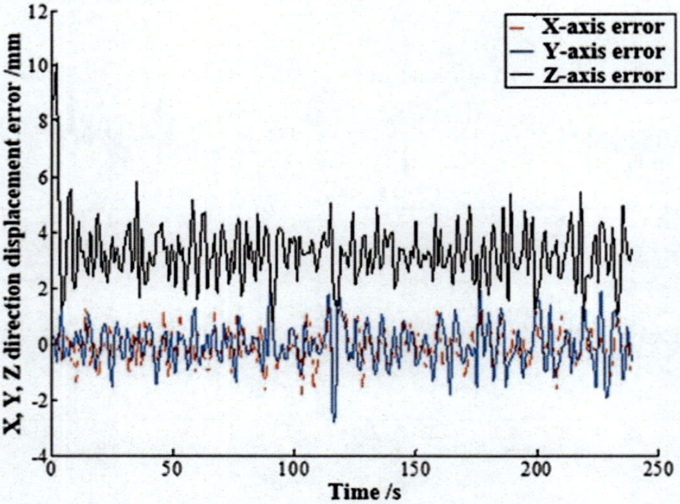

Fig. 10.33 Error analysis of the downward linear motion of the CDPR along the *X*, *Y* and *Z* directions

10.4 Astronomical Observation Experiments

The final scale model of the FAST feed support system is adopted to carry out the astronomical trajectory experiments. During the astronomical observations, the effects of vibration control of trajectory compensation of the fine-tuning Stewart

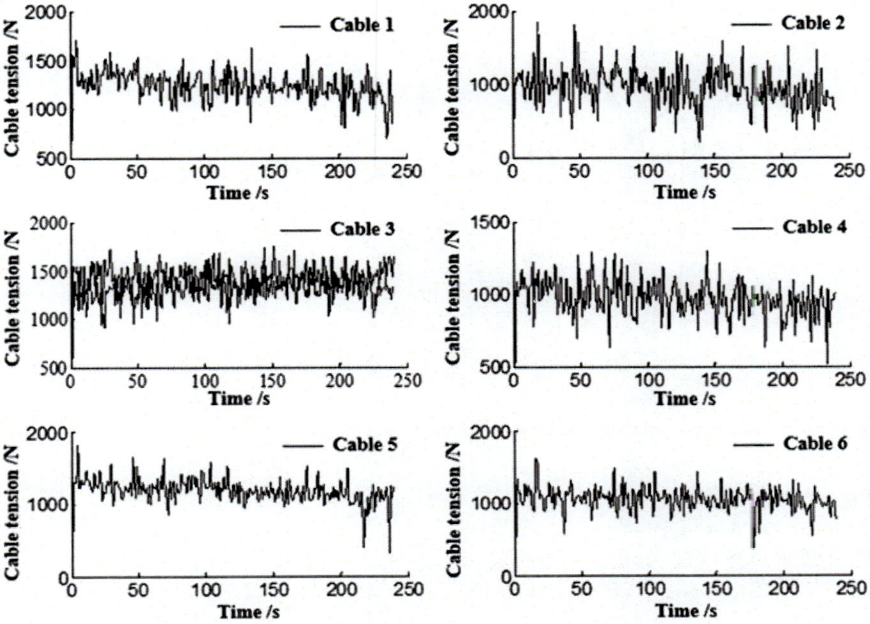

Fig. 10.34 Experiment 2: Cable tension curves of the CDPR with downward linear motion

platform are verified. It is mainly investigated whether the terminal trajectory accuracy of the feed support system can meet the astronomical observation requirements. The global coordinate system is established at the spherical center of the active reflector, as shown in Fig. 8.2. The experimental trajectory is shown in Fig. 10.35. Firstly, the feed support system moves from the initial point G_0 (0, 0, −10.51) m to the geometric midpoint G_1 (0, 0, −9.60) m of the signal receiving surface with a speed of 10.98 mm/s. Then, it runs to the center point G_2 (0, 1.34, −9.51) m of the astronomical trajectory with a speed of 4.47 mm/s. Finally, the feed support system carries the receiver along the astronomical trajectory with a velocity of 4.47 mm/s to G_3 (4.79, 0.34, −8.31) m and G_4 (−4.79, 0.34, −8.31) m in turn, and returns to G_2 to complete the whole trajectory. The experimental trajectory corresponds to a maximum zenith angle of 30°, and the total time is 5000 s, including 4500 s for the astronomical trajectory. The measured wind speed is 1.6–3.3 m/s.

Through the positive kinematic solution, the actual terminal trajectory of the feed support system can be derived from the pose data of the cable platform measured by the total stations as well as the motion parameters of the A-B rotator and the fine-tuning platform from the Turbo PMAC card. As shown in Fig. 10.36, theoretical and actual trajectories are compared, and the theoretical trajectory given in the experiments completely coincide with the actual trajectory.

The terminal trajectory errors of the feed support system are calculated and shown in Fig. 10.37. In the trajectory section from G_0 to G_1, the terminal errors are relatively

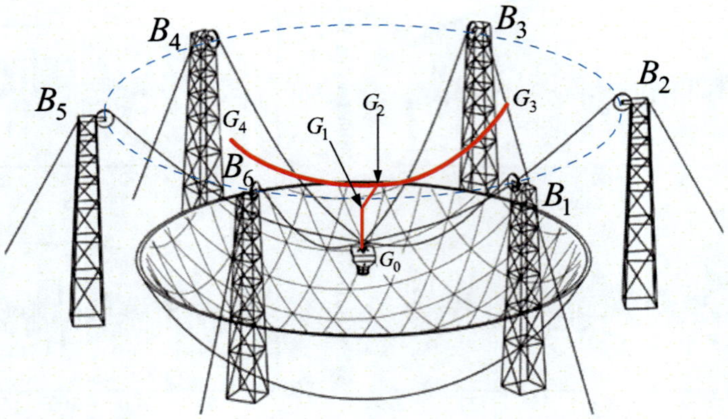

Fig. 10.35 Terminal trajectory of the astronomical observation experiments

Fig. 10.36 Actual and theoretical terminal trajectories of the feed support system

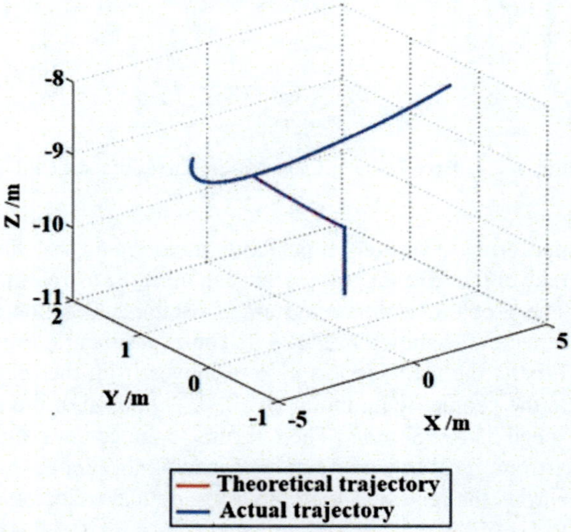

large due to the lack of the vibration control. Subsequently, the terminal errors are relatively stable at the speed of 4.47 mm/s (higher than the required observation speed of 3 mm/s). The terminal position errors along the direction of the coordinate axis of the global coordinate system are shown in Fig. 10.37a, which are kept below 5 mm, and the terminal errors along the *Z*-axis direction are the smallest. During the whole motion, the RMS error of the terminal trajectory of the feed support system of the 40-meter scale model of FAST is 1.872 mm, and that of the astronomical trajectory section is 1.684 mm, which satisfies the requirement of the astronomical observation position accuracy of RMS 2 mm.

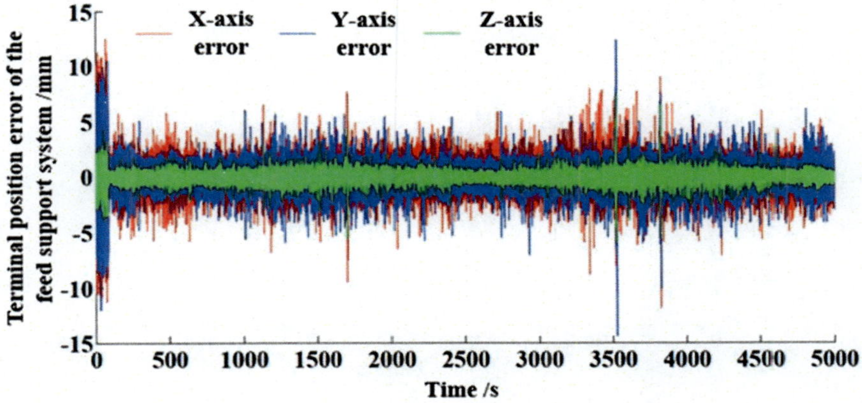

(a)

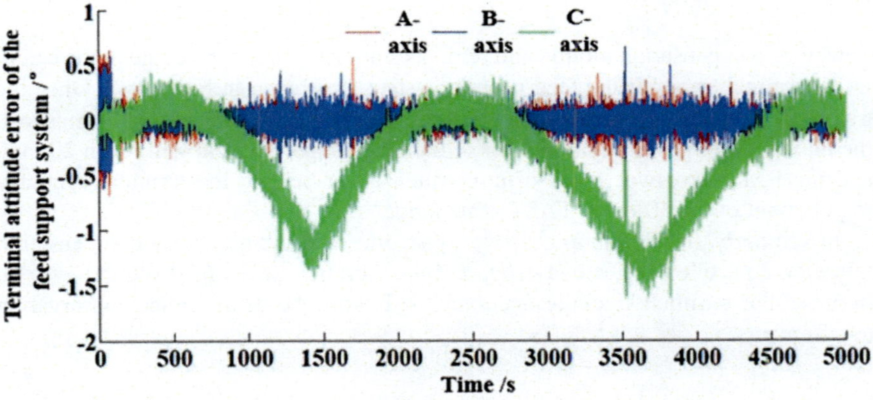

(b)

Fig. 10.37 Terminal trajectory errors of the feed support system. **a** Terminal position errors of the feed support system; **b** terminal attitude errors of the feed support system

The terminal attitude errors of the feed support system are shown in Fig. 10.37b, which are described by the RPY angles. In the base coordinate system, the RMS errors of the terminal rotation angle are 0.045° for the *A*-axis, 0.044° for the *B*-axis, and 0.40° for the *C*-axis, respectively. It should be noted that the *C*-axis rotation in the coordinate system of the base platform is the self-rotation of the moving platform. The angle error of the *C*-axis does not affect the astronomical observations, and are not controlled in the trajectory compensation vibration control. The accuracy of the terminal *A*-axis and *B*-axis rotation angles of the feed support system meets the attitude accuracy requirement of RMS 0.2°.

In order to quantify the role of the fine-tuning Stewart platform in the astronomical trajectory experiments, if the fine-tuning Stewart platform does not perform

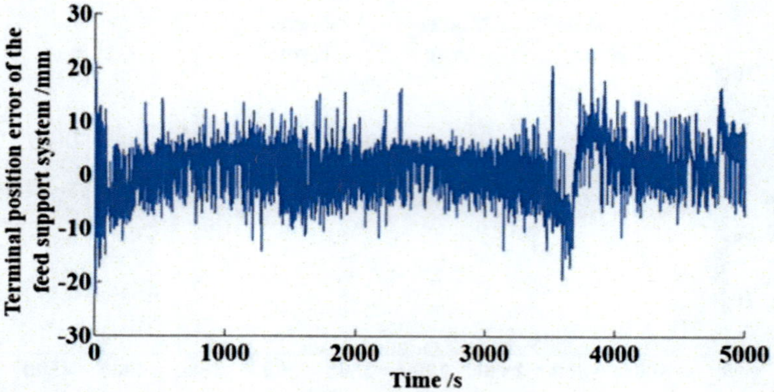

Fig. 10.38 Terminal position errors of the feed support system when the fine-tuning Stewart platform is stationary

trajectory compensation motion and remains stationary in the initial state, the calculated terminal errors of the feed support system are shown in Fig. 10.38. Once the trajectory compensation vibration control of the fine-tuning Stewart platform is lost, the maximum terminal position error of the feed support system will reach 20 mm, and the RMS error will reach 6 mm, which is far beyond the terminal accuracy requirement of the 40-meter FAST scale model.

In summary, the fine-tuning Stewart platform with the trajectory compensation vibration control can effectively suppress the vibration of the feed support system, improve the terminal tracking accuracy, and meet the astronomical observation requirements.

10.5 Conclusion

This chapter presents the engineering practice and experiments of the final scale model of the FAST guided by the theory in this book. Firstly, the mechanical structure, control system and control software of the 40-meter scale model built at the Miyun Observing Station of NAOC are introduced. After that, semi-closed-loop and full-closed-loop control experiments are conducted with the scale model to verify the contents of this book, and the force characteristics and control accuracy of the cable-driven parallel robot are further investigated. Finally, the scale model system is used to conduct the astronomical observation experiments to test the effects of trajectory compensation vibration control of the fine-tuning Stewart platform. The experimental results reveal that the fine-tuning Stewart platform with the trajectory compensation vibration control achieves the terminal accuracy of RMS 1.684 mm, which well meets the astronomical observation requirement of RMS 2 mm of the scale model.